From the
PRAIRIE
to the
PACIFIC
A Blue Angel's Journey

Praise for From the Prairie to the Pacific

"From Vietnam combat to the *Blue Angels* to commanding America's Flagship, Captain Gil Rud has lived life to the full, a life of rollicking fun and increased responsibility. You'll laugh at his antics in college, flight school, and combat squadrons—and ponder lessons in human nature and the treatment of others he offers in each chapter. This fun book is honest, insightful, and unforgettable... strap in!"

—Kevin Miller
Author of the *Raven One* trilogy

***

"From farm boy to naval aviator to *Blue Angel* and carrier captain, Gil Rud's life is a fun, funny, moving, and exciting adventure! A must read!"

—Nicholas A. Veronico
Author of *Hidden Warbirds: The Epic Stories of Finding,*
Recovering, and Rebuilding WWII's Lost Aircraft

***

"*From the Prairie to the Pacific* is a remarkable memoir that chronicles the journey of a young man from rural North Dakota to the cockpits of Navy jets and eventually to the bridge of the carrier, USS *Constellation*. Gil has also managed to capture the essence of the Naval Aviator during one of the most transitional times in the history of the United States Navy. With superb attention to detail, he takes the reader through three tumultuous decades of peace, war and stunning technological advances.

Joining the Navy in the early years of the Vietnam war, Gil experienced the pressure and thrill of Navy flight training culminating with the ultimate challenge of flying combat sorties at sea. With insight and humor, he reflects on the experiences and emotions that he faced as a young pilot, husband and father. His success took him to command of an A-7 squadron and finally the pinnacle of aviation accomplishment, commanding the world-famous *Blue Angels* during their successful transition to the advanced F/A-18 *Hornet*.

Gil Rud has a wonderful ability to make the reader feel that *From the Prairie to the Pacific* is an intimate conversation between two friends over a beer. His Navy shipmates will describe his literary effort as 'Well done, sir.'"

—John F. Schork, Captain (USN Ret.)
Author of more than 10 novels and former A-6 *Intruder* pilot

"If Carl Ben Eielson from Hatton, North Dakota, is considered the greatest aviator from the state, then a gentleman who grew up 14 miles away in Portland is a close second. Most people will know Gil Rud because of his time as the Boss of the *Blue Angels*, the high performance and acrobatic arm of the US Navy.

Certainly, Gil goes through his days with the *Angels* and the pride that went with it. But what this book is about is an author's honest assessment of himself and the ups and downs of life. With each passing word, especially early in his career, Gil peels a layer away at a time revealing a young man who had his faults but also got the job done.

He doesn't skirt the details of nights drinking with his buddies; high school, college and military. Through it all, with ensuing maturity, he maintains his allegiance with his family, friends and country. He flew missions in Vietnam, but didn't ask for credit. He performed night landings in a fighter jet on an aircraft carrier in the middle of an ocean; instances that best could be described as harrowing.

Settling into a family lifestyle didn't come overnight, yet his three children grew up to be very successful athletes and people.

When it comes to a book, this one is the Boss of a military man and how Gil Rud overcame the obstacles that constantly followed him."

— Jeff Kolpack
Sports writer for *The Forum of Fargo-Moorhead* newspaper and author of several books including *North Dakota Tough* and *COVID Kids*.

"With humility and a brilliant application of self-deprecating humor, Boss Rud's memoir is an eye-opening tale of a young man's evolution through a storied naval career. Leveraging "farm boy common sense" to navigate the challenges that come with 5,600 flight hours in combat and leading the world's most renowned flight demonstration squadron, this text not only provides a unique perspective into naval aviation but also serves as a blueprint into the intangible qualities that make a great leader. Boss Rud's account of his career will not disappoint with incredible story telling and exclusive insight."

—Ryan Nothhaft
Grandson of the first *Blue Angels* leader, Butch Voris

From the
PRAIRIE
to the
PACIFIC
A Blue Angel's Journey

GIL RUD
Captain, USN (Ret.)

ISBN: 978-1-943492-99-2 (Hardback)
ISBN: 978-1-958407-00-4 (Soft Cover)

Book design by designpanache

ELM GROVE PUBLISHING

San Antonio, Texas, USA
www.elmgrovepublishing.com

Elm Grove Publishing is a legally registered trade name of Panache Communication Arts, Inc.

Contents

Foreword
by Rear Admiral Garland Wright, USN (Ret.)

MY MOST ENDURING memory of Gil Rud came from the very first day we met. Gil was the Commanding Officer of USS *Constellation* (CV-64), and I was just checking aboard as his Reserve Unit CO. The experience involved some casual conversation, a few cups of coffee, and fire in the ship's engineering spaces. It also provided a leadership lesson that I will never forget.

We should start by saying that fires at sea are not a good thing. A ship, after all, is just a big fat metal container filled with often humorless flammables. Aircraft carriers like *Connie*, for example, can carry up to two million gallons of aviation fuel, and more than 2,000 tons of ordnance. So a fire, no matter the size, can quickly ruin an otherwise perfectly nice day.

Dialing 911 doesn't do any good. There are no pointy-helmeted-water nozzle-jockeys to come riding in on cool red trucks to save the day. It's completely up to the crew. And that's why by necessity, every Navy sailor is a trained firefighter. When the normally routine dangers associated with operating a warship at sea turn into the kind of wide-eyed, teeth-sucking, adrenalin-driven escapade that shipboard fires can so rapidly provide, then it's the ship's damage control teams who determine if that day's routine will include loud noises and lifeboats.

So back to that first meeting. Coffee had been poured and we were just starting our conversation when Gil said to me, "Gar, I want to thank you for coming by to introduce yourself. There is nothing more important than building a strong team from the get-go. But let me apologize for one thing. I normally wouldn't do this, but if that phone rings (pointing to the shipboard handset on his desk) then I am going to have to take the call." He left it at that, and I didn't think anything of it. The meeting went very well, and just when we were finishing, the phone rang. Gil quickly picked it up and I remember him saying things like, *What's your take on the situation?—Let me know what else you need—I trust you and your team*—and stuff like that. After he hung up he said to me, "Gar, during the time that we've been sitting here, our damage control teams have been fighting a fire in our engineering spaces. It was serious, but I knew they could handle it. We have trained them very well, they are well led, and I trust them to do their job." Then he said, "I could have called down there every few minutes to check on them, but that would have only added unnecessary, energy-robbing pressure. They know when to call me. You gotta trust the people who work for you."

Now it's one thing to talk about the importance of trust. It's another thing to act on it. Gil had just demonstrated what it was like to be cool, calm, and professional under pressure—and he let his team do their job without interference or second-guessing. I thought to myself then, and I think now, this is the kind of leader that I want to work for. This is the kind of leader that I always want to be.

If you ask Gil, he will tell you he never intended to write a book about leadership. He'll say that writing this book was simply about capturing his life's journey, giving thanks for the many opportunities he's been provided, and sharing some fun stories that happened along the way. I find it both interesting and telling that Gil doesn't think of himself as an exceptionally good leader. And for that matter he doesn't consider himself exceptionally good at anything. But truly great leaders can't help themselves. They just exude it. And often the most interesting, engaging, and inspiring leaders are the ones who are authentic, who naturally connect to others, who don't take themselves too seriously, and are genuinely funny. That summarizes my friend Gil Rud.

So a few remarks about his book.

Other than the foreword, this is one of the most amusing, entertaining, and delightful reading experiences you'll find. Gil is a natural storyteller and a gifted writer. His book proved to be exactly what I hoped it would be—clever, funny, and written in a way that I could imagine myself in the situations that he so artfully describes. His chronicles about the trajectory of his life, growing up on a small North Dakota farm, joining the Navy, meeting the many challenges of learning to fly Navy jet aircraft, and ultimately serving as the Boss of the Navy's elite Flight Demonstration team (aka the *Blue Angels*), provide a wonderful and uniquely delightful accounting in which his wit and his personal and professional life are intertwined.

There are lots of stories that will make you laugh out loud. Among my favorites are the accounts of his adventure-filled days as a young junior officer and "nugget" squadron pilot. Typically, these tales include several other interesting, party-prone colleagues. But of all the many characters introduced in this book, it's evident that Gil's own best character is himself. His chronicles about the trajectory of his life also include many poignant and touching stories. He does a superb job of describing the difficult life choices between having a military career and supporting a family that anyone who has ever worn a uniform can identify with. But I think my absolute favorite parts of the book are where he describes the many action-filled moments in his life as a Navy pilot and the literally dozens of times that quick thinking, calm, cool decision making saved the jet and his life. Gil's descriptions of the time-compressed pressure that often accompanies envelope-stretching flight in a Navy jet aircraft are perfectly captured.

You will thoroughly enjoy the privilege and the pure fun of spending time with Gil Rud. But I know first-hand that he has many more highly entertaining stories, anecdotes and adventuresome details than can't possibly be squeezed into a single book. (So Gil, if you are reading this—please get going on writing your next one right now!)

Finally, to the reader: buckle up—you are in for a highly entertaining ride!!!

Gar Wright
RADM, USN (Ret.)

Preface

"ON THE RUNWAY, WIND CHECK, maneuver, diamond burner loop with a left turn out, lets run them up, thumbs up, off brakes now, burners ready now." And the diamond is rolling for the start of another *Blue Angel* airshow with a farm kid from Portland, North Dakota leading in the #1 F/A-18 *Hornet*. Day dreaming? Nope, this was for real, and it was one heck of a journey for this extremely long-shot prospect to go from a one-room country school to the flight leader of the *Blue Angels* and the commanding officer of the aircraft carrier USS *Constellation* (CV 64). It was a rough and often hilarious ride with some incredibly good luck along the way to just stay alive.

1. Family Background and the Early Years

I WAS BORN SEPTEMBER 3, 1944, at the hospital in Mayville, North Dakota. I joined my sister Linda who was four years older. We grew up on the family farm, which was located on the Middle Fork of the Goose River, 10 miles west of Portland. Portland was considered to be our hometown although as a child, I very rarely went to town. Instead, our social center was the Bang (pronounced Bung) Lutheran Church, located 3 miles from our farm. It was named after a church in Norway that many of the early pioneer families had been members of prior to immigrating to the United States.

To better understand my background, I will share some family history. My father, Theodore C. "Ted" Rud, was a second-generation farmer of Norwegian heritage. My grandfather, Gilbert Rud was a Norwegian immigrant pioneer, who originally started the farm. Gilbert and his six brothers heard of the land that was available in Dakota Territory. In 1878, Gilbert and his older brother, Halstein, left their families behind and traveled from Zumbrota, Minnesota, to Fargo, North Dakota. From Fargo, they walked the 70 miles to the Middle Fork of the Goose River. They each then laid claim to 160 acres of land in the same general area. Eventually, six of the seven brothers started farmsteads, all within a five-mile radius. The hard part lay ahead, as the rules involved with the claim required building a structure and spending the winter before the claim could be verified.

Winters in North Dakota are simply brutal. They involve lots of snow and extreme cold. Temperatures can reach 40 degrees below zero, and in January, rarely get above zero. Dangerous blizzards sweep the prairie with little warning. Temperatures preceding a blizzard are often quite pleasant, lulling people into venturing out to visit neighbors, gather wood, or even travel to town for supplies. Suddenly, a few snow flurries drift from the darkening sky. The temperature begins to fall and the wind starts to pick up, almost always out of the northwest. Within a few minutes, a beautiful winter day can transform into a killer blizzard.

Imagine heavy snow being blown into huge drifts by hurricane force winds. Visibility is extremely limited and at times non-existent. If a person was caught even a short distance from shelter, they often got lost and froze to death. Despite this dangerous challenge, my grandpa, Gilbert, constructed a sod-based cabin into the side of a hill. He then spent the winter there, as well as part-time in a nearby cabin owned by the Odegaarden family. Gilbert's cabin was built facing south on the north slope of the river-valley, which provided protection from the worst of the winter weather. The river was about 50 feet lower than the cabin so there was no danger of flooding. Other pioneers built closer to the river only to be inundated by spring flooding, which could expand the river from 20 feet wide by 5 feet deep to a raging torrent that was almost a half-mile wide, covering the entire valley.

After surviving the winter, the Rud brothers traveled back to Zumbrota to get the rest of their families. This was a one-way distance of 380 miles. In the spring of 1879, the entire Rud family packed up all their belongings, joined an immigrant wagon train and headed back to the North Dakota farms in time to plant the first crop and begin their new life as pioneer farmers.

My mother, Clara Helena Leland, was also of Norwegian descent. The Leland family came to America from Norway, circa 1900. My grandfather, Herman Leland, was a fisherman operating from an area in Norway known as Lofoten Island. At the turn of the century, Norway was the poorest country in Europe. The Leland's were scratching out a living in an area that was close to the Arctic Circle. Grandpa Herman returned from a particularly hazardous fishing trip and decided it was time to head for new opportunities in America. He and Grandma Leland packed up their meager belongings and the two oldest children, Hennie and Johnny, and sailed to America. Unlike the Rud brothers, they came too late to gain pioneer-land ownership. Instead, they became sharecroppers, renting land from a large land baron, Norman Brundsdale, who eventually became the governor of North Dakota. Their home farmstead was only 2 miles from my grandfather's, so they were neighbors.

The family name, Rud, or Ruud is very common in Norway. It refers to a clearing in the woods where a farmstead would normally be located. If there were a farm and people living there, then it would be a family name attached such as Paulsrud, Johnsrud, etc. The spelling varies between Rud and Ruud. It was not changed when my ancestors moved to America. Unfortunately, Rud is pronounced Ruud, but of course in America folks tend to mispronounce it, add letters such as Rudd, or Rude, etc. You cannot imagine how many times that I have corrected people on the proper pronunciation. The other three family names are Leland, Sondrolie and Reise. The Lelands kept the same name in America, however the Sondrolies changed to Lee, and the Reises to Rise.

Unfortunately, this zeal to become Americans included discouraging our generation from learning Norwegian. We were encouraged to learn only prayers and songs. They were very conscious of what they referred to as a "Norwegian accent," which they thought would hinder us in getting jobs. Of course, since they would talk in Norwegian when they did not want us to know what they were talking about, we quickly picked up on off-colored jokes and the various cuss words.

Our farm was located 10 miles west of Portland, 13 miles east of Finley and about 14 miles south of Hatton, North Dakota. So, it was indeed, in the middle of nowhere. All the local pioneers were of Norwegian heritage and were practicing Lutherans. A priority for the community was to build a Lutheran Church. Bang Church was founded by the Rud brothers and a few other families of homesteaders and served as the social center for the entire community. It consisted of both a church and a community hall, which was used to host church-affiliated events and Sunday school for the children.

Because life as a pioneer farmer, especially in the Dakotas, was so challenging, it was essential that they supported one another as a community. Planting and harvesting of crops this far north required precise timing to stay within the confines of the short growing season. Harvest was especially daunting, requiring dozens of men to complete the task on each farmstead. This was accomplished by an unselfish attitude of "help thy neighbor and he will help you." According to my dad, they also had a whole lot of fun working together as a community. If a farmer was injured or became incapacitated by illness, the neighbors would step in and plant/harvest his crop for him. This practice still exists today and is a hallmark of life on the farm in North Dakota.

As I mentioned, my dad and his friends had a lot of fun working together. They were also not above pulling some tricks on each other. Halloween offered an excuse to do just that. One of the older farmers in the area scoffed at the foolish pranks that my dad and his friends would pull so they decided to really get under his skin. He had a wagon full of wheat sitting in his yard ready to take to town to sell. They unloaded the wheat from the wagon into grain sacks. They then hoisted the empty wagon onto the top of his barn. Once it was situated on top of the barn, they carried the sacks of grain up the barn roof and refilled the wagon. The next morning, he went out to milk his cows, only to find a wagon full of wheat on top of his barn.

My dad and his buddies were hiding in the trees and had the opportunity to watch the old fellow scratch his head at the sight of the wagon on his barn roof. He then went up to examine more closely to discover that it was full of wheat. He went from head scratching to "it must have been the Devil" mode. At that point, my dad and his buddies emerged out of their hiding spot to accuse the old fellow (he was known to partake of a drink or two) of somehow driving his horse drawn wagon, full of wheat, up onto his barn roof. Eventually they calmed him down, explained how they had accomplished the feat and proceeded to unload the wheat, lower the wagon, and fill it up with wheat again. It was the ultimate Halloween prank.

My dad was an accomplished ski jumper when he was a young man, competing around North Dakota, Minnesota and in Manitoba, Canada. He related as to how he was entered into a competition as a 16-year-old (not afraid of anything) kid against some older guys in Canada. Most of them were of Norwegian descent, and of course he spoke fluent Norwegian. He was the first to jump. The jump was a big one built on the side of a hill that eventually ended up on a frozen river. Of course, the landing area was on the slope of the hill, and then you eventually stopped on the river. Now there were several choices for how high you started on the jump. This was predicated on how fast the jump was. The slower the jump surface was, the higher you started from.

Now my dad could jump a long way, but your competitive score was based on both distance and form. His form not being a strong suit, my dad knew that he needed a particularly long jump. As he headed toward the highest start point, this older Norwegian warned him that the surface was fast so he should start lower. "Aha!" Dad thought. He knows my jump will be long, so he is trying to talk me into a shorter one. Well, my dad proceeded to the very top. Feeling quite cocky, he started his run. "Boy this is fast; I am going to get a really long one." He said that it felt wonderful until he saw the landing area disappear behind him. Yep, he landed flat on the frozen river, broke both skis, but fortunately no bones. It was definitely the longest jump of the day, but you must stand up to qualify. As he recalls, the audience was very appreciative of my efforts, but the veteran Norwegian jumper just laughed and gave him the "I told you not to start there."

My dad did not like school and like most other farm boys of his era, he dropped out after his freshman year in high school. He might have stayed in school longer, but his mother insisted that he attend an agricultural high school in Fargo where he did not know anyone and did not fit in. Despite dropping out early, he was an avid reader and very knowledgeable about both history and current events. He did not serve in either WWI or WWII. He was 15 when WWI ended and 39 when WWII began. Although never mentioning it to me, I realized later that this failure to participate in either conflict was a burden for him. So much so, that I found out from one of his friends that my dad started savings accounts for all his friends that were serving in WWII. It was all done anonymously and he would never admit to being the mystery benefactor. This fellow told me that they knew it was Ted and that it was not a small amount of money. Later, my dad was very proud of my service in Vietnam.

***

FIRST PERSONAL MEMORIES, CIRCA 1948–1950: Growing up on the farm was a great experience. I was a freckle-faced, redheaded, little fellow with big ears, and in fact, closely resembled the character "Opey" from *The Andy Griffith Show* (no offense intended to the great actor and eventual producer/director Ron Howard).

Since our immediate family consisted of just my sister and me, and she was four years older (if you ask her, she will tell you that I did not exist), much of my playing was done alone. This situation allowed me to develop a great imagination. As per the Kenny Rogers hit song *I am the Greatest*, I played entire baseball games by simply tossing the ball up and taking a swing. If I hit it, I was a great hitter and it was a home run. If I missed it, I was a great pitcher and it was strike three.

The farm was a great place to play hide-and-go-seek. I remember crawling under the large grain storage structure that we had on the farm. It had a drive-up dump area that was used to off load the grain into a pit. At the bottom of that pit was an electrically operated elevator that could direct the grain into one of several bins located in the storage structure. The structure was built upon a foundation of cemented-together rocks that kept the bottom about 3 feet off the ground. This also allowed access for maintenance on the elevator. It was nice and dark under there and the perfect place to hide from my sister and some cousins.

Unfortunately, I crawled in a bit farther than I should have and fell into the pit. It was probably only about 5 feet deep, so no injuries, but also no way to crawl out. Maximum panic set in. What if nobody finds me? What if somebody starts up the elevator? What if there are ghosts under here? "HELP!!!" I was only 5 years old at the time, so to me it seemed like hours that I was trapped under there. My dad came to the rescue as always (probably within a few minutes), with one of several lectures I was to receive on farm safety. The good news was that I was certainly the winner of that hide-and-seek game.

Life on the farm was simple, but full of adventure. For instance, from June until September, I rarely put on a pair of shoes. Our feet were so tough that we could run on gravel roads and even step on sharp objects with limited damage. We were basically self-sufficient raising our own fruit and vegetables with a massive garden of potatoes, corn, tomatoes, carrots, etc. June berries: chock cherries and gooseberries grew wild along the river. We also had an orchard with apple trees. My mother and my dad's sisters were all expert at canning, so our basement was full of mason jars to ensure a balanced diet that lasted through the long winters.

We had around 100 leghorn chickens that provided lots of eggs. Of course we consumed those eggs, but also sold them to the local creamery. We usually milked around 10–15 cows of mixed breed. The best milk producers were the Holsteins, but the Guernsey's provided a more cream-rich product. We had a separator machine, which allowed us to sell the cream. The milk we consumed and also used it to feed the pail-fed calves. Our bulls were always Herefords, and we usually maintained a herd of about 100 Hereford/mixed breed (offspring of the milk cows and the Hereford bull) beef cattle. We also had some domestic ducks and a litter of feeder pigs that we purchased each spring. Because of the inherent risk associated with small grain farming, diversification provided by the various animals ensured that a bad crop would not result in a total financial disaster.

In addition to small grains such as wheat, barley, soybeans and flax, we raised crops that could be fed to the cattle. Corn, alfalfa and oats provided everything needed to keep our animals fed. No need to purchase expensive supplements in those days.

My dad loved baseball and played outfield on a local country team. He bought me my first baseball glove, which was flat as a pancake (taught me how to use two hands to catch the ball) when I was five. For whatever reason, I could only throw sidearm and of course, left-handed. My mother taught me to write right-handed because there were no left-handed desks in the one-room school.

My parents, Ted and Clara Rud, at their wedding. October 14, 1939.

2. Growing Up in the Country

THE COUNTRY SCHOOLS were divided up by what township your farm was located in; ours was in Primrose Township. There were three schools in the township and the one closest to us was Primrose #2. It was a one-mile walk along a dirt road, or a one-mile ski in the winter. I was always early and my sister was always late. We never walked together. Marian Thompson was my first-grade teacher, a very sweet lady who still managed to maintain discipline while teaching all eight grades.

We had a weird teacher during my second-grade year, Ms. Plant (pronounced "Plunt"). She had a younger brother named Kenny and the two of them lived at the school (very odd). She had no discipline, which we students thought was awesome. Then after about a month of chaos, she and her brother simply vanished. My mother took over for the rest of that year (she had a one-year teaching degree from Mayville State Teachers College), so that was a bit strange. Talk about not getting away with anything. Then we got Mrs. Vern Hanson for my third-grade year, followed by Mrs. Avis Grandalen, who was an awesome teacher for 4th through 8th grade. I was so lucky to have her as a teacher and mentor. She prepared us for high school and beyond, and she also brought her daughter Patty with her to add to my class. Patty was very pretty and lots of fun. Initially there was another boy in my class, but he was learning disabled and dropped back, so from fourth grade through eighth it was just me and three very smart girls: Carol Peterson, Mary Leland (my first cousin) and Patty Grandalen. *I pretty much finished last at everything.*

The dirt road that we walked, rode bikes or skied on to school went through an area that we called "The Haunted Anton Gilbertson Woods" and "spooky old house." My sister Linda was the best ghost storyteller ever so of course she made up all sorts of grisly fables about this area. I distinctly remember hauling ass through there for fear of being nabbed by ghosts.

Like most kids, my best memories of grade school were centered on what went on during the three recess periods. With all eight grades being represented, we are talking games with participants ranging from ages 6–14. As a 6-year-old, I was exposed to some rough games, teasing, and I suppose what is now called "bullying." Of course, that bullying was mostly accomplished by the girls who were all bigger than I was. Probably the most common game was something we called "Break-Through." Simple enough rules, which consisted of a single player standing in the middle of the playground, with the rest of the students lined up parallel across one end of that playground. The object of the game was for the student body to all run to the other end of the playground during which time that single player in the middle would attempt to tackle someone. And yes, I mean *tackle,* like in football. If he or she accomplished that tackle, then there would be two against the student body, until finally the entire student body was now available to tackle the last two players. They usually targeted one of those two, but sometimes they would go after both. If they tackled both, the game would end in a tie. If not, the survivor would be the big winner. Of course, considering the violent nature of the contest, inevitably there would be injuries. Parents expected and generally tolerated an injury, but if you ripped or damaged your clothing you were in big trouble. With only a couple of pairs of pants (normally blue jeans) meant to last the entire school year, mine ended up with numerous patches.

The only playground equipment we had included a swing set and a manually activated "merry-go-round." The older kids, as they attempted various atrocities disguised as "playing with the little kids" abused both devices. They would put us on the merry-go-round and get it going at a speed that would eventually get you dizzy. Then one of two things would occur: you would be ejected from the ride, or you would hang on and get sick. The swing event was really a competition consisting of which 8th grader could get their little kid higher. That was quite fun, although like most things we did, totally unsafe. The other ridiculous thing that we attempted as older (7th and 8th grade) kids, was to see if we could pump the swings so high that we would loop over the top of the swing set. Fortunately, no one ever accomplished this feat, since it would probably have resulted in a major injury or even a fatality.

Bicycles were a big part of our lives as well. Since there were no paved roads anywhere near our farms/school you would expect that we would have dirt bikes. Except there was no such thing as a dirt bike, so we rode regular bikes on the dirt and gravel roads. My cousin, Donnie Leland, who was two years behind me in school, got this awesome English-style racing bike with the skinny tires. Unfortunately, it did not perform very well in loose gravel, resulting in several spectacular spills. We did construct jumps in the ditches as a precursor to the BMX racing done today. I still have a couple of scars from those events.

The Grade School Fall Carnival was an event that we all looked forward to. It was set up as sort of an open house with various booths to raise money for the school. All sorts of games were set up with prizes for the winners. One year, when I was in about the 3rd or maybe 4th grade, I had the honor of overseeing the jellybean jar. It was a very large glass jar filled with jellybeans. The object of the game was to guess the number of beans in the jar. You did this by buying a ticket and then writing your guess as to the number of beans by your name on a sheet of paper that I carried on a clipboard. It involved some salesmanship. I carried the jar and clipboard all around the carnival until a certain deadline time.

The next phase was to empty the jar and count the beans. Just as they were announcing last call for the jellybean competition, I was carrying the jar and clipboard down the stairs to the basement when I tripped and dropped the jar on the cement floor. "Nooo!" Jellybeans everywhere. I still distinctly remember thinking that this was indeed the end of my life! Of course, the adults were laughing hysterically and I was mortified. I hauled ass for the door and started to run home. My dad caught up to me and, stifling a laugh, assured me that they were already at work counting/eating the jellybeans. It was one of the first—but certainly not the last—failures in my life.

***

AND THEN THERE WERE CARS, PICKUPS, TRUCKS AND TRACTORS: We drove all of them of course, starting legally at the age of 12 with a farm permit, but actually at a much younger age. When I was 10 years old, we had two John Deere tractors. The big one was a 1939 John Deere G with a two-cylinder, 34-horsepower engine, capable of pulling a three-bottom plow. The other one was a 1938 John Deere A with a two-cylinder, 26-horsepower engine, capable of pulling a two-bottom plow. We also had a 1954 Minneapolis Moline Model UD, four-cylinder, 44-horsepower engine, capable of pulling a four-bottom plow.

The two John Deere tractors were started by manually grabbing and turning a large flywheel, which was difficult to do, especially with a cold engine. I did notice that when the engine was warm, it started much easier. I was always pestering my dad to be allowed to drive the tractors by myself. He rather smugly stated, "When you can start one of the John Deeres, you can drive it by yourself."

Cleaning snow from the farmyard using John Deere G equipped with snow-removal blade.

One hot summer day when I was 10, our hired man, Bennie Lee, came into the farmyard with the Model A and shut it down near the house so he could wash up and come in for dinner. Dinner was what we called the noon meal. The engine was warm from working in the field, so I decided to see if I could start it. I opened the petcocks on each of the two cylinders, gave it a little throttle, grabbed the massive flywheel and turned it. The engine fired right up. Unfortunately, I did not let go of the flywheel, so it threw me headfirst into the dirt. My mother was watching all this and, I am told, became quite alarmed by the potential disaster taking place in the front yard. I dusted myself off, puffed out my meager chest, got on the tractor and drove it around the yard. This was a big step in my young life, as I was now a contributing member of the farming team.

My dad was a firm believer in minimum fall tillage to prevent wind erosion during winter and early spring dust storms. This attitude came from his experience with the terrible drought and subsequent dust storms of the 1930s. Most farmers plowed their fields in the fall, which left them nice and black and ready for seeding in the spring with minimum preparation. Since we only disc plowed our fields in the fall, that process left lots of straw as top cover against wind erosion. In the spring, it necessitated a process called "pony pressing" to seed the new crop.

Pony pressing was accomplished by attaching a packer and a seeder to a plow. We would use all three tractors with my dad leading driving the G, my sister on the A, and me in the rear with the Moline U. The two John Deere tractors pulled their pony presses at about the same speed, which was a bit faster than my tractor. Hence, I was always last. No problem, except my sister could not drive straight, or, more accurately, did not give a hoot whether she drove straight or not.

Since a farmer's reputation is often judged on how straight he seeds his crop, this situation was unacceptable to my competitive nature. Normally, each tractor would simply follow in the lead furrow created by the preceding tractor. In my case, I needed to concentrate especially hard to straighten her awful work. I soon discovered that her seeming incompetence was her way of eventually being relieved of any farming responsibilities. By the way, my sister was a very attractive young lady, known to draw the attention of the boys in the area while out in the field. This resulted in one of them running off the road while trying to get a better view. I also began to realize why some of these older boys were befriending little old Gil.

***

My Best Friend Was Richard Fugleberg: Since his last name was so difficult to pronounce, he picked up the nickname, "Fugie." Fugie lived just 3 miles up the road and we were only six weeks apart in age. Although he lived close, it was in another township, so we went to different one-room grade schools. We did go to the same church and Sunday school, sang in the choir and eventually acted as ushers for the Sunday services.

Sunday school, during the school year, was held an hour before church services. In the summer, instead of every Sunday, we had a consolidated summer session that lasted for two weeks. How do you keep a bunch of kids under control and interested in boring Bible stuff when it is absolutely beautiful outside? It is certainly not easy, especially for a hot-tempered, redheaded, preacher's wife by the name of Dorothy Ree.

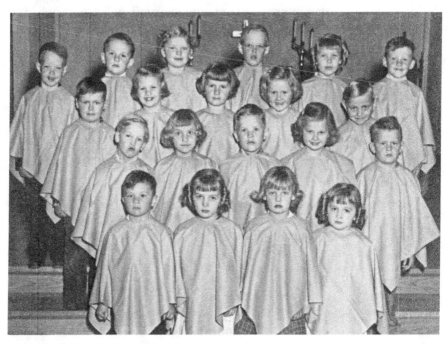

Bang Sunday school 1st - 3rd grade choir members. Starting with the back row, left to right: Allen Odden, Richard "Fugie" Fugleberg, Carol Peterson, David Odden, Phyllis Thykeson and Gil "Opie" Rud. Third row: Tommy Sparrow, Arlene Rud, Kathy Thykeson, Mary Leland and Vernon Thompson. Second Row: Marlan Groven, Volborg Thompson, Lowell Thykeson, Lorraine Leland and Paul Rud. Front row: Donald Leland, Patricia Rygg, Diane Groven and Susan Rygg.

Now Fugie was pretty much your average farm kid… no, not really. He was full of energy and had a heck of a time sitting still for more than a few minutes. I don't remember exactly what the offense he committed was, but it was provocative enough to cause Mrs. Ree to slam a broom into the floor so hard it broke the handle. Now that by itself was no big deal, however, she then threw him out of the summer session. Oh my, it was 10 o'clock in the morning and he sure as heck was not going to go home to what would be a most unwelcome reception. So, he headed for the haunted "Anton Woods" to hide out until the end of the school day. The woods was only about a quarter of a mile from the school, so we visited with him over the noon hour and brought him a sandwich.

I do not recall the outcome of this situation, but he was back in school the next day (more than likely with a pretty sore behind).

***

4H STANDS FOR HEAD, HEART, HANDS AND HEALTH: This is an organization that plays a big role in the life of farm children. The purpose is to help youth acquire knowledge, develop life skills and form attitudes that will enable them to become self-directing, productive members of society. To us it was a wonderful opportunity to get together, play Break-Through and even flirt a bit with the girls. We did have animal or crop-related projects that culminated in a once-a-year competition known as "4H Achievement Days." This event rotated between the Steele County towns of Hope and Finley. Both had special areas and buildings used to house and show the various animals.

My first memory of Achievement Days involves my sister, Linda. Linda was supposed to be raising a Hereford steer that she named "Lucifer." She named him and I think that is the last time she saw him until it came time to compete at Achievement Days. In other words, this was not an important affair to her. Part of the event involved a well-attended parade of the competing livestock through the town, which in this case was Hope. Since Lucifer had never really seen a halter or even been outside of a pen, this parade offered him a chance at FREEDOM!

The parade seemed to go relatively well for the first few minutes, with the local boys paying much more attention to Linda than to Lucifer. Suddenly, Lucifer realized that the only thing between him and a great gallop to freedom was this pretty little lady. He took off like a bat out of hell. Linda held on for maybe a few seconds before dropping the halter, putting her hands on her hips and letting the damn beast go. And go he did, creating chaos in downtown Hope. The good news was that every teenage boy in Steele County took up the chase, hoping to get Lucifer back to the damsel in distress. They eventually did, although as I recall, it took a team effort.

Although we did not have a history of raising pigs, I talked my dad into purchasing feeder pigs for my 4H projects. I experienced varying degrees of success until the spring of 1958. That is when Dad and I went to an auction and purchased a litter of incredible pigs. As I recall, they were "grade" hogs, in other words, not purebred. I think they were a cross between Yorkshires and

Chester Whites. We trucked them home and started feeding them ground oats and milk, nothing special or exotic. To compete in the market hog category, the pigs had to be born in a certain time window and ideally reach a market weight of between 180 and 220 pounds by the September Achievement Day competition. These pigs literally grew like weeds. I picked out the two I thought would show best, we loaded them up and headed for Hope.

Since I had two hogs to show, I needed to enlist the help of another 4H boy to show one of them. I got a good friend, Merle Evanson, who was two years older than me to help me out. The judge was an animal science professor from North Dakota State University, Dr. Johnson. He was not only an awesome judge, but also a colorful character. He stopped by my pen a few minutes before we were going to enter the show ring. "Holy cow!" he said, "Do you have proof of when these pigs were born?" I did and he left just shaking his head. I thought to myself that this is probably not a good sign.

I am estimating that there were probably 20+ hogs being shown in the market class. There is no real way to lead a pig, so you use a cane-like device to sort of guide the animal in a manner that makes it easy for the judge to see and evaluate him. There are three levels of awards given out in the competition. A white ribbon is awarded for third place, a red ribbon for second place and a blue ribbon for first place. Several hogs are placed in each category. Finally, two hogs are chosen from the first-place group as the overall Grand Champion and the runner up Reserve Grand Champion. As part of this process, it is not unusual for the judge to send the white ribbon hogs, followed by the red ribbon hogs into a corral so that he can get a better look at the finalists.

Merle and I did not last more than about five minutes before the judge motioned us to get our hogs into the corral. "Crap, this is embarrassing!" He started out giving white ribbons, then red ribbons and finally blue ribbons. He congratulated all the contestants, then sent them and their hogs back to their respective pens. "What the heck is going on?" I thought.

Then I heard the judge say, "Now boys, bring out those two massive beasts. Ladies and gentlemen, I put these two hogs in the corral early because I was afraid that they were going to eat the other pigs." This statement drew a huge laugh from the crowd, embarrassing me even further. The judge went on to say, "I am absolutely amazed, because these two are the same age as the others, the biggest of which was a little over 220 pounds. These hogs weigh

270+ pounds and they are perfect examples of a great market hog. They are not purebreds either. "Son," he said to me, "What did you feed them?" I replied, "Just ground oats and milk."

Another burst of laughter from the crowd. "How can this get any worse?" I thought. He then looked at my hog very closely and said, "What happened to his tail?" His tail was just a stub instead of the usually curly tail. I replied, "His brother here bit it off when they were little." Now the crowd is really getting into it, and I am starting to sense something positive. The judge looked at me and said, "The brother here must have known it would come down to this, so he enhanced his own fortunes by biting off his brother's tail. Here you are son." After handing me this massive purple ribbon that said "Reserve Grand Champion," he turned to Merle and said, "Congratulations, this hog is the 1958 Grand Champion!" *4H can be pretty lucrative from a business perspective since these two hogs brought me a premium price that ended up paying for a full year of college. Of course, college then was only $240 a year.*

Every spring, my parents would purchase baby chickens that we would place in a small brooding coop with heat lamps to keep them warm. These were actually replacement chickens for our flock of leghorns. One year, the purchase included a couple of domestic ducks, which are not good for anything other than entertainment and grasshopper control (they had ravenous appetites and would spend all day chasing down grasshoppers and crickets).

These ducks would wander all over the farm and especially enjoyed swimming in the river. I decided that I would enter them in the poultry competition for 4H. I played with them all summer, but never fed them anything. They just lived on bugs and whatever else they could scrounge up around the farm. I brought them to 4H Achievement Days, entering them in the poultry contest. Not too much competition in the domestic duck class, just me and one other kid. He had obviously fed his ducks well because they were fat and waddled around like a fat duck will do.

Mine on the other hand were beautifully feathered, but probably did not weigh even half what his did. Oh, by the way, domestic ducks are supposed to be raised for human consumption, so fat is good and feathers are just for fluff. For whatever reason, the judge never did weigh the ducks, he just looked at them, was impressed by how tame and wonderfully feathered mine were, (they both strutted around and spread their wings to show off). He gave me blue ribbons, and the other kid red. I swear that my two ducks both smiled at me when I placed the blue ribbons on their cage. *We never did eat them; they eventually died of old age.*

***

HORSE RIDING ADVENTURES: During my grade school years, horses became very popular. All my friends had horses. My dad had two beautiful Belgian workhorses, "Boots" and "Sue," but no riding horses. He referred to riding horses as "broncos," and to him, they were useless animals that would need to be fed. Just because I did not have my own horse, did not mean that I did not go riding with my buddies. The problem is that when you do not have your own horse, you get stuck with the nastiest animals available.

My favorite was Fugie's older brother Harlan's very spirited stallion. He was one of the faster horses in the neighborhood, but of course he had a mind of his own, so I had virtually no control over him. I distinctly remember one instance, in which I had some trouble getting him saddled up and ready to go. This left us a good distance behind the other riders who were heading off into a wooded area along the river. Harlan gave me the advice to just let him go, and that he would follow the other horses and take the shortest route to catch up.

OK. Big mistake. He took off like he was coming out of the starting gate at the Kentucky Derby with me hanging on to the saddle horn and whimpering like a little girl. His shortest route to catch up included through every low hanging tree branch; he jumped over whatever object was in his path and through every mud puddle until we caught up. This cowboy lost his hat (never did find it either), ripped a hole in my precious blue jeans and was left with some rather impressive scratches on various parts of my body.

Another horse of note was Abner Knutson's bit reigning nag (most decent riding horses react to neck reigning). My friend Merle had this beautiful quarter horse and I borrowed the nag. So, what do you think might be the dumbest thing that a 12-year-old kid could decide to do with this pitiful excuse for a horse? Oh, I know—let's play cowboys and Indians and go for a night gallop down a dirt road. And let's do it bareback (no saddle) just like the Indians used to do.

I swear, I remember this like it was yesterday. I was at a full gallop (fortunately that was not very fast on this piece of crap) when suddenly a jack rabbit leapt out of a bush alongside the road. The horse disappeared from under me, moving sideways like an NFL running back avoiding a tackler, while I continued straight ahead—now horseless. I managed to tuck my head and roll as I hit the dirt at probably 20+ mph. Then, (this is not supposed to ever happen), the horse came back to his original track and put a hoof in the middle of my back. Consequence: one ruined white t-shirt with a hoof print on it, and some more injured pride.

Now for one final horse memory, and my last 4H Achievement Day. I had not planned on participating in this event. About a week prior, I got a call from an older boy, Arlo Thykeson. Arlo was the owner of a beautiful Palomino gelding that he had planned to enter in the show and riding competition. He had a conflict so he asked me if I would show the horse for him. I said, "No thanks because I am a terrible cowboy and will only embarrass the horse to say nothing of myself." He laughed and said, "Don't worry about that, just do it and have a good time." Then he added, "As you know, I have probably the best-looking horse in the county, so he is a bit of a chick magnet too." Hmm! I wonder what that means? "OK I will do it." Now, in addition to the horse being impressive, Arlo had the most beautiful saddle and bridle too.

The contest was quite simple. We walked the horses around the arena (really just a field near the grain elevators in Finley), and lined them up facing the judge. She then had us dismount and mount back up. The next step was to back the horse up. As I mentioned, this was a great horse with one big flaw—he would not back up when whoever was riding attempted this maneuver. Instead, he would rear up on his hind legs like "HI HO Silver" and the Lone Ranger. Divine intervention. A bolt of lightning and loud thunderclap ended the competition. The lady judge said, "OK, we need to get you folks out of this storm. The good news is that I have seen enough to award the Grand Championship to this beautiful Palomino and his rider, who by the way performed a perfect mount and dismount."

Apparently, being a sissy cowboy paid off since I had done the mount and dismount exactly by the book, versus the normal manly jump on and off. As the other competitors stared at me, I was embarrassed to accept the award, turned the horse toward the barn and hoped that I would not fall off on the way back. I later gave the ribbon to Arlo. He laughed heartily and said, "Keep it kid, and thanks for showing the horse for me. How did the chick magnet work for you?" Blushing, I answered, "Great, I guess."

***

BULL RIDING: Fugie had some cousins who lived on a ranch in Wyoming. Pretty much every summer he would go out there and participate in their roundup and branding of their cattle. They were real cowboys, even participating in rodeos, and Fugie got pretty darn good at being one too. So much so that when he got back, we decided that he would show me how to ride a bull. What could possibly go wrong? OK, so we needed a bull. Well, we had a 2,000 pounder, but it would be tough to get up on him (he might also kill us) so we opted for a smaller one.

When one of our milk cows had a calf, instead of that calf nursing from its mother, we would put the calf in a pen with others like him and pail feed him milk. The pen was good-sized, but not big enough to attempt any rodeo riding. So, we got a halter and led the half-grown bull calf out into the yard. It would have been a lot smarter to lead him into a fenced-in pasture of course, but he was so docile that we decided the unfenced farmyard would work just fine. We put a rope (I think we used a milking strap) around his belly just like the rodeo riders do. As I mentioned, he was docile as can be and more interested in eating grass than paying attention to us.

Fugie got on the bull's back, wrapped his right hand into the strap and said, "I am ready, let him go!" I dropped the halter, Fugie spurred him in the ribs and said, "Yahoo! Ride 'em cowboy!" Absolutely nothing happened. He just continued to eat grass. Fugie spurred him and hollered some more. Suddenly, the bull calf raised up his head, realized that he was free of the pen and took off, bucking like a real Brahma bull. Fugie hung on for dear life, but ever so slowly the strap he was hanging onto began to slide until he was underneath the calf, dragging along through the only mud puddle in the farmyard, where he fell off. Rising out of the mud, Fugie said, "Did I make eight seconds?" By now, the calf was out of the yard heading at full gallop for places unknown. My dad walked up and very calmly said, "OK, cowboys, now let's see how good you are at lassoing, because you have the rest of the day to track down that calf and get him back into the pen." As I recall, it took us most of the day to do just that.

***

DANGEROUS EVOLUTION: We had a corral in our yard that was used to trans-

34

fer cattle from the barn area to the feedlot. It was also used to load cattle into trucks for a final ride to market. It had a special chute area that we used for the local veterinarian to give shots and to de-horn and castrate the year-old beef bulls. This was necessary to keep the Hereford heifers and cows bred by only one purebred Hereford bull. De-horning and neutering the bull calves also kept them from fighting and injuring each other. At the end of the chute, there was a gate made from iron pipe. It was designed so that the animal would attempt to get through it. Once his head was in the gate, I would slam the pipes together and drop a pin in a slot to hold the animal in the gate. Once the veterinarian had completed his procedures, I would pop the pin, the animal would back up a bit to free himself, I would open the gate all the way and he would head out of the corral into the summer-river pasture.

Normally, this worked great until we tried it on a half-Hereford, half-Holstein yearling bull. This fellow was much larger than the others and was already mature enough to be considered a no-nonsense bull. They prodded him along to get him to the pipe gate. Suddenly he charged the gate. Bam! He hit that gate with me on it and ripped the entire gate loose. I was still hanging on to the gate, getting a wild ride. Finally, I jumped off, ran, and climbed over the fence to watch as he shook off the gate. He now began to sniff the air and uses his front hoofs to paw dirt just like a full-grown bull.

At this point, he started heading toward a group of highly prized, pure bred, Hereford heifers. The next thing I knew, my dad showed up with his WWI Springfield 30-06 rifle, and quickly ended the potential love life of this maverick animal. *As a farm kid, you get exposed at an early age to the sometimes violent and seemingly merciless actions required in managing a successful farming operation.*

***

SKI JUMPING: I mentioned earlier that my dad was a ski jumper. With that background in mind, let's fast-forward to sometime just after Christmas circa 1956. My buddy Fugie got a brand-new pair of skis with binders and all that fancy stuff. I had a pair of old cross-country skis with no binders. So, we cut up an old rubber tire inner tube and used those strips to slide around my boots.

Dad made the little jump out of hay bales and then instructed us to prepare the landing area. "You have to pack the landing area really well to make it safe." Oh yeah—who at the age of 12 has the patience to pack a landing area? So, Fugie goes first. Wow, he really soared through the air and then he landed, ski tips first, into the woefully inadequately packed snow. Both skis stayed exactly where he landed and he continued nose first down the slope. Luckily, this is the same guy who would eventually be inducted into the Mayville State University Football Hall of Fame, so he survived with probably the first of a half dozen or so broken noses.

***

CIGARETTE SALESMAN: Growing up in the 1950s meant that most of our adult role models were smokers. This included my dad. When I was about 8 or 9 years old, he quit smoking. I mean he just quit, "cold turkey" as he described it. Up to that point, he had been smoking a brand of cigarettes called Raleighs. He ordered them by mail, getting several cartons delivered at a time. He kept these in a small desk, opening a new carton when he needed to. When he quit, he still had a desk full of cigarettes.

Of course, I took note of this. I overheard some older kids at the church hall asking each other if they had any cigarettes. "Aha," I thought, believing that I had a solution to their problem. Imagine an eight-year-old kid showing up at the next get together with not a pack, but rather a carton of cigarettes. I then sold them for a huge profit, at 25 cents a carton. It did not take long before one of these kids got caught smoking. An angry father then inquired, "Where did you get these cigarettes?" "I bought them from Gil Rud," muttered the kid. Of course, that soon got back to my dad. Much later in life, he laughed about it, saying that he was glad that I was selling them rather than smoking them, at least at that point.

***

FIELD OF DREAMS: Our social life, especially pre-high school, consisted mostly of church events and 4H. As we got a bit older, Fugie, Dennis Braaten, Dale Evanson, my cousin Donnie Leland, Tom Sparrow, Johnny and Lloyd Jemtrud all got into cars. I enjoyed driving cars but was not into working on them. Vern Thompson (who was a year behind me in school) and I were much more into sports, especially baseball.

My dad constructed a ballpark of sorts in the valley pasture below our farm. He used telephone poles and chicken wire fencing to construct a backstop. We had plenty of room for a full-sized baseball field although it was pock marked with gopher holes and of course, since it was still an active pasture, cow pies (piles of cow shit).

My dad would throw batting practice to me using three baseballs. He would stand quite close to get some speed on the ball. When I hit them, I needed to keep track of where they went, because I would then need to run out and bring them back. If Vernon were available, then either he or I would act as a fielder. This gave the two of us a big jump on the other kids, as we both became starters on the high school team as freshmen. Vern and I also pitched batting practice to each other. Vern got so used to batting against left-handed pitching (me) that he ended up absolutely owning the best pitcher in our league, Donnie Halvorson from Hillsboro, who was a lefty and good enough to eventually play baseball for the University of North Dakota.

I also had a pitching mound set up in the farmyard with an old mattress set against one of the outbuildings as a target. My dad measured off 60 feet 6 inches from the mattress to the mound, which was the legal distance for a real baseball diamond. One day when I was in 7th grade, Johnny Nelson, an insurance salesman and athletic booster for the town of Portland stopped by the farm to talk to my dad. I was throwing balls against the mattress. Johnny came over and said, "We have a Little League team in Portland and we practice on Tuesday evenings. Why don't you come to practice next Tuesday?" I did and was surprised to find that they were using 45 feet versus 60 feet 6 inches. Needless to say, I was quite effective.

***

YMCA CAMP: When I was eight my dad's sister, Rosie, enrolled me and my friend, Dale Evanson, who was a year older, in a week-long YMCA Camp. The camp was located at Cormorant Lake in Minnesota, about 100 miles from the farm. The camp was for 8–13 year olds, so it was quite an adventure for this little farm boy. Before attending the camp, Aunt Rosie taught me how to swim in the river by the farm. As I recall the swimming lesson consisted of demonstrating how to "dog paddle," and then tossing me in the river. It was a very effective teaching technique, unfortunately, I am sorry to say that the dog paddle is still my best swimming stroke.

The camp was my first time away from my parents for any length of time and turned out to be a great confidence builder. I admit that I was mortified when they drove away at the start of that week, but eventually I settled in and had a blast. Rosie was an amazing lady. She was college-educated and a veteran of two years teaching school in Nome, Alaska, in the 1930s. No way to get there except by airplane or dog sled. She had a huge influence on me as to what I might accomplish in life.

***

INTERACTION WITH FAMILY: Two of my dad's three sisters, Rosie and Christine, graduated from college. Aggie, being the oldest and the sweetest of my aunts and uncles, helped raise the younger children so she was not afforded the opportunity to go to college. She was just so kind and always looking to help the family. For her sacrifice, she inherited 320 acres of land that I eventually purchased from her after my return from Vietnam in 1972.

Also, part of the family was my dad's cousin Peter Lee. Peter's mom died when he was a baby, so he was raised by my grandmother and considered to be a son and a brother to the rest of the family. Peter also graduated from college with a degree in electrical engineering. He and his wife Audrey had a huge family. Their oldest children, Patty and Barbara, were fraternal twins, a couple of years younger than me. They would pair-up with another cousin, Rachael, to provide this farm boy some city slicker victims for various farm boy ploys, including, "Please hold this fence down for me while I cross it." *Bzzzt!* "Oh yeah (snicker), it is electric!" Despite my unappreciated antics, we did have a lot of fun together.

***

FAMILY FISHING TRIP AND A CLOSE CALL WITH MOTHER NATURE: Dad took a fishing trip every summer to Lake of the Woods with his local farmer friends. When I was about 10 years old, he took the family to Baudette, Minnesota, which is located on the Rainy River next to Lake of the Woods. It absolutely ruined me for fishing because it was what is best described as "arching rod country." In other words, you tossed a line in the water and bang, you got a bite, and your rod would arch toward the water—and we are talking good-sized walleye and northern pike.

Lake of the Woods is quite shallow so storms can quickly turn calm water into nasty/dangerous waves. We had rented a small boat with a little five-horsepower motor to fish from. That little engine was perfect for trolling, but not so good to get out of the path of a fast-moving thunderstorm. Whether an old wives tale or not, the fishing really is awesome right before a storm hits. We waited a bit too long to head for shore and got caught in a nasty thunderstorm with probably 40+ mph winds and driving rain.

Since there was no way for us to make it back to the Rainy River and the safety of the resort, my dad headed for a nearby island. He could not see more than a few feet, so he had me up in the bow looking for the shoreline. Not being smart enough to be afraid, I thought it was an awesome adventure as I guided him to a sandy spot where he drove the boat ashore. Shortly thereafter, the resort owner showed up in a large powerboat to rescue us. He chewed my dad out for not avoiding the thunderstorm and said that he was fully prepared to hunt for bodies instead of survivors. That got my attention.

***

FIRST KISS: At the age of 12, my baseball playing buddy, Vern Thompson and I attended the Red Willow Bible Camp. As I recall, there were four large cabins for boys and the same number for the girls on the other side of the camp. The counselors told us to pick a cabin and set up our bunks. The cabins were named Matthew, Mark, Luke and John (good Lutheran heroes all). John was the deepest in the woods, so Vern and I headed for John. Weird, nobody else showed up in that cabin. It was a couple of days before the counselors found that just the two of us had this whole cabin. They finally assigned us to one of the others and poor John remained uninhabited.

A Bible camp it might have been, but it was really my first exposure to girls that I had not grown up with. My second Cousin Arlene, who was, and still is, a great friend, lined me up with this amazing girl, Marjorie Odegaard. She was intimidatingly beautiful, but like the words of the country song, "She don't know she's beautiful." She was also sweet and even a bit shy.

Now the big deal at the camp was to complete a swim across Red Willow Lake. The lake was a half of a mile across, so it was no small feat, especially when your primary swimming stroke is the dog paddle. I have no idea what came over me as we watched some older boys lining up for the swim, but I turned to Margie and said, "If I make this swim, will you give me a kiss?" She smiled that wonderful smile and said, "Yes I will—and it will be a good one!"

I was so excited I almost fainted. And then, reality sank in. What if I don't make it and die before I get my first kiss? Fortunately, they had a boat that took us to the other side of the lake where we started the swim. That same boat then monitored our progress and picked up the folks who were not able to make it. Well, they asked me several times if I wanted to be picked up, but I steadfastly refused. Why were they asking me? Because everyone else had already finished the swim, and little old "in love" Gil was still dog paddling the last few yards to the beach. I came out of the water coughing and sputtering, exhausted. And there she was, my angel. "Oh, my goodness," I thought, "Am I dead? Is this heaven?" No, but that sweet kiss is something I will never forget as long as I live. She was from a different town, so although I saw her a couple of times, she had a boyfriend, and I was way too shy to pursue a relationship. *Thanks Margie wherever you are.*

***

FARMING ADVENTURES: Farm work was divided into basically three main events. Spring planting, summer haying and fall harvest. We always had about 40 acres of alfalfa, which is a perennial type of crop that makes for a nutrient-rich food for livestock. The growing season in North Dakota allowed for two crops to be harvested each summer. The process involved cutting the crop, raking it into rows, and using a baler to produce individual hay bales that could then be stored and fed to the cattle over the winter months.

The Case baler that we used was tractor-towed by my dad and powered by a Wisconsin air-cooled engine. It had a wire tie system that required two folks sitting one on each side of the baler to manually tie the bales. This was normally the job of my aunt Rosie and aunt Aggie. These bales were bound very securely, and unfortunately were also very heavy, weighing over 100 pounds each.

Once the hay was made into bales it was then my job to drive the 1940 International truck with a bale grabber to pick up the bales. This device was attached to the left side of the truck. It had a trigger that initiated the loading sequence. This required me to accurately strike the bale. The claw would then grab the bale and hoist it alongside the box to the point where our hired man, Bennie Lee, would snatch the bale with a hook and stack it in the truck box. We would continue loading until he filled the truck box to as much as seven bales above the box confines. *As with most farm operations, this one was fraught with risk, required precision, and was another precursor for me to eventually fly formation and land on aircraft carriers.*

Starting in August, we began the harvest phase of the small grain crops. I always thought that we had plenty of money. I was never hungry, and I had what I considered to be nice clothes to wear. This was reinforced when in 1949 my dad purchased the very first self-propelled combine in our neighborhood. Having grown up and successfully farmed through the great depression, he paid cash for the combine. In fact, he never bought any piece of machinery that he did not pay cash for. He refused to be in debt for any reason. Of course, his method of not ever purchasing anything on credit meant that we made the most of older machinery, which you fixed yourself.

The combines were an exception. In addition to the first self-propelled John Deere model 55 in 1949, he also purchased the first John Deere Model 95 in 1957, and finally a revised model of the 95 in 1962. He specifically purchased this latest model because it was not as tall and therefore fit into the machine shed on the farm. We would have to let air out of the tires on the earlier model to get it through the doors of the shed. These combines were his pride and joy. I operated everything on the farm, except for the combine. He reserved that responsibility for himself.

Following the purchase of the last combine, my dad was on the first round of a 40-acre field of oats just to the north of our farmstead. Normally, we would off-load the grain directly from the hopper of the combine to a truck while on the run. In this case, we were very close to the trees, so he kept the off-load auger in the folded position. When the hopper was full, he repositioned the combine so that I could pull the truck alongside. The next step was to unfold the auger so that it would off-load into the truck.

Dad stood on the access ladder to the combine while I stood in the truck box. The auger was manually lowered using a large nylon line with a pulley. The nylon line had a wooden handle on it for my dad to grip. The combine included a large flywheel with belts connected to various moving parts while the combine was in operation. This was still running while he lowered the auger. After safely lowering the auger, my dad was retrieving the nylon line when the wooden handle caught in the flywheel. During the retrieval, the nylon rope had looped around his thumb so when the rope caught it literally amputated his thumb. His thumb landed on the combine ladder. He looked at the severed thumb and said, "Well now, that came off quick and easy."

I picked up the thumb, gave him my shirt to put over the stub, and we headed for the farm. Harvest is a family affair, so my mother was near the grain elevator waiting for me to bring the first load home. My dad said to my mother, "I cut my thumb pretty good, so Gil is going to take me in for a few stitches. We will be back in a couple of hours." Then he had me stop by the house and get a mason jar and some formaldehyde to put the thumb into in case they wanted to re-attach it.

We traded the truck for the car and headed for the Mayville Hospital, which was about 15 miles away. He was getting pretty pale, so I was hauling ass down the highway and laid on the horn as we went through Portland at a high rate of speed on the way to the hospital. He said, "Hmm, it appears that you have some experience driving these speeds." (I was 17 when this happened). We arrived at the hospital, and I ran in ahead of him announcing to the first nurse I saw, "My dad cut off his thumb in a farm accident!" She said, "You mean your dad cut his thumb?" I quickly produced the jar with the thumb in it, and she said, "Oh my goodness, he certainly did!"

Dr. Vandergon, our local family physician looked at the situation and said, "Well Ted, I can send you to Fargo and they can attempt to re-attach the thumb, or I can just stitch up the stub." My Dad said, "We just started harvest, so I don't have time to go to Fargo, just stitch it up." This was one tough dude! We drove back to the farm, jumped in the truck and drove back to the combine. He missed a total of about three hours of work. That evening, my mother asked him how many stitches he got. "Oh, quite a few." I thought about this incident many times as I went through various challenges throughout my life. *"Heck, just stitch it up and get on with the job."*

We had another close call during harvest. We had a 16-foot-tractor-pulled swather that was used to cut the grain and place it in a row for later pickup by the combine. We used to put it lengthwise on a trailer to move across the bridge over the Middle Fork of the Goose River. Loading and unloading of this machine was a nasty affair.

My dad was unhitching the trailer from our 1940 Dodge pickup when the swather somehow pinned him between the trailer and the pickup. I saw what was happening and jumped in the pickup, pushed in the clutch (it was in first gear) and my dad manually pushed the pickup away from himself. He was one powerful Norwegian. Being 11 years old at the time, I became very emotional at almost losing my dad. "Are you OK?" I stammered. He said, "Oh yes, I am fine. Thanks for your quick-thinking son, I was having a tough time pushing that pickup with it locked in first gear."

The final phase of harvest involved corn silage cutting and hauling. We partnered with a neighbor, Orion Klath, for this phase of harvest. Orion's father, Albert, passed away when Orion was only 18 years old, so Orion successfully took over the operation with close advice from my dad and the help of a WWII veteran, Lloyd Hanson, who acted as his hired man.

Corn can be harvested three different ways. You can combine it and end up with kernels, you can harvest it and get the cobs, or you can make use of the entire corn plant by cutting it for silage. We normally used the silage option. We shared ownership of a single row International Harvester corn chopper. It is hard to imagine now, but that silage cutter only took one row at a time. We used two trucks, alternating to keep the chopper going all day. We drove the trucks within a few inches of the wheel of the old Minneapolis Moline Diesel U tractor that my dad used to pull the silage chopper. The chopper utilized the power takeoff of the tractor to run it.

The trucks we used included a 1940 International 1.5 ton that dad bought used from the Farmers Union Oil Company. Although it had lots of miles on it, his theory was that their mechanics took very good care of their trucks. Orion had an awesome International truck, which he bought new in 1948. It was a beautiful machine. Two tons with a 5-speed transmission, which included a high/low range for each gear or essentially a 10-speed transmission. The steering wheel was enormous and I had to sit on a pillow to reach the various controls, so I normally drove our truck. Both trucks were fitted with a large wooden sideboard that rose well above the normal height of the truck box. This sideboard was on the right side of the box and was used to deflect the blown silage into the box. *As the box began to fill up, my dad would motion us to move forward or backward on our position alongside the machine to level out the load. It was a great precursor to understanding and properly following signals that would later be given to me by taxi directors and Landing Signal Officers when landing aboard aircraft carriers.*

The fields were normally a half-mile in length or a little less in some cases. With one truck, we were able to fill up it up with a total of one mile of cutting.

This meant that after finishing up a half-mile row, we would follow the cutter until it started on a new row in the opposite direction. At that point we would get the truck back into position as fast as possible, so as not to waste any silage. *This challenging action helped prepare me for aircraft rendezvous and other relative motion skills like underway replenishment of ships that I relied on in the Navy.*

On completion of the mile of cutting, the full truck would be driven to the farm for off-loading. This was normally a short distance, usually less than a mile. Upon arrival at the farm, the silage was dumped into a trench silo, blown into a vertical silo, or stacked into a well-packed pile. If it was a pile, then a neighbor, Otto Thykeson, would use his D-6 Caterpillar bulldozer to blade the silage and pack it. As the pile got bigger, it would involve backing the truck at high speed, as far up the stack as you could get, hoisting the truck box to near vertical, then going back down the stack with the silage spilling out of the box. It was great sport for a 12-year-old farm kid.

***

COUNTRY BASKETBALL TEAM: Edgar Overbe was teaching school in Enger Township, which borders Primrose to the immediate north. He was pretty old at this time but had been a basketball coach. I believe we were in 7th or maybe 8th grade at the time, and he decided that we country kids should have a basketball team so that we would be better prepared to play high school ball. Of course, we had no gym, but we did have barns with large haylofts for feeding milk cows. By this time, many of the milk cows were sold off, so we had two barns, Fugleberg's and Lyle Erickson's converted into basketball courts.

The problem was the incredible amount of dust that accumulated over years of storing hay for the cattle. When we finished a practice session, we all looked like we had been working in a West Virginia coal mine. Coughing and spitting like crazy. Edgar finally got us a game against Hatton in their gym. He got us some Farmers Union uniforms that were designed for much older kids. They were red and they were made from some extremely uncomfortable material, which I believe may have been wool. The pants had to be cinched up at the waist and the tops were so big that the bottoms of the under-arm straps were about waist high. Of course, they laughed at us.

Hatton had a really good team with kids who later became high school stars. Since they were much bigger than our motley crew, they built a substantial lead by half time. So, they sent all their 8th graders to the locker room and started the second half with just 7th graders. Well, by pressing on defense, bounce passing and shooting free throws underhanded, we caught up to them and they had to re-dress their 8th graders for the last quarter in order not to lose. Some of our guys went on to play for Hatton High School, with most of us playing for Portland High School. That day was in some ways the start of a rather heated rivalry. *By the way, as a 50% free throw shooter in high school, I should have stayed with the underhand method.*

***

I Hope He Makes it through the Night: In the spring of 1957, I was a 12-year-old 7th grader. My dad bought a brand-new 1957 John Deere 70 diesel tractor, and I was determined to be the first to take it into the field. It was still early April and I was just getting over the flu bug. The weather was cold, windy and wet as I took the new machine out to dig up a few acres of old corn ground. Of course, I froze my ass off, but I was too stubborn to admit defeat, so I finished the field before heading for home.

I am sure that I already had the pneumonia bug, but with this foolish exposure to the elements, it really took hold. I woke up that night unable to breathe properly. I was a skinny little shit, so I literally was putting my fingers under my rib cage and pulling on both sides to breathe. My dad, who never showed any sign of panic or alarm over anything, wrapped me in a blanket and carried me to the car; he and my mother drove me to the hospital in Mayville.

Events are a bit fuzzy here, but I do remember them putting me on oxygen. By this time, I was running a very high fever, which I think was over 104 degrees. I had the feeling that I was drowning. Since I was allergic to Penicillin, I was in deep shit. They used another drug, which was not as effective. They must have also given me something to sleep, because I was drifting in and out of consciousness when I overheard one of the nurses say, "If he makes it through tonight and we break the fever, he has a chance to survive."

"Aww Geez," I thought. I am really in trouble now. If I die, I am sure to go to hell, because according to my Lutheran upbringing, as a typical 12-year-old boy, I had broken at least five of the 10 commandments, so I was screwed. Not only that, but my parents would find my magazines with photos of Marilyn Monroe and Jane Mansfield that I had under my bed. I remember wishing

I were Catholic so I could confess all this evil and have a shot to at least get sent to purgatory, which I assumed was a better option than Hell. *As you may have discerned from my thought process during this crisis, my knowledge of Christianity and the after-life options was not all that impressive.*

Obviously, I made it through the night with my fever breaking the next day. I was in what I would describe as an intensive care room with an elderly gentleman from Hatton, by the name of Mr. Beck. He was attempting to recover from a massive heart attack, while I struggled back from life-threatening double pneumonia. The good news was that we were both baseball fans, so we had some great discussions regarding that topic. Also, he had a huge family so there was always somebody visiting.

One day, we decided that we would plug in his radio so we could listen to some music. Since neither of us was supposed to get out of bed at this point, I decided to reach over and plug in his radio. In the process of doing this I accidently hit the "Holy shit" (STAT) alarm on his bed, which resulted in nurses and doctors arriving with all sorts of nasty-looking stuff to save him from what they presumed to be another heart attack. He was laughing so hard that he almost did have a heart attack.

After about four days, I felt fine with no fever and no breathing difficulties. Unfortunately, to be safe, I had to remain on bed rest for another six days. I was not supposed to get up for any reason, even to go to the bathroom. So, I was supplied with a bedpan. I did not consider this an option, so of course, I got up and regularly went to the bathroom located right there in our shared room. When the nurses would ask me if I had used the bedpan, I would say; "No ma'am, I don't have to go yet." Pretty soon they showed up with an enema machine. These crafty and wonderful nurses (Mrs. Elwood Brend, Bonnie Nelson and my aunt Francis Lee) knew what I was doing of course but delighted in the enema ruse to get my attention. *I guess that I came pretty close to dying that first night, and I will have to admit that there was no bright light or beautiful singing angels, or any other sign of a positive after-life so I was very relieved to wake up the next morning.*

3. The Bang Bombers Go to Town

SEPTEMBER 1958: Fugie and I headed into town for the first day of high school in his 1932 Model A Ford. Arriving on the outskirts of the huge metropolis of Portland, North Dakota, population 641, we started down the small hill that led to downtown. Fugie said, "I wonder if all the city people are up yet? If not, they soon will be." He then retarded the spark on that old Model A, and it backfired its way into town. That announced the arrival of what folks would eventually refer to as the "Bang Bombers," aka Richard Fugleberg and Gil Rud. The "Bang Bombers" was the title of a church affiliated tumbling team that we had been members of.

Now we should have been better prepared, but our older siblings had already graduated. So, we showed up with lunch buckets. Why was this a big deal? Because lunch was served to everyone in the school cafeteria, so we looked like what we were, a couple of hick farm boys.

It was bad enough being freshmen, but this arrival drew the attention of the older townies. They rolled in hot, with various insults until the Otteson boys showed up. These two farm boys from a neighboring parish just to the south of Bang, called Perry, came to our aid with the pronouncement, "Richard and Gil are our friends and anybody who messes with them will answer to us!" Carl and Selmer did not play sports due to farming duties, but they were two nasty, mean fighters that not even the football players would dare to mess with.

By the way, I mentioned that the population of Portland was 641; as I recall that consisted of 636 Norwegian Lutherans and five Irish Catholics. The Irish Catholic family was the McLeods. Mr. McLeod was the Northern Pacific railroad depot agent. One of his boys, Glen, was a year older than me. His name will come up often, as he was a close friend and a gifted athlete.

The high school consisted of a basement area where all the student lockers were located. The boys and girls bathrooms/shower areas and associated sports locker rooms were also located there. Since there were no girls' sports in those days, their locker room hosted the visiting team's boys for basketball games.

The main floor consisted of the basketball gym and a stage for the band, choir and plays to be held. The gym also had two tiers of seats on one side for fans to watch the games. A row of chairs was placed along the opposite side of the gym, right next to the out-of-bounds line on the basketball court. The basketball court had an awesome wood floor that was state-of-the-art for 1958.

The second level consisted of individual classrooms. The third level was dominated by a large assembly area, which could hold all the roughly 100, 9–12th grade students. We were each assigned a desk in the assembly/study hall. These were assigned by class and alphabetically within the class. Seated directly behind me was a townie by the name of Robert Seaver. We eventually became great friends and teammates, but we had a rough start. He was much bigger than me. I was a little squirt at about 5 feet 5 inches and 110 pounds I had not even changed voice yet. So, he began to pick on me. Super irritating. I guess this was the portend of future behavior from this hot-tempered Norwegian. In any case, I challenged him to a fistfight behind Art Hagas' pool hall in downtown Portland. Fortunately, just as we squared off to go at it, one of the older boys stepped in and stopped it (probably kept me from getting my butt kicked). The good news was that he never bothered me again, and I seemed to gain some respect, despite my small size.

As if starting high school was not traumatic enough, the expectation was that if you did not suffer from a debilitating disease, or were restricted by farming duties, you were expected to play or at least dress for football. With a total of approximately 50 boys, we still played 11-man football against schools, most of which had at least twice as many students to pick from. Despite the relatively small enrollment, the Portland Pirates were very competitive at all sports.

As a freshman, I only played in one game. I can't remember whom we were playing, but we were leading by something like 44-0 when my name was called to go in as a 5-foot-5-inch, 110-pound defensive tackle. I was not the least bit thrilled to get into the game; rather, I was petrified. The ball was hiked, the quarterback began to hand it off when he was hit by one of our blitzing linebackers and fumbled. By this time, I had been literally pulverized by some

200-plus-pound offensive lineman, when suddenly the fumbled football hit me on the helmet. I grabbed it, and I was immediately piled on by what seemed like everybody on the football field. I survived and mercifully never played another down during my freshman year.

***

BASKETBALL: Next came basketball season. I need to provide some background on Class B basketball in North Dakota 1958–62. This was the number one form of entertainment, and it was amazing to see how many people came to games. The games were normally on Tuesday and Friday nights. If a fan wanted to watch a game, they needed to be in their seat before the start of the junior varsity (B Squad) game, which was normally 6:30 p.m. The B squad consisted of 9th and 10th grade players only. A few of the 10th graders were also members of the varsity, but no 11th or 12th graders were allowed on the B Squad. It was a "up or out" system, in that if you did not make the varsity as an 11th grader, your basketball days were over.

With the popularity of the sport, it was a lot of fun as a 9th grader to play before sold-out and raucous crowds. And there were great rivalries between the schools, especially for us country kids. Our farm was on the border of three school districts, so I had friends/enemies that I had grown up and gone to church with, who now attended either Finley or Hatton high schools, versus being a Portland Pirate. We were also located only 2 miles from Mayville, which had a high school roughly twice the size of ours.

As a sophomore, I played mostly "B" Squad, but dressed for varsity and finally got into a game against Finley. I was so excited to play that I managed to foul out in just one quarter (8 minutes) of play. Since that was the only time I played, coach Pederson (who was also an outstanding high school science teacher and mentor for our class) started out the next year showing statistics for the past year. I had a 20-foul-per-game average. Just to remind me, he said, "Rud, this is basketball. Football is over."

During my sophomore year, we had a very good varsity basketball team led by Doug Eiken. We had three players who could dunk the basketball, but that was considered to be showing off, so they never did it during warm-ups or in games. That is until an away, out of conference game in a town called Enderlin, which was 100 miles from Portland. Coach Pederson decided that this was an opportunity to intimidate the opposing team, so he had Doug Eiken, Dan Anderson and Glen McLeod all dunk the ball during warm-ups. It worked, as I recall us charging out to a huge lead and winning big. We should have gone to the state tournament that year; unfortunately, we were upset in the first round of the district playoffs when our star player, Eiken, got injured.

Although I played very little, my junior year was very memorable. We had a 6-foot-7-inch center, Dan Anderson. He was not only tall, but also a deadly shooter. Complementing Dan was Glen McLeod, who at 6 feet 1 inch could dunk the basketball and was both a great ball handler and scorer. Ole Stavedahl filled out the front line at 6 feet 2 inches and was a great rebounder and defender. Rich Strand and Bob Heskin were the starting guards and Fugie also played a lot. I was the backup center. This meant that I faced-off in practice with big Dan, who eventually grew to 6 feet 10 inches, starred at Augsburg College in Minneapolis, and even played a few years as a professional in the American Basketball Association. As a junior, I was 5 feet 10 inches tall so the only way I could score against Dan was to shoot a hook shot. As a senior, I grew to 6 feet, and using that hook shot I was able to average over 10 points per game against centers much taller than me.

During my junior year, that 1961 team was the best in the history of Portland High School. I believe that we only lost two games during the regular season, and one of those was to our coach's alma mater, Hatton High School. Of course, his buddies from Hatton rubbed it in, big time. When it came time for a rematch to be played in Portland, it was the only time that I ever heard our beloved coach say, "Boys, no mercy tonight. We get ahead, we continue to pour it on!" We did, winning by over 30 points. We also had a very formidable zone defense that frustrated smaller/faster teams.

The Class B Tournament format consists of a district, regional and finally state tournament. We absolutely roared through the district tournament and easily won our first-round regional game against Sargent Central. Next was Ellendale. They matched up well against us and they employed a strategy of double-teaming our big man, Dan Anderson. It was effective, however, Glen McLeod stepped up and scored 30 points, and many of those shots came from what would be considered three-point range these days.

Next, we played Edgeley, easily beating them 89-68. By the way, these games were played in Fargo before huge crowds. Toward the end of the championship game, a bunch of us bench players got a chance to play. The other team also sent in their reserves, so it was a wild and woolly few minutes of run and gun. I remember stealing the ball from one of their guards. There was no one between me and the basket, so it was going to be an easy layup. Just before reaching the basket, I noticed Ronnie Nelson coming down the other side of the court. Ronnie was a senior who never got to play. Instead of taking the layup, I passed the ball to him and he scored what I think was his only basket of the year. After the game, Coach Pederson, who I greatly respected, came up to me and said, "Gil, I am so proud of you for passing that ball to Ronnie versus taking the shot." *This was a lesson in teamwork and more importantly in doing the right thing that I tried to emulate throughout my life. Don't get me wrong, I did not always do the right thing, but at least I tried.*

Now it was time for the big stage: on to the first round of the state tournament in Minot. What an experience that was. Most of the town of Portland traveled the 200 miles to Minot, ND, as it was only the second time in the history of Portland High School that the basketball team had made it to the state tournament.

1961 Portland Pirates. *Left to right: Eric Strand, Jens Strand, Ron Nelson, Bob Heskin, Richard Fugleberg, Ole Stavedahl, Dan Anderson, Glen McLeod, Richard Strand, Gil Rud, Roger Erickson, Bryan Kringlie and Robert Seaver.*

The state Class B basketball tournament is the most well attended sporting event in North Dakota. It was on television as well, so you can imagine the pressure. Up to this point we had been shooting lights out, making almost half of our attempted shots. This was the last of the first-round games, and we were playing New Town, which is located close to Minot. In fact, they had played their regionals at the same venue as the state tournament. In other words, a bit of a home court advantage.

We stayed close for three quarters, but you can't shoot 19 percent and expect to win. We got beaten 63-44. Oh, by the way, I got into the game at the very end, and got fouled on the last play of the game. The clock had expired, so all the players left the court, yet I still had to shoot two free throws. Of course, I would rather have been anywhere but alone on that court in front of thousands of fans and statewide television. And yes, I missed them both. If that were not embarrassing enough, I then ran over one of our cheerleaders in my haste to get off the court. All of this caught on television. *Not my finest hour, but there would be other failures to follow, all of which prepared me for a challenging future as a naval aviator.*

We were embarrassed and dismayed at our performance, but that is not the end of the story. Now into the consolation round, we came back and beat Rock Lake and then New Rockford to earn the 5th place trophy. This was the best finish ever in the history of Portland High School, and considering the reception we got returning to Portland, you would have thought that we won it all. Great memories.

***

BASEBALL: Finally, baseball season arrives. Although I played all four sports, baseball was the only one that I was good at. Portland had probably the best pitcher in North Dakota. Doug Eiken was 6 feet 3 inches tall and about 210 pounds. He could throw the baseball at least 10 mph faster than any other high school pitcher. The problem was (as Bob Ueker described in a famous baseball movie), "he was just a bit on the wild side." In 1959 we did not have batting helmets. Therefore, facing Doug in practice was scarier then landing on an aircraft carrier at night. We all longed for the relative safety of playing in games where the other team would have to face his fastball. By the way, Doug became a college ace, got signed by the Philadelphia Phillies and made it to Double A before sustaining a career-ending injury.

Having practiced for several years in our cow pasture field, I was both a good hitter and outfielder. I was the only freshman to start on the baseball team. Since I was at the time also the shortest player with decent speed, I batted lead-off. Our first game in 1959 was against our super rival, Mayville. As I mentioned earlier, Mayville was only 2 miles from Portland, on the other side of the Goose River. It was a college town with Mayville State University located there, and it was about three times the size of Portland. They had a beautiful baseball field with the dark green wooden grandstands and real player dugouts. My dad bought me a really good glove, spikes and a bat. The problem was that I did not like the bat as well as the one I had used for years in the pasture. That pasture bat was a 36-inch Joe Cronin. Joe last played in 1945, so basically the bat was the same age as I was—14. It was about 3 inches longer and certainly heavier than a fellow my size should have been using, but I loved it. Oh yeah, it had also been broken and repaired several times by my dad who was, in addition to being a farmer, a darn good carpenter.

I walked the first two at-bats and scored both times. The third time up, the pitcher grooved one and I connected. I suppose I should embellish just a bit here and say I got a home run. Not quite. However, it was a high and very deep fly ball that the center-fielder caught at the fence. After the game, I was walking to meet my dad in the outfield parking lot. As I was placing my bat and glove into the trunk, either the opposing coach, or maybe he was the umpire, Red Soholt, a former star player, said to my dad, "Ted, I don't want to see that darn bat in the hands of your kid ever again!" My Dad said, "Sorry about that Red, I have been using it to hit him fly balls and did not realize he was going to use it in the game." Red responded, "Yeah, well there is no way in hell that a little kid like him can hit a baseball that far." Unbeknownst to me, my dad had corked the bat so he could hit me more realistic fly balls.

I was a bit of a smart ass, and always competing, so I said, "You know, dad, if you had used just a little more cork, I could have hit a home run." I then got the LOOK! This was a very cold stare that was every bit as effective as a slap or punch—which was not something my father ever needed to do. Just the LOOK! By the end of that high school season, and for the entire American Legion season, I became the cleanup hitter. I batted cleanup for all four years, which was certainly my claim to athletic fame.

***

FOOTBALL: At the beginning of the 1959 football season, I had grown to about 5 feet 8 inches and 140 pounds. I was way down the depth chart, so I opted to miss the first game to show my pigs at 4H Achievement Days. I came back to practice to find out that my cousin Jim Rud, who was a senior offensive and defensive end, had been injured. I was one of four or five kids sent to the end of the practice field to field a punt and attempt a return against the whole rest of the team. This rather bizarre situation turned out to be a tryout to replace Jim as a starting end. I caught the punt, and out of sheer fear, managed to avoid a few tacklers, returned it a couple of yards without getting killed and became the starter for that Friday night.

I was one of three sophomores that started with one junior and seven seniors forming the rest of a very good football team. My whole goal was not to screw up. We were playing Larimore, another school at least twice our size. It was their homecoming so there was a large crowd and about an hour before the game a huge thunderstorm dumped over 2 inches of rain on an already soaked football field. As the kickoff approached, I was so nervous, I nearly threw up. There was an aspect of fear, but it was not in anticipation of potential injury or getting the shit kicked out of me, because the shit-kicking part at least was inevitable. It was fear of failure, screwing up and letting my teammates down. They set up to kick off. As an end, I was back, able to potentially field a short kick off, but really to block for the returner who was behind me.

The field literally had standing water with the rest of it rapidly turning to mud. Hank Knudson, a senior running back, said to me, "If the ball comes to you, just let it go and block somebody. I will pick it up and return it." The guy kicked off and instead of the ball doing the usual end over end movement, it spun sideways and of course came right to me. Crap! Hank said, "You have to take it, just don't fumble." I caught it and immediately got clobbered by half of the kickoff team. By the time they all got off me, I initially thought that I had lost my left eye. As it turned out, my helmet had been twisted around (it was too big of course) so that I was looking out through the ear hole of my helmet with my right eye only and had a mouth full of mud. But I did not fumble, and I was ALIVE.

Now we had the best athlete in North Dakota, Doug Eiken, and in addition to being a running back, fullback and sometimes QB, he was an awesome punter. He not only kicked it a long way, he also had unusually long hang time. So, during his first punt of that game, I took off down the field to cover the punt returner. I might have been small and lacked skill, but I was reasonably fast. I remember distinctly the sensation of getting past the last blocker and having a clear shot at the returner. One problem: the ball had not arrived yet. No "fair catch" signal so kept on going. Catch — Hit; a simultaneous event that caused him to fumble and one of our guys to recover. What a feeling to be congratulated by my heroes, the older guys on the team.

Another memorable event from that year was the first play from scrimmage in a game against the best team in the conference, Hillsboro. They were known to have a good passing game, so we were prepared to put pressure on the quarterback. I attempted to do that, got knocked on my ass by their offensive tackle, started to get up, and was promptly run over by this massive, scary fullback by the name of Breen. The good news was that he was carrying the ball on a draw play that had fooled all my teammates. While he was running over me, he tripped and fell, which amounted to a tackle for a loss. I got congratulations all around, and comments like, "Hey kid, how did you sniff that one out?"

The next football season, my junior year, was a tough one. We lost seven seniors and had only two returning juniors to play their senior year. One of them, Glen McLeod, was a very good player and was going to be our QB. Unfortunately he was also an awesome basketball player, and was recovering from a back injury, so he decided to forgo football in order to be ready for basketball. I believe that he might have also looked at his offensive line, and thought, "I am going to get killed." This left us with only one senior, Bob Heskin, who was a strapping 145-pound fullback. It was a long year with only two wins, although we did manage to stay close in most games.

Near the end of the season, we had a game scheduled with a newly formed consolidated school in their first year of 11-man football. Since it was their first year, they were playing the teams in our conference, but those teams were limited to not using any seniors. They had beaten all of them, and then showed up to play us. I am sure they looked at our abysmal record and thought, "We are going to kick some butt today." What they failed to realize was that they were up against our first string, sans Bobby Heskin. We absolutely kicked their ass. After a sack, I was helping their incredulous QB off the ground, and he said, "Are you sure you are not playing any seniors?" I smiled and said, "We don't have any seniors." His reply, "Oh shit, now I understand."

***

ART HANSON'S HOME FOR COUNTRY BOYS: Due to the nasty weather, road conditions and our busy after-school sports schedules, Fugie, Dennis Braaten and I all moved into town to share one bedroom at Art Hanson's for the months of January and February. We had meal tickets to the only restaurant in town, The Strand Café. They would punch our meal cards for breakfast and dinner. We ate lunch at the high school. I still have fond memories of those huge stacks of pancakes. There were two older boys, Kenny Thykeson, and Orin Johnson, who also stayed at the Hanson home.

One night, we found this old alarm clock that had a huge bell-like affair on it. We set it for 2 a.m. and put it under those boys' beds. When we were being too noisy, old Art would poke a broom handle into the kitchen ceiling, which was right below our rooms. Well, this alarm clock went off and those two guys were cussing up a storm trying to find the thing! Art was banging away with his broom handle, and we were laughing loudly. That almost got us thrown out of our winter home.

***

HOCKEY: It was not a high school sport back in those days, however, that did not keep us from playing on a frozen pond in Portland, or at Island Park in the big city of Mayville. The one in Mayville had a fence around it. There were no real rules and certainly not any protective equipment. Checking resulted in some spectacular falls on the rink that had no boards. Whoever showed up and either did not have any, or forgot their skates became the goalie. This was normally not that dangerous, as most of us did not know how to lift the puck. However, Earl Larson figured out how to lift the puck and he fired one at the goal, which was being guarded by Rick Aamold in his overshoes. Unfortunately, it caught him right in the mouth and took out his front teeth. Ouch! The bigger (and richer) kids all had hockey skates. I had a pair of old figure skates, with serrated toes and super sharp blades, which allowed me to stop on a dime, and run on the toes. I avoided many a check with those skates.

***

FROGMAN PRACTICE TEACHER AND A MANLY CATCHER'S MITT: As I men-

56

tioned earlier, Mayville State University was located only 2 miles from our high school. Since it was primarily a teacher-oriented curriculum, we got the benefit of having their Physical Education majors' practice-teach and help coach our teams. One of those coaches was Rip Rorhagen. Rip was a former Navy Frogman (predecessor to the SEALs), and a left-handed pitcher from Fertile, Minnesota. He taught me to throw at hitters who dug in, and to warm-up throwing the ball all over the place to make them a bit apprehensive. This was no problem because I was so darn wild anyway.

My dad had a catcher's mitt, which was a "Bill Dickey" model with a hole in the center. He wore a leather work-glove under it. We always kept it in the trunk of the family car in case we had an opportunity to play catch. Earl Larson forgot his catcher's mitt at one away game, so he used my dad's. I threw relatively hard, and my control was such that he probably caught 120 pitches in a seven-inning game. His hand swelled up something awful and he wondered aloud how tough my dad was to use this darn thing as his primary catcher's mitt.

***

TRACK AND FIELD: As a four-sport letterman, I went from one sport right into the next, throughout the school year. The exception to this was the simultaneous seasons of baseball and track. For example, I pitched an entire baseball game, and then rushed over to the college track to run in the mile relay on the same day. I remember smoking a cigarette (I started smoking during the summer between my junior and senior years) on the way to the track meet. I arrived just in time to run the first 440-yard leg, handing off to 6 feet 5 inch Dave Odden, running second. Ron Holkesvig, who had just won the mile run, was the third runner, and Richard Strand, our fastest guy, ran the anchor. We not only won the conference mile relay championship that day, but we also set a school record that lasted for 10 years!

We had only eight guys to compete in the district track championship at the University of North Dakota the next week. The good news was that all of us were capable of scoring points. Ole Stavadahl was the high jumper and Richard "Fugie" Fugleberg was the pole-vaulter. Fugie did not have any real training, but he still managed to break 10 feet. The problem was that he often performed the vault without executing the final portion, which was to land on your feet in the pit. Instead, he would go over like an arrow with his feet pointing straight up, and then come down the same way, landing on his back/head. To protect himself, he would wear a wool beanie (which offered no real protection).

At one meet, the track coach from Cooperstown tried to get Fugie disqualified because after making the winning vault, he landed outside of the landing pit (ouch). The pit supervisor simply pointed at the cross bar, which was still in place and said, "The crazy kid made it over the bar, so he is the winner." Ron Holkesvig was the miler, Richard Strand and I were both entered in the 440, and a great hunting buddy of mine, Larry Nordbo, was entered in the javelin. Larry had been in a nasty car accident his freshman year, so he did not participate in anything other than throwing the javelin.

Larry was super smart but had a huge attitude. He was in fact, a self-described "hood." Up to this meet, he had refused to wear, what he considered to be, the sissy track shorts that were part of our track uniforms. Instead, he competed in a pair of jeans. This getup did allow him to sneak up on competitors. I remember one of the guys from Hillsboro (perennial track champions) asked me, "Who is the little shit in the blue jeans? He just out- threw me by 20 feet." Duane LaChapple was entered in the sprints, and Dave Odden as a member of the mile relay team. Originally, I was entered in the mile run, but convinced the coach that I had no chance of placing in that event (plus I hated it). I convinced him to enter me in the 440-yard dash instead. There were three heats of eight runners in each heat. Since I was the last one entered, and they already had eight entrants filling all eight lanes, the starter asked the runner in lane eight if he would share the lane with me. The starter then explained that whichever one of us got the lead; we could use the whole lane.

Fortunately, my running style for this event consisted of sprinting the first 110 yards, striding the next 220 yards (full speed but with longer steps), and attempting to sprint the last 110 (this often resulted in little more than a survival shuffle for the last few yards). For whatever reason, this last now nine-man heat was loaded with two of the best runners, with the other two top-competitors already having run in the earlier heats. During this race, we all stayed in whatever lane we were assigned. To compensate for that, the starting blocks were staggered, with the ideal options being the inside lanes where you could see all the competitors. The worst lane of course, was number eight, which was mine. I was just happy to be in the race, so I took off like a rabbit being chased by a wolf and easily got the lead over the other runner in my lane.

I broke into my stride after 110 yards and really felt good. As I turned the last corner and headed down the final 110 yards, I started my sprint and realized, "Holy Shit! Nobody has passed me; I am in first place." I managed to stay there until the last 10 yards when my buddy Rich Strand and another good runner from Cavalier or Pembina, I can't recall which one, finally passed me. I finished third in the heat. The amazing thing is that the three of us were all faster than any of the 16 runners in the previous two heats, so Rich finished first and I finished 3rd in the meet. My finest track hour by far. For you track experts, my time was 54.9 seconds, which translates to a 54.6 for 400 meters. Not bad for a guy who was smoking a pack of cigarettes a day. And yes, I threw up. Richard Strand, the winner, who was a smoker, also threw up after every 440. If we didn't, then we felt that we had not tried hard enough. *This attitude came in very handy later in the challenges I was presented with at Aviation Officer Candidate School.* By the way, with only eight participants we still managed to finished 2nd in the district meet.

1962 qualifiers for the state track meet. Left to right, back row: Duane Lachapple, Richard Strand, Gil Rud, and Larry Nordbo. Front row: David Odden, Ole Stavedahl, Ron Holkesvig and coach Ron Haedt.

***

WATER SKIING: By the time I was 16, I had arrived on the party scene. Nothing too wild, but we did drink some beer and literally everyone smoked so that was not a big deal like it is today. During the summer months, we spent a good deal of time at Golden Lake, which was located just a few miles from our farm. My dad built a 14-foot speedboat that was designed like a hydroplane. It was ultra-cool, especially when outfitted with a 55-horsepower Mercury outboard engine. Fugie also had a ski boat, so we became pretty good at water skiing. So much so, that we decided to build a water ski jump. Thank goodness that project never came to fruition, or we would have more than likely been maimed for life.

The ultimate summer destination was Detroit Lakes, Minnesota. Detroit Lakes was a teenager's dream for the Fourth of July weekend. We would rent some affordable cabins on the lake and then fill them with as many kids as we could fit through the door. The town had a large pavilion at the edge of the lake, and they booked the best rock bands in the area. These bands performed nightly for what was usually at least a 4-5-day time frame.

Between my junior and senior years, I towed the ski boat to Detroit Lakes, and we had a blast. I partied heavily in the evenings, but never mixed alcohol with driving that boat. On Saturday evening, we loaded the boat on the trailer, and headed to the pavilion for one last night of celebrating. On Sunday morning, I headed back to Portland just in time to make a scheduled American Legion baseball game with our traditional rival, the neighboring city of Mayville. July in North Dakota can be warm, and this was one of those days. It was 100 degrees and humid. I was the starting pitcher, so I took one last drag on my Camel cigarette, grabbed my glove, and headed to the bullpen (just a strip of grass down the right field line) to warm up.

Nursing one heck of a hangover, all I could think of was, "Take it easy or you will throw up." Never pitched a better game. I had perfect control, excellent velocity and as I recall, 10+ strikeouts. I think I now know why some of the major league pitchers back in those days partied so hard. My dad rarely attended a game, not because he wasn't interested, but because he had ulcers and he would get too nervous if he watched. Well, he attended that game, which was perfect to see his little boy at his best.

***

HUNTING: I was not much of a deer hunter. I did accompany my dad when he hunted with a shotgun slug. Next to bow hunting, this is the most difficult method of getting a deer. You must get within about 40 yards to have any chance, so you must sneak up on the deer (we did not have any deer stands). We just hunted on our own land, which included, at that time, approximately 90 acres of trees and pasture along the Middle Fork of the Goose River. According to my dad, the worse the weather was, the better chance we had of sneaking up on a deer. This was one of those days. Very windy and although no snow yet, it was cold. He knew a spot where the deer liked to go on days like this, so I soon found myself crawling alongside him toward the top of this ravine that sloped down to the river. We lay on the ground as he scoured the area for any sign of a deer. Suddenly, he placed his shotgun on a fallen log and fired. I don't see a thing and there is no reaction from the area he shot into. He looks at me and smiles, "Got her." I was incredulous, "Got who, where?" Dad leads the way down to the ravine and there in a nice little spot shielded by bushes is a beautiful White-Tailed doe. The slug went right through her neck, killing her instantly.

One more deer hunting memory I have is from the opening day (it was usually just a three-day rifle season) in 1961. I was doing some late season disc plowing when Fugie pulled up to my tractor on his 1938 Harley 74 motorcycle. He was not a serious deer hunter, but he did have a 30/30-saddle rifle slung over his shoulder. I had some coffee with me, so we sat down by the wheel of the tractor to watch all the hunters driving around looking for deer. Suddenly, Fugie said, "I am bored, watch this." Since a statement like this from him was usually followed by some type of mayhem, I watched as he fired up the Harley. He roared off toward the nearest ridgeline overlooking the river. He then grabbed his rifle, fired off several rounds, and then roared off again, repeating this action before coming back to finish his coffee. We then watched, laughing, as hunters in vehicles of all makes and models tore down the country roads and across fields looking for the deer that of course, did not exist.

I never actually hunted deer myself. What I did enjoy hunting were game birds, including ducks, geese, pheasants and partridge. The first shotgun that I had was an old single-shot 16-guage that had been stored in the garage and not used for many years. The firing pin was so dull that it would not set off the cartridge until the third pull of the trigger. Yeah, I know, not the safest or most efficient weapon, but it was mine.

Opening day of pheasant season. We had every variety of domestic animals imaginable on our farm. We raised chickens, ducks, pigs, beef cattle and dairy cattle. We also had many varmints inhabiting the area (they loved to prey on chickens who were on the bottom of the food chain), and game birds of all sorts. Throughout that summer, a particularly audacious Pheasant cock would often perch on a fallen log near our corral. I had a plan: if he is still around on opening-day, I am going to get me a pheasant. As you may recall, I mentioned that my shotgun firing pin was so dull that it would take three trigger pulls to actually fire. Sunrise, opening day, pheasant cock perched on his tree stump. *One-two-three, blast.* Down went the pheasant, off his mighty perch. That was when I got my first sportsmanship lesson from my dad, "Son, when we hunt these game birds, it is traditional to get them to fly before you shoot at them. What you just did is called 'pot shooting.' We are not that hungry."

***

DUCK SEASON VS. FOOTBALL: Unless you have played football, it is difficult to explain what one feels like on Saturday morning following a Friday night game in which you have played every down on both offense and defense. Now that I am old enough to experience some age-related muscle aches and stiffness, I will compare it to that. Unless I live to be 100-plus, I will never be as sore as those Saturday mornings. I literally could not get out of bed without sissy squeals of anguish. And then, it was, don the hunting gear, grab a lunch box full of peanut butter and jelly sandwiches, grab the shotgun (I had graduated to a 12-gauge pump) and jump in the truck with my hunting buddies. It was usually still dark at this point, and cold as hell. The good news was that most of my hunting buddies had played in the same game the night before, so no one was moving pain-free. We were not "duck blind" hunters. We went to the numerous sloughs and river bottoms where we knew the ducks flew in and out of and usually ended up jump shooting them vs. waiting in a blind.

One of my friends, David Seaver, had this awesome double-barrel ten-gauge shotgun. It was enormous and with double aught buckshot, he could bring down a Canadian Honker or Snow Goose from exceptional heights. On one day, we started by bagging a couple of mallards. David just winged his, so he did the old "snap his neck" routine and stuffed it in the back of his hunting jacket. We spotted a flock of snow geese feeding in a sheltered spot in a muddy field. We began crawling slowly toward the flock attempting to get within range. Suddenly, I hear this "quack, quack" and the thunder of a double-barrel ten-gauge going off with both barrels at once. I turned to see a mountain of mud, blasted free of the earth, and David in mid-air flying backwards to end up flat on his ass in the mud. Meanwhile, the mallard, with his neck at a 90-degree angle to his flight path was flying off, along with our now fully alerted flock of snow geese. I laughed so hard I too ended up in the mud. Dave was now up, reloading his ten-gauge, and hauling ass after the mallard. He never did catch it.

***

OPENING DAY OF GOOSE SEASON, CIRCA 1961: The whole gang headed off to Rock Lake, North Dakota for opening day. And yes, it was a school day. We hunted all day and then headed back to make football practice. You could probably get by missing school, but not football practice. Of course, the Superintendent, Dusty Rhodes, got suspicious when a bunch of boys failed to show for school. As it was the same time frame as corn harvest, it would not be unusual if some of us farm boys were missing, but, alas, there were townies gone as well. Superintendent Rhodes called my dad, "Mr. Rud do you know that Gil is not in school?" "No, I did not know that." "Well, do you have any idea where he might be?" "Oh ya, I would betcha that those boys are all up in Rock Lake hunting geese." We were all met in the locker room as we put our pads on and invited to spend two hours after football practice in detention. Small price to pay for a great day of hunting, and no, I was not in trouble at home. My dad simply asked, "Did you get your limit?"

***

SUMMER EVENINGS ON THE FARM: One of my very close friends, and a high school classmate, was David "Haftor" Wiggen. Haftor was Dave's dad's name and it seemed natural to us to use that handle as a fitting nickname for the oldest boy in the family. Although a townie, Haftor spent the summers working for one of our neighbors, Howard Holkesvig. Since the two farms were only 2 miles apart, we spent many evenings together fishing in the Middle Fork of the Goose River. We had a minnow trap that we hung from the bridge just below our farmstead. Of course, rather than catching minnows, we had that trap full of beer, which stayed relatively cold in that minnow trap. My mother would often ask, "Where have you boys been?" The reply was pretty much always, "Down at the bridge, fishing." What we were really doing was smoking cigarettes and drinking beer. Technically, because it was a minnow trap, we were indeed fishing even though the catch was a bottle of Hamm's beer that was advertised as being, "From the land of sky-blue waters."

This is the Portland High School Class of 1962. Starting in the back row left to right: Mary Leland, Ole Stavedal, Gloria Eide, Richard Fugleberg and Ronald Holkesvig. Third row: Larry Nordbo, Shirley Amb, Paul Haugen, Patty Grandalen, Robert Roholff, Cheryl Holkesvig and Richard Strand. Second row: Mollyann Beck, Carol Peterson, David Wiggen, Vicki Strand (Valedictorian), Dennis Braaten, Phyllis Thykeson, Duane Lachapelle and Mary Mjogdalen. Front row: Arlene Rud, Gil Rud, Cheryl Aldrich, Robert Seaver, Sandra Carlson, Carolyn Viken, David Odden and Kersti Oja.

4. Off to Party in the Big City of Fargo, North Dakota

THERE WAS NEVER ANY DOUBT in my mind that I was going to attend North Dakota State University (NDSU). What was still a bit up in the air was what major I would pursue. My Aunt Rosie and my Uncle Peter, who were both graduates of NDSU, pushed me toward pursuing a degree in engineering. Peter was a very successful electrical engineer and Rosie an established educator. I also had a good friend, and second cousin, Jim Rud who was starting his junior year in civil engineering. The problem was that I did not particularly care for, nor was I, proficient at math. I checked out the engineering curriculum and decided that I would be struggling for four years at something that I was not really interested in doing. What put the nail on the head of the engineering coffin though was Freshman Orientation Week. I had so much fun and spent so much money that I could not afford to purchase the slide rule and other engineering related equipment. So, I used that as an excuse to major in agricultural economics, commonly referred to as Ag Econ. Later, I learned that many struggling engineering majors referred to engineering as "Pre Ag-Econ."

I did not live in a dorm. Rather, I lived with my two maiden aunts, Rosie and Aggie. They lived in a small two-story, three-bedroom house located at 726 12th Street North. This was to be my home for almost five years. It consisted of a very small bedroom with a little desk and a bathroom shared with my Aunt Rosie. I spent very little time there as I was hanging out at the fraternity house or with friends who had apartments.

It was a 12-block walk to the campus, which was no big deal until the winter hit. Then it was a 12-block run in below zero temperatures. I did not have a car until my senior year. The car was a 13-year-old Dodge 4-door sedan with fluid drive. Not exactly a chick magnet. One night, while parked on 7th

Avenue, near my aunt's house, a drunken driver struck my car at a high rate of speed. Since the whole rear end was smashed in, the insurance company considered it a total loss. Wow, OK I thought because now I can get enough money to

726 12th Street North, my college residence from 1962–66.

buy another car. Not so fast. The Blue Book value was a whopping $55. So, my brother-in-law, Dave Kringlie, who was a trained auto body repairman, cleaned the car up enough for a one-time trip back to the farm. From that point, my dad, and a talented neighbor farmer by the name of Truman Thykeson, cut off the back part of the car and attached a Ford pickup box (the pickup had been wrecked on the front end). We called it a "Dodge Arrow," and it survived many more years as a field-spraying vehicle.

***

COLLEGE SPORTS: Having spent the last four years playing football, basketball, baseball, and track, it is not easy to suddenly give it all up. Although I was not recruited by anyone to play anything (should have been a strong hint), I was not about to give up without at least giving it a try. So, I walked on to the NDSU Bison freshman football team.

This was the first year that the athletic department decided to heavily recruit to boost a weak varsity football program. In 1962 as a freshman, you were not allowed to play on a varsity team. As I entered that locker room for the first time, I was intimidated by the size and in many cases the age of the players. In those days, the active-duty armed forces had football teams other than the academies. So, some of these guys had spent four years in the Marines and were just now starting their freshman year. As I recall, that first day, there were almost 80 players. After the first practice we were probably down to 60, and at the end of the season there were a total of 34 of us remaining. I was a backup defensive end, which meant that I got lots of practice time against a very formidable offense. This included several players who went on to be members of the first Division 2 national championship team, three years later.

Our freshman backfield consisted of quarterback Billy Hanson, fullback Rich Mische, and a stable of running backs that included Ron Hanson and Rudy Baranko. Mische became a starting fullback on the varsity. Ron and Rudy became starting linebackers on the varsity. Gene Gebhards was eventually an All-American offensive guard. In other words, I got my butt kicked daily. We only played three games: Concordia, University of North Dakota and South Dakota State. I got in for a couple of series against Concordia and actually got a sack. I chased their QB all over the field and finally dived and put two fingers in the back of his low-cut spikes that tripped him up. It was not exactly a "thump your chest" type of sack.

Although I did not get into the game, the experience of beating the University of North Dakota was awesome. Why? Because it had been eight years since either a freshman or varsity Bison team had beaten them. We strutted around campus proudly after that one.

My decision to end college football aspirations was an easy one. During a practice toward the end of the season, I was playing defensive end. The offense started what we called a "student body left" play. I was the right defensive end. What I faced was a 210-pound fullback and a 205-pound halfback leading the blocking for another 205-pound halfback (both became starters on the varsity as linebackers). In this case, the best option for me was to take out the lead blocker so that the linebacker behind me could get the tackle. The last thing I remember was leaving my feet to _____?

I awoke via the aid of smelling salts with the entire team standing around me, and coach Gentzkow saying, "Welcome back kid. You have been out for several minutes." "Wow," I said, "this is so embarrassing, I can't believe I got cold-cocked by a damn running back." "Oh no kid, it was the pulling guard, 6 foot 3 inch 250-pound Gebhards that got you. How do you feel?" "I feel fine, coach," I replied. In actuality, I had a headache for several days, but I kept right on playing, as a stubborn Norwegian will do. I did, however, decide that this would be the end of my football, and it was. *Oh, by the way, this incident never found its way into my application for flight training along with other potentially disqualifying incidents. I will have a lot more to say on flight physicals later.*

The 1962 Bison freshman football team, which began the season with over 60 players, and ended up with these 34. I am kneeling in the front row, number 71.

***

A WILD NEW YEAR'S WEEKEND IN MINNEAPOLIS — RIDING THE THIRD ENGINE TO WISCONSIN: I was visiting my good friend and high school football teammate, Richard Strand over the holiday break in 1963. Rich was attending Augsburg University where he was a member of the football and track teams. He lived in a house full of college athletes near the Augsburg campus. There was not a whole lot going on the night before New Year's Eve, so we started looking for adventure (did I mention that the two of us together often elicited bad behavior?) Rich had gotten a summer job through the Augsburg athletic department's contacts working as a fireman for the Soo Line Railroad. Since they now had only diesel locomotives, the fireman was just a carryover position that the labor union managed to hold onto. In other words, he had a wonderful time riding these massive freight trains with three or four locomotives pulling 100+ railroad cars, and he got to know most of the crews. He said, "Let's sneak into the Theo Hamm Brewing Company in Saint Paul and jump the locomotive when they slow down to pick up the case of beer that is always left on the dock next to the tracks." I was more than a bit nervous about this plan of action, so I asked, "Isn't it a bit risky, maybe even slightly criminal to sneak into the brewery?" "Oh, I did not really mean, 'sneak,' since one of my buddies who flunked out of school is working there and he has a key we can borrow."

So, we are off on another well planned, albeit idiotic, adventure. We get into the brewery and onto the loading dock easily enough, and sure enough there is a case of Hamm's beer sitting on a small extension of the dock that is reachable from a passing locomotive. Rich pointed down the track to where a train is coming. He began waving at the lead engine and held the case of beer over his head. The train slowed to maybe 5 mph and blew its whistle (which scared the shit out of me). Rich explained how to jump on board the locomotive (there is no way to do it safely). The engineer leaned out of the cab and said, "Hey Rich, how are you doing? Do you guys want to ride with us to Altoona, Wisconsin?" "We sure do, will the Rail Haven Bar be open when we get there?" "It sure will, Jane opens up at 8 a.m. and we are due to arrive at 8:30 to switch crews, so we will all be heading to the Rail Haven for a night cap, so to speak."

Somehow, we got aboard without getting run over, and Rich introduced me to the engineer and the rest of the crew. Since it was now about midnight, and it was cold as hell, the engineer had one of his crew lead us back to the third locomotive where we could find a spot to warm up and even get a little shuteye. I suppose I should mention that there was a bit of alcohol fueling this adventure, so it wasn't long until we were soundly sleeping.

As we were pulling into Altoona, one of the crew informed us that a railroad inspector was waiting at the station. "If he spots you guys, the engineer will be in big trouble, so you have to jump off right now." Now wide-eyed, I asked, "OK, but how fast are we going?" "Oh maybe 10 mph and there is a huge pile of snow along the tracks for you guys to jump into." So, we jumped into the snow bank, which would not have been a big deal if we had been going 10 mph. At our real speed of closer to 20 mph, it was a nasty experience as we went rolling along in the snow. It did not help that the engineer and crew were now laughing heartily and pointing down the street from where we had landed. "The Rail Haven is just a block or so that way. We will buy you guys a beer when we get there."

We brushed ourselves off and headed for the Rail Haven. Like all structures in the North Country, it had a small mudroom that you first entered, with a second door to get into the bar. This was necessary so that all that cold air didn't come in with the arrival of every customer. That second door consisted of a giant photograph simply entitled "The Man." It was a photo of Vince Lombardi, and Rich then explained that the owner of the bar was Jane Thurston. Jane was the sister of Fuzzy Thurston, a famous Green Bay Packer lineman.

It was 8:30 a.m., but the place was rocking with all the night shift railroad employees enjoying a nightcap. Jane also served food, and Rich was a charmer so although pretty much broke as always, we enjoyed a couple of beers and a great lunch. We played some pool, watched a football game until about 3 p.m. when we decided we better get back to Minneapolis for New Year's Eve. "Wait, we don't have a car." No problem, we will just hitch hike. Jane helped us out by giving us a huge cardboard cutout of the Hamm's Bear (remember the jingle, "From the land of sky-blue waters") to which she taped a sign that said, "Students to Minneapolis, and we have beer to share!"

We had been standing by the highway for just a few minutes when this car pulled over to pick us up. It was a rather strange sight as the car had a bunch of tin cans and whatnot dragging behind. Then we noticed the "Just Married" scrawled across the back window. As it turned out, their motivation was to have someone reasonably sober drive them to Minneapolis where they could start their honeymoon. We pretended to be reasonably sober and, in comparison to this happily married couple, we were. Rich drove and I shared some beer with the newlyweds, who spent the remainder of the trip making out in the back seat. We managed to get to the New Year's Eve party safely, and our new friends joined us for the celebration before officially embarking on their honeymoon. Thus ended a crazy New Year's holiday for a couple of small-town boys in the big city.

***

ALPHA GAMMA RHO FRATERNITY: As a freshman, I paid little attention to Greek life on the campus. During my sophomore year, I noticed how many of my classmates were sporting fraternity jackets.

Alpha Gamma Rho, a great organization that helped to shape my life, and provided life-long friendships. My picture is second from the left in the bottom row.

A great friend (we are still great friends today) and teammate from Portland, Roger Erickson, who was a year behind me in school, joined the Alpha Gamma Rho (AGR) fraternity as soon as he arrived on campus. I had several other friends who were also members, so during the spring quarter of my sophomore year I became an AGR pledge. Alpha Gamma Rho is a social-professional organization with almost all its members majoring in an agricultural-related field. After becoming an active member in the fall of my junior year, I considered moving into the fraternity house, but decided that I could not afford to do so. Even though I did not live there, I did spend a great deal of time studying and hanging out at the house. It was, and still is today, a wonderful organization that helped me to learn a great deal about leadership, ethics and integrity. *If I had not become an AGR, I very much doubt if I would have overcome the many challenges involved with my Navy career.*

***

COLLEGE ACADEMICS: I would describe my college academic performance as mediocre, at best. I did well in subjects that I liked, and not so good in those that I was forced to take to graduate. Those of us who majored in agricultural economics were all assigned an advisor. My advisor was Dale Anderson. At the time, he was still working on his Masters (he eventually earned a Doctorate). He was not that much older than I was, had a great sense of humor, and became flustered by my choice of courses each quarter. There were three avenues to pursue in Ag Econ. They consisted of business, production, and science. Dale thought I should pursue the science option, which included lots of advanced chemistry courses. The business and production options on the other hand, included several more liberal arts-type courses with lots of girls in the classes. Duh! The science option was not going to happen.

My overall grade point average ended up a whopping 2.86 with 4.0 being straight A's. However, my grades in agricultural-related subjects were pretty good, in fact, good enough to earn me the status of being selected into Alpha Zeta, an honorary professional society for students in the field of agriculture and natural resources. As part of my induction, my advisor, Dale Anderson, was my sponsor and introduced me. This was a big deal, so my parents were in attendance. Dale introduced me with the following comment, "I just want to say that I have done some research, and I found that Gil Rud is the lowest overall grade point ever accepted into Alpha Zeta." He did this with a big smile on his face and then added, "He is also the most difficult student that I have ever had the opportunity to advise for course selection, as I don't believe that he has ever totally accepted my recommendations."

It was then my turn to say a few words. I looked directly at my mom and dad, and said, "I may be the lowest GPA ever accepted into Alpha Zeta, and my good friend and fraternity brother Jerry Bergman here may be the highest (he was brilliant and had a 4.0 GPA), but upon graduation, when we start looking for jobs, both of our resumés will say that we are members of Alpha Zeta." My dad gave me a smile and a knowing "thumbs up." *Ironically, 47 years later, Dr. Dale Anderson and I would again share a stage at North Dakota State. We were both receiving Alumni Achievement Awards, and he told the exact same story, adding that he didn't care how successful I had managed to become, I was still the most difficult student he had ever attempted to mentor. Thank you, Dr. Dale Anderson, for your patience in getting this stubborn Norwegian a degree.*

***

LOVE LIFE DURING COLLEGE: Not much to discuss here. I had lots of lady friends, but most of the interaction was in groups versus actual dating. Of course, I was in love with several totally out-of-my-league ladies, none of which I attempted to date. If I didn't try, then I had not been officially rejected, which was good for my morale. When it comes to girls in my life, I should probably go into some detail about my second Cousin Arlene and her awesome hairdresser friends/roommates. They lived in a huge apartment above a furniture store in Moorhead. Most of these ladies hailed from Mahnomen, Minnesota, and they hosted some great parties. Many of them went by nicknames, but all of them loved to have a good time. I spent a lot of time there along with my former high school buddies from Portland.

Fargo, and its sister city Moorhead, were great party towns. There were three colleges: North Dakota State University, Moorhead State University and Concordia College, plus Dakota Business School, Josef's School for Hairdressing, and St. Luke's School of Nursing. Since it was a relatively young and active nightlife area, there were lots of bars to party in. Unfortunately, the legal drinking age was 21, and I looked even younger than my age of 18. Not to worry, I acquired a very effective fake ID. It was a driver's license from a 28-year-old, 5 foot-6 inches, 145-pound man. Since at the time, I was only 18, 6 feet tall and 190 pounds, how did I make this work? Easy, the guy was handicapped with a walking and driving disability, all clearly noted on this driver's license. Entering a bar, I literally dragged my right leg behind me in a rather pitiful shuffle. Worked pretty much every time except when I got a bit juiced and dragged the wrong leg on my way to the men's room. Luckily, a friendly bar maid warned me that the bartender was calling the cops. I quickly healed from my disability and hauled ass out the back door.

Finally, during my senior year at NDSU, I got up the courage to date a wonderful lady, Beth McLaughlin. She was a sophomore and was a member of the sorority across the road from my fraternity. She was a cute, feisty, redhead that loved to have a good time. And we did. So much so that although we did not work out as a couple, we stayed friends and eventually I introduced her to the love of her life, and husband of 50 years, John "Lightning" Davison. I will have more to say on him later. There were a few other girls that I hung out and partied with including Karen Fosse, Nancy Avery, Mona Brandhagen, Diane Smith (student nurse), and a singer from the Five Spot Bar in downtown Fargo. For the life of me, I cannot remember her name, but she sang renditions of Petula Clark, and she was super sweet. Turns out, she was underage, and a student at Concordia Lutheran College in Moorhead.

By the way, I had a thing for singers, and Mona Brandhagen was also an awesome singer. She would do gigs at private clubs. One night she called me, because she was starting to feel uncomfortable at a private club in Moorhead. I was a good-sized fellow and had no sense when it came to backing down from a scrap. So, I headed over to the club. It was run like the old speak easy bars with a little window that they could look through to see who wanted to come in. Mona told the owner that she would only continue to perform if they let me in. So, they did. Not sure what I would have done if there was trouble, but like I said, too stupid to be scared, and certainly not about to turn down a request for help from a beautiful lady.

One of my adventures with Nancy Avery also involved a colorful fraternity brother of mine, Byron Bollingberg. Byron was an awesome dude, but not much of a student. At the time of this adventure, he was a 6th year senior with a 1.98 grade point average. He had so many credit hours, that to get to a 2.0 and graduate, he would need something like 20 hours of straight A's.

Anyway, it was during a holiday break, when the fraternity house was officially closed. Of course, we had keys, and we decided it would be a great idea to show our dates the house. While we were making some peanut butter and jelly sandwiches in the kitchen, we were literally accosted by this outraged, old-fart alumnus. He began calling our girlfriends' harlots and other awful names, as he demanded that we leave the house. Of course, with my Norwegian temper, I threatened to kick his ass if he touched any of us. Turns out, he was a big shot in the fraternity hierarchy. When school came back in session, Byron and I had to go before some sort of disciplinary hearing at the first fraternity meeting. Luckily, the rest of the brothers considered this alumnus to be a basic asshole, so we got off with some sort of double-secret probation.

Karen Fosse, a western North Dakota cowgirl, and I had some great times together. I say this tongue-in-cheek since we were a combination that sometimes elicited bad behavior. Herb Johnson's Barn was a favorite spot for folks to dance and party. It was located on a farm near Arthur, North Dakota, about 30 miles from Fargo. It really was a barn that they converted into a state-of-the-art dance hall.

Since they booked the best rock and roll bands in the area, it was a great destination for many of the young folks. It also attracted some "bad guys" who enjoyed stirring up trouble in the sizeable parking lot that also served as the BYOB bar. One evening, Karen and I were drinking beer and doing a little smooching in the car when we noticed some disturbing behavior. Three guys were hanging out near the exit of the dance hall. They seemed to be physically picking on heavily inebriated guys, and then hitting on their girlfriends. Since at this point anyway, both of us were relatively sober, we decided to do something about this.

"We need a plan. OK, let's go in the other entrance and then come out the exit where they are accosting folks. I will act totally inebriated, hang on to you for support, and appear as though I am afraid of them. I will let them push me once, and then I will take the big one while you nail the closest one of the other two with a kick in the shins." It worked to perfection with one small hitch. While working over the big guy I managed to break my left wrist with one too many punches. Fortunately, all three of them hauled ass, and I am sure they never picked on another drunk. The wrist issue required a visit to St. Luke's Hospital emergency room where my sister, Linda, was on duty. She was quite used to my adventures. She just shook her head and said, "Will the other guy be showing up soon?"

***

BABYSITTING: My sister Linda and her husband Dave lived very close to me, so it made sense that I would be available to help them out by watching their two boys, Jim and Dale, who were just a year apart in age and very active. Occasionally, I would mix a sitting event with more than a study session. I am sure you have heard about how a guy with a puppy can be a positive influence on a potential date. With that in mind, imagine the effect of a guy babysitting two adorable (well cute anyway) little boys might have on said ladies. Unfortunately, these sitting sessions, when fueled with a few beers, would often turn a bit raucous even ending up with, "Has anyone seen the boys?"

***

GETTING MY PRIVATE PILOT'S LICENSE: During my senior year at NDSU, I started to get serious about how the situation in Vietnam might affect my future. There was absolutely no doubt in my mind that I was going to serve. I was enjoying the sixties culture as much as anyone my age; however, that did not include drugs or the anti-war movement.

Having grown up watching and marveling at the skill of local crop dusters, I was very interested in flying. I also loved farming. In early 1966, my dad and I had a long talk about what my plans were after graduation from college. He was 62 at the time, and we were farming 880 acres, and feeding about 100 Hereford beef cattle. My dad owned 320 acres and rented 320 acres from my aunt Agnes and an additional 240 acres from a neighbor, Christ Lunde. I say "we," because I spent every summer working on the farm.

My major was agricultural economics, which included many courses on animal/crop science. I felt well prepared to step into the farming role as my dad's partner with the goal of eventually taking over the operation. I shared this vision with my dad. Being a man of few words, his response was, "No, I don't think that will work." I was a bit surprised, so I asked him, "Don't you think that I would make a good farmer?" "No, I don't think you would make a good farmer because this is dry land farming (no irrigation) so we rely on God for providing good enough weather for us to make a living." Since I had a reputation as a bit of a "wild child" who had been smoking, drinking and generally raising hell since the age of 16, he added with a wry grin, "I am afraid that God does not owe you much good weather."

Now it was time for the sage advice that would forever shape my life. "Son, I know that you are interested in flying." He then handed me a coupon that he had cut out of a magazine. "Here, take this to Kundert Aviation in Fargo. This coupon and $5 will give you a one-hour introduction flight in a Cessna 150. If you like it, I will give you $700, which you earned working last summer on the farm. That should be enough to get a private license, and maybe a shot at flying in the military." Old Norwegians do not hug, but his extra-firm handshake was as close to expressing love for his son, as he would ever muster.

I wasted no time in making use of that coupon. I cashed it in on March 27, 1966, for my introductory flight. It was awesome! My initial flight instructor was the owner of the business, Chuck Kundert. He was a no-nonsense veteran of WWII. I believe that he flew either B17s or B24s and survived harrowing missions over Germany. He was a great instructor, but he absolutely did not tolerate me being unprepared for a flight. His criticism of my piloting skill/lack of skill was quite harsh, preparing me for what was to come as I worked my way through naval aviation training. He emphasized the use of common sense to survive challenges, which saved my butt many times over the next 30 years of flying.

I had found my true love, but I had a problem. I was due to graduate in a couple of months, and I needed more flight time to get accepted into a military pilot training program. I decided to delay graduation by dropping a five-hour Agricultural Engineering course, which left me short of the credits needed to graduate. My plan was to use this extra time to complete my private license and then apply for a pilot training program.

Not so fast. The Steele County Draft Board was keeping a close eye on folks in the 2S (college deferment) status. When I did not graduate on time, they declared me 1A and drafted me into the U.S. Army with orders to report to Fort Leonard Wood, Missouri, as a Buck Private. I appealed and was honest about the reason for not graduating, which was a good thing, because they knew that I had a B going in that course when I dropped it. They gave me one more quarter to complete graduation and to enroll in an aviation officer program.

I used the rest of that spring and summer to fly as often as possible, attaining a private license in October 1966. I experienced a couple of adventures during solo flights. On one flight, I was having a wonderful time, to the point where I completely lost track of where I was. Suddenly, I got that shot of adrenalin, also known as a cold shot of piss to the heart. "Shit! I am lost." Don't panic; just drop down low enough to read the name of the town on their water tower. It was Ada, Minnesota. "Now, just head west to highway 75, turn left, and follow it to Moorhead, and then to Hector Field." *This "common sense" approach to flying and navigation later came into play during my first night aircraft carrier qualifications.*

On a solo cross-country flight to Jamestown, North Dakota, I got caught in a snow squall. I had only rudimentary instrument training, so this was not good. I was heading from west to east toward Fargo when I entered the squall. As a commonsense farm boy, I knew that the weather almost always moves from west to east. I managed to make a 180-degree turn very carefully on instruments, and soon emerged from the storm. I then gingerly followed the storm front into Fargo. After landing, my instructor, who had called the Jamestown Flight Service Station to find out that I had taken off just as the storm hit (without warning), expressed great relief that I, and more importantly, his precious C-150, survived.

As soon as I got my private license, I began the search for a military pilot slot. Here is some background I will share that is relevant to this process. NDSU was a state university, which in those days meant that it was mandatory for all incoming freshman to participate in Reserve Officer Training Corps (ROTC) for two years. The choices were either Army or Air Force ROTC. Since I had always been interested in flying, I chose Air Force. I loved the academic portion of the course but did not care for the drill stuff. In other words, I did not take it seriously, and it was only worth one credit hour.

The academic portion of Air Force ROTC was interesting, and I did very well. Unfortunately, the drill portion took place on Fridays. During the spring quarter of my freshman year, I was a member of the Bison Freshman Baseball team, and we played most of our games on Friday afternoon. I assumed that we would automatically be excused from the ROTC drills. Not so fast, I had neglected to fill out the necessary forms validating an excuse for missing the drills. The result: I got a D in Air Force ROTC for that quarter. No big deal I thought because it was only a one-hour credit course. As it turned out, it was a huge deal!

I decided to start my quest for a pilot slot with the local Air Force recruiter. He was initially impressed with my flight time until he saw my college transcript. "You got a 'D' in Air Force ROTC?" I began to explain the baseball conflict when he suddenly stood up, pointed at the door to his office and said, "Don't let it hit you in the ass on the way out!" *Well, that certainly did not go well.*

I stopped by the fraternity house to see Ron "Odd Job" Kofoid. Ron had mentioned that he was in the Navy Aviation Reserve Officer Candidate (AVROC) program and had completed the first part of a two-step process to attain a Navy commission with a pilot slot in naval aviation. Although it was too late for me to enter that program, he described the Aviation Officer Candidate (AOC) program as one that I might be able to qualify for.

He recommended that I start with the local Navy recruiting office, which was conveniently located on the second floor of a very popular bar called the Flame Lounge in downtown Fargo. Senior Chief Lightning, who I think was an Aviation Boatswain's Mate, was the recruiter. I decided to be up front with the issue of the "D" in Air Force ROTC. The Senior Chief looked at my private pilot's license and said, "I don't give a shit if you got an 'F' in AFROTC! You have a pilot's license, and we are killing our pilots off at a pretty good rate. Take this test. If you pass, I will get you a physical and you will be on the sunny beaches of Pensacola, Florida in a couple of months." Thus began my career in naval aviation.

***

ADVENTURES AT SOUTH SIDE TEXACO: During the fall quarter of 1966, I had a very light academic load, so I started looking for a part-time job. My fraternity brother, Lloyd "Boomer," Well was working for his brother-in-law, Ron Greenwood. Ron owned a Texaco gas station located in South Fargo near the new 94 interstate. Ron's station included three maintenance bays and a full-time licensed mechanic. The station had several gas pumps on two separate islands. Due to its ideal location and superb reputation, Ron had more business than he could comfortably handle. So, he welcomed me into the family of employees. He especially liked my background as a farm kid, and I became the prime operator of the stations 4-wheel-drive, Willys Jeep. The Jeep had an 8-horsepower alternator in the box for boosting batteries to start cars, and a snowplow attached to the front bumper.

I graduated in December of 1966 and went to work full time, while I awaited acceptance into a military flying program. I typically opened the station at 6:30 a.m. Of course, it was still dark and cold as hell. One morning, before officially opening the station for business, I decided to change the oil on the Jeep. To do this, I needed to position it so that I could use one of the three lifts that we had to work on vehicles. The night crew had already placed it in almost the appropriate position. I just needed to adjust it a few inches backward. Since I was alone, I needed to move the jeep, while also checking the position of the hoist. The Jeep was always hard to start in the mornings, which required pulling the manual choke (suppresses air and supplies more fuel) all the way out to give it enough gas.

I decided that the best way to move the few required inches was to place the Jeep into reverse. While still standing outside the vehicle, I decided that I could lean through the driver's side door and move it by just using a couple of clicks on the starter. VROOM! That baby roared to life, now at full throttle due to the position of the choke. WHAM! The Jeep backed into the roll-up glass doors just as I managed to use my left hand on the clutch and my right hand on the brake to stop it. I shut the engine off and slowly stood up to find that I was literally inches from being cut in half by the broken door. Instead of fear of death, my fear was that I would be fired from my job.

The boss, Ron Greenwood, arrived a few minutes later to see the destruction this incompetent employee had caused. I will never forget what he said, "Are you OK kid?" I stammered, "Yes sir. I am so sorry, and I will pay for the door." (At $1.50 an hour, it would have taken me several years to do that). He laughed and said, "Kid, we all make mistakes, and that is why I have insurance." He then gave me a hug and said, "We can fix the door, but thank goodness we don't have to sew you back together." *This was a valuable lesson in leadership and empathy that I would later attempt to emulate when someone who worked for me in the Navy made a mistake.*

Following this close call, I became the most hard-working and diligent employee at the station. I became skilled at plowing snow from not only the station, but also from customers' driveways. It was common practice for us to drive customers to their job location, bring their car back to the station, complete the work, and then pick them up after their workday ended. Like I mentioned earlier, this really was a full-service station. On one occasion, I drove a professor from NDSU to the campus, took her car back to the station, and completed the service on the vehicle. I then drove back to the campus at the end of the day and brought her back to the station. I looked younger than my 22 years, so on the way back she said, "Son, you seem like an intelligent young man; have you ever considered going to college?" "Well, yes ma'am, in fact, I graduated in December, but being classified 1A for the draft, this is the best job I could get."

I loved the job, but it was arduous, and after a 12 hour-shift, I was worn out. At the time, I was dating two girls (no, they were not aware of each other). One was a little nurse by the name of Diane Smith. She was great fun and a very feisty lady. One evening, I was supposed to pick her up at 8 p.m. for a date. I got home about 7 p.m. and decided to take a combat nap prior to the date. Big mistake. I woke up at midnight. Since there was no such thing as a cell phone, I was in deep shit.

The next day, I was back at work when I spotted her marching across from her car to the door of the station. I quickly asked Don, our licensed mechanic, to cover for me as I busied myself under a car to change the oil. I said, "Don, tell her that I am out on a call with the Jeep and won't be back for a couple of hours." She walked into the maintenance bay and said, "I am looking for Gil Rud!" Don, kind of giggling, said, "Wow, I can't imagine why a beautiful lady like you would be looking for that guy?" She said, "Because he stood me up last night!" "No way! What a foolish boy he is. You will find him hiding under that car in the third bay." Diane then proceeded to chew my ass out in front of Don, the boss, and another employee. I had been called a lot of names before this incident, and I have been called a lot of names since, but these really stung. So much for getting any letters from her while I was in the Navy.

Now on to the next bungled relationship. After the loss of my 1953 Dodge to the drunken driver incident, I purchased a 1955 Chevy from my girlfriend, Beth McLaughlin's dad. He ran a Chevy dealership in Halstad, Minnesota. Now, at one time, a 1955 Chevy was a real chick magnet. Not this one. It was an 11-year-old, four-door sedan and had a six-cylinder engine that ran great up to 60 mph. At that point, it would begin to lose 5 psi of oil pressure per mph. The car also had a defective gear linkage that resulted in it getting stuck in first gear. This would require stopping, opening the hood, and manually pulling the linkage back to neutral.

One day, Beth showed up at the Texaco Station. She asked to borrow my car to make a quick trip home to Halstad, which was about an hour's drive. "The keys are in the ignition, and the car is ready to go." A customer then asked me a question, which distracted me, and before I could show Beth the first gear issue, she was gone. She ended up driving about 20 miles in first gear. Not good. However, her dad did fix the linkage before she came back. She was only a sophomore so although we continued to date, we broke up before I left for the Navy. Bottom line — no support letters for me during Navy training from girls other than my mom and sister. I know, pitiful.

Most cars in the 'sixties in North Dakota were equipped with head bolt heaters. This was an electric heater that kept motor oil from congealing to the point where you could not get the motor to turn over when it was really cold. The problem was that you had to have this heater plugged into an electric outlet, so unless you were parked in your garage or driveway, they were of no use.

There was a new Holiday Inn hotel just a short distance from our station. It had conference facilities that often were used to host large groups for business events, or parties. During one of these large events, a blizzard struck with corresponding temperature drops and wind chill down to 40 degrees below zero. Most of the cars in that Holiday Inn parking lot would now not start without a boost to the battery. No problem, we had the Jeep and the big alternator to provide that service.

I had already put in a twelve-hour shift when the boss approached me about staying on to man the Jeep until midnight. He said, "Gil, if you stay on duty, I will pay you time-and-a-half." (This raise in pay translated to $2.25 an hour, versus the usual $1.50). Then he sealed the deal by handing me a pint of Peppermint Schnapps to keep me warm. I ended up boosting 40 cars in four hours at $5.00 per boost. The boss made $200 minus my salary of $9 and whatever the pint of Peppermint Schnapps cost him. I can say that I certainly did not feel the cold and thank goodness this was before "DUI."

During this same storm, I got a call from a fellow who had a brand new 1967 Ford parked in an alley in downtown Fargo. When I showed up, I discovered that he was extremely upset that his brand-new car would not start. I put the jumper cables on, and the engine turned over, but it would not start. I gave a little shot of ether to the carburetor to assist light off, but still no engine response. He was furious at the car, not me. Then he grabbed my can of ether and sprayed a huge amount into the carburetor. He got back in to attempt another start when I said, "Stop!" I then took cover behind the Jeep, and he hit the starter. BOOM! The breather cover blew off and ricocheted off the walls of the businesses in the alley, ending up a good 100 feet away. I walked up to this now in-shock customer and said, "Give me back my can of ether, and also that will be $5 for the boost." *Lesson learned: Do not get frustrated with the performance of a mechanical machine; it will get you back in the end.*

***

My First New Car: Following official acceptance into the Aviation Officer Candidate program/assignment to Class 20-67, I decided that I needed to purchase a new car. It was a 1967 Pontiac Tempest convertible. It was eggshell white, with a black top and red interior. It also had a red boot for the top when it was down. I did not have any money to speak of, so I shaved the purchase price by having a manually operated top, and a three-speed floor-mounted shift versus a four speed. It also had bench seats versus bucket seats. Still, it was pretty darn cool. I finally had me a real chick magnet. And I got Florida license plates even though I was still in North Dakota. The first time it got above 32 degrees, down came the top. Total price as I recall was $2,700. I borrowed the money from Aunt Rosie and paid her back at the interest-free rate of $100 a month. I was in debt for the first time, but I never missed a payment.

5. An Officer and a Gentleman

THE 1982 MOVIE *An Officer and a Gentleman* starring Richard Gere, offers a pretty accurate depiction of Aviation Officer Candidate School (AOCS). I had a check-in date of Sunday, May 23, 1967. I drove my 1967 Pontiac Tempest convertible to Chicago, checked my road map, and found to my amazement that Pensacola was almost straight south of Chicago. It was even in the Central Time Zone. This is when I realized that even though I was heading to Florida, it would be the far western edge of the state. I got the "gouge" (inside information) from fraternity brother, Ron Kofoid, who, as I mentioned earlier, had completed the first session of Aviation Reserve Officer Candidate (AVROC) training, that I should not check in until the very last minute.

I arrived in Pensacola on Friday the 21st and hooked up with LT Glynn, whose younger brother Pat was a fraternity brother of mine. LT Glynn had also graduated from North Dakota State a few years earlier, and he was now serving on shore duty as a Naval Flight Officer. He and one of his friends showed me around the area, including the beautiful and enticing Pensacola Beach. "Wow, this place is going to be awesome."

On Sunday, I drove through the front gate at NAS Pensacola at about 11 p.m. *Never in my wildest dreams imagining that in 19 years I would be legally flying inverted at 200 feet in a* Blue Angels *jet over that same gate.* I got directions to the Aviation Officer Candidate Check-In site, drove there, and parked my car. I locked it up and headed for the building. It was not hard to find because of all the unpleasant noise emanating from inside. I carried my suitcase through the entry and was greeted quite pleasantly by what turned out to be a candidate officer. He asked if he could help me with my suitcase. He then grabbed it and threw it across the floor, screaming, "Get up against the bulkhead maggot!"

"OK, but what is a bulkhead," I asked. I don't remember much of that evening other than I had the most difficult time calling myself "Aviation Officer Candidate Rud." For some reason I kept calling myself "Air Officer Candidate Rud." Things went quickly downhill from there as I received special attention for being the last one to check in. We eventually got assigned to rooms in Poopyville, which was a nasty old WWII barracks that was used to house the new class for indoctrination. It was called that because we were outfitted in these awful coveralls and tin-pot helmets called "poopy suits."

Assigned by alphabetical order, my roommates were Max Tea, Joe Uhrig and Chris Visher. Chris was a long-haired hippie from Reed College, a liberal arts school in Oregon. He had arrived early and made the mistake of checking in on Friday. This meant that he spent the entire weekend getting harassed while the rest of us were hanging out on Pensacola Beach. Max was a football player from the University of Utah, and Joe was a graduate of Ohio State. Joe and Max were both married, and had their wives set up in apartments in town. Chris and I, both being single, became great friends and liberty running mates. The hippy from Reed College, and the farmer from North Dakota State, made quite a pair.

Staff Sergeant Borce was our drill instructor (DI) for that first week of indoctrination. He was from Brooklyn, New York and a native of Puerto Rico. The combination of New York Street speech and Hispanic accent meant that only one of us could understand anything that he said. Steve Rochford, who was also from Brooklyn could both interpret and perfectly mimic the DI. Of course, understanding exactly what a DI says is not a requirement, because his meaning becomes obvious.

Having just enough experience from my two years of mandatory ROTC, and being in pretty good physical condition, I managed to stay in the middle of the third row during marching drills and go pretty much unnoticed. That was until Tom Wurzbach and I were assigned to clean the head (bathroom). Somehow Tom had been able to stash a pack of cigarettes in one of the stalls. We each lit one up and were getting by with this ruse by blowing the smoke into the commode and flushing it away. The DI heard all this flushing and assumed someone had got dysentery, or a couple of poopies were up to no good. He snuck up on us while we were smoking, flushing and giggling like a couple of schoolgirls. This did not end well of course, as we were made to do all sorts of extra physical training (PT), beginning with the chewing and swallowing of our cigarettes.

We finally completed hell week and proceeded to Battalion Three. There we were introduced to our class drill instructor (DI), Gunnery Sergeant Pennington. I am not positive of this, but I believe that we were Gunny Pennington's first AOC class. I really can't put into words my first impression of this man, but I am going to give it a shot. He was immaculate in every respect. A true recruiting poster Marine, he had just returned from Vietnam where he had been badly wounded. Unlike DI Borce, who simply scared the shit out of everyone, DI Pennington was a man who commanded respect, and one whom you desperately did not want to fail. He also had a sense of humor, and a grin that you could detect when he was enjoying a particularly silly screw-up by the class, or one of us individually.

Of course, there were many of these incidents. In one case, we were doing a drill with our M1 rifles, which included snapping the breech shut at the DI's command. One of my classmates (who eventually had a very successful career as a helicopter pilot) snapped his breech shut, but unfortunately his finger was in the way and got pinched/stuck into the rifle. This was an unbelievably painful situation, but in a sick way, also very humorous. Especially when Gunny Pennington took advantage of the situation to prolong the pain and misery of the candidate. I don't remember the Gunny's exact words, but it was something like this, "Candidate ___, that breech slam sounded pretty mushy to me." He then checked his immaculate uniform very carefully (while the unfortunate candidate's finger was still trapped in the breech) and stated, "You better hope that finger of yours did not squirt any blood onto my uniform or you will be buying me a new one!" Finally, he said, "Most sane candidates would have removed their finger from the rifle by now so get it out and get your ass to the dispensary to have it amputated so this doesn't happen again!" This was of course followed by his famous grin of amusement.

In the AOCS program, there were three areas that a candidate was graded on: Academics, Physical Fitness and Military Bearing. If you met a certain criteria, you would be awarded a piece of tape for your name tag. Academics was blue, physical fitness was white, and military was red. A few folks earned all three. One of my roommates, Joe Uhrig, earned all three and together with top peer ratings from the rest of us, he became the Regimental Commander:

the only candidate officer wearing five bars. I was horrid at military bearing, but pretty good at physical fitness, finishing near the top of the class in the cross-country course and the obstacle course. Unfortunately, the physical fitness grade also included swimming where, as previously mentioned, I excelled in the dog paddle only.

During one of our numerous swimming tests, we jumped off a 15-foot tower, which simulated abandoning ship. We then had to swim underwater to the end of the pool, simulating avoidance of oil and fire. After jumping off the tower, I decided that I could get to the end of the pool faster by going all the way to the bottom and using my hands to pull myself along the bottom to the end of the pool. I did this successfully. I popped up at the end of the pool to be met by Gunny Pennington. He said, "Holy shit farm boy! You can't swim a stroke and you still jumped off that tower." Without thinking I said, "Screw you, I made it didn't I?" That cost me 100 push-ups, but it also changed the way DI Pennington treated me the rest of the way to eventual commissioning. By the way, I could swim well enough to pass all the swim tests and stay out of "sub-swim," which often led to failure in the program. But, since swimming grades counted 50 percent of Physical Fitness, the white tape was not to be.

***

ACADEMICS: Our class was a mix of all sorts of college majors, with most being technical in nature. I am proud to say that with a degree in agricultural economics, I achieved academic excellence and the corresponding blue tape for my name badge. How did I do that? I am a superb guesser. Most of our exams were multiple choice/true and false. I used common sense to eliminate answers rather than know the right one. For instance, in our math and physics exemption exam, there was a time limit of 60 minutes. I figured if the math problem had two even numbers in it, then the answer would also have an even number. Some lucky coin flipping was also involved, but most importantly, I answered all the questions. Because if you did not finish the test, the problems you did not answer were counted as wrong answers.

***

CANDIDATE OFFICER WEEK: In Class 20-67 there were 27 of us left out of 32 that made it through hell week. I have no idea how many started hell week and dropped out before that was over. As I recall, not all the other five dropped on request (DOR'd); a couple of them were injured and dropped back a class. During Candidate Officer Week, our class was together with an AVROC class that was finishing up their second summer session, so we ended up with a very large number of candidate officers. The rank/number of bars you were bestowed with for this week depended on how your peers rated you, and your military performance during the previous three months.

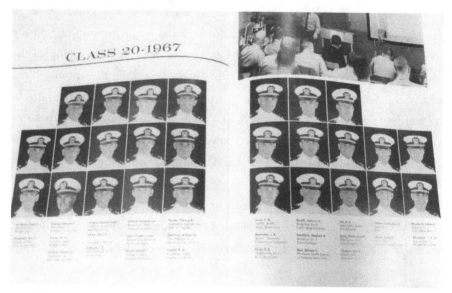

The 27 graduates from Aviation Officer Candidate Class 20-1967.

I never volunteered for anything, and I was sorely lacking in military bearing/shoe shining/brass shining/room locker and personnel inspection performance. My peers were not impressed with my performance either, as my priority was to get through the bullshit and get on with the flying.

The result? I was a one-bar man, self-assigned as the regimental commander's driver (and we did not even have a vehicle). During that week, we had a get-together at the ACRAC (officer candidate) Bar. Gunny Pennington joined us, and one of the candidate officers asked him who he thought would end up flying jets. His answer to this "brown noser" was, "Well, it sure as shit is not going to be you." He then named all six of us that ended up in jets. "The jet jocks will be Uhrig, Gellman, Visher and Ewald." (All four of these guys were outstanding performers throughout AOCS.) "Also, it will include the skylarking, no-load, one-bar assholes, Rud and Roach, who don't give a shit about being a Naval Officer but will end up as jet jockeys." *He was right on target.*

6. Snakes and Survival Training

PREFLIGHT: We had been commissioned Ensigns and now joined the Naval Academy and NROTC graduates for the rest of the training required to earn our Wings of Gold. Part of the preflight training we went through prior to getting into the cockpit involved survival training. The first part involved water survival. We were deposited in Pensacola Bay with our survival kit, which included a life raft. It was a very realistic scenario since we had to inflate our life vests and extract the survival equipment, including the raft. We then inflated the raft, crawled in, and began the wait for rescue by helicopter. Since I was a heavy smoker, I had placed a pack of Lucky Strikes into a waterproof bag. It wasn't long before the other smokers (most of us smoked in those days) noticed me lying back in my raft puffing away. Most of them would have paid $10 for a cigarette, but I was a nice guy and provided a smoke to all who successfully paddled close enough.

The next phase of survival training involved simulated jungle survival. During the classroom part of this training, one of the instructors brought in his pet boa constrictor. It was a very large, nasty looking creature that he swore was a perfectly tame pet. We were lined up in vertical rows of desks, so he started this snake crawling over the first guy in line and on down the row. When it arrived at the Marine sitting in front of me, he leapt out of his seat, pulled out his survival knife and said, "You better get that damn snake away from me right now, or I will cut his ugly head off!" The instructor quickly gathered up this monster and placed him in a cage.

Next, it was off to the survival training area located east of Pensacola in a swampy, desolate area of a neighboring Air Force base. We were only given water and told to find our own food from whatever we could scrounge up in the swamp. As we were scouring the swamp for something edible, we came across a very large black snake. This city kid next to me asked, "Is this snake poisonous?" I said, "No, he is not poisonous." Before I could say anything else, this fool grabs the snake. It quickly reacts by coiling around his arm and then biting the shit out of him. He pulls it off and screamed, "Oh my God, you said it was not poisonous!" "Lucky for you it is not poisonous, but it is meaner than shit, and often kills the poisonous snakes." By the way, we killed, fried and attempted to eat the black snake. Did he taste like chicken? Hell no. He tasted like snake, all rubbery and nasty.

Since we are talking snakes here, one more episode that occurred during AOCS comes to mind. One of my classmates, Jim Watts, was from Tyler, Texas. He was a great storyteller, and one evening he related the following from when he was growing up in East Texas, "Me and my Daddy were out fishing and we passed under a cypress tree. Suddenly a giant water moccasin drops right into our boat. That is the day that Daddy and I both outdid Jesus as we literally walked on water to that shoreline while that nasty snake had our boat all to himself." You must imagine this being told with a very slow East Texas drawl.

7. Real Flying Finally Begins

VT-1 AND T-34B EXPERIENCES: It was a good thing that I had some prior flight time, because I certainly was not God's Gift to naval aviation. In fact, I would describe my entire training command performance to be average at best. The prior flight time did give me the confidence to know that I could fly and most importantly, that I would love every minute of it.

My initial instructor had a background in the multi-engine P-3 community. He was very much "by the book." Initially impressed with my ability to talk on the radio and land the aircraft, he gave me some good grades. On our second flight, he opened the brief by asking me why I had not shared that I had prior flight time. My answer was that unlike my competitors from the Naval Academy and NROTC, I had personally paid for my training, and that I should be graded on how good I was compared to the other trainees, versus a higher standard that he was now going to set. No dice.

Oh well. Since he was also a city boy, I could always bullshit him about emergency landing options. One time, he pulled the power and said, "You just lost your engine, where are you going to land?" I replied, "In that soybean field." He started looking for a soybean field, having no idea what it is, which gave me time to find a suitable place to land. After we made a low approach and Wave Off, he said, "So that is a soybean field?" *It wasn't, but a little fib got me an "above-average" grade.*

I flew most of my flights in the T-34 with the same instructor. However, toward the end of the training, I had an opportunity to fly with a fellow who had flown A-1 *Skyraiders* in Vietnam. He was awesome. He showed me bomb runs, low-level navigation, fighter-type maneuvers, and all sorts of great flying. I decided that I wanted to fly A-1s, and I told him so during the debriefing of our flight. He then informed me that the A-1s were being retired and that if I enjoyed that type of flying, I should attempt to become an A-4 *Skyhawk* pilot. This was great advice that I certainly took to heart.

At the completion of the T-34B syllabus, the entire flight class (probably about 20 pilots) assembled in the ready room to receive our orders for the next phase of training. The Marine Captain in charge of our flight read off the orders, which were either to jets at NAS Meridian, Mississippi or to propeller-driven aircraft at NAS Whiting Field located just east of Pensacola in Milton, Florida. Our flight and ground school grades had been combined into a total score, which had a certain cutoff that had to be met to get into the jet pipeline. I made that cutoff. The captain said, "Ensign Rud, VT-7 NAS Meridian." Then he addsed, "I will give you these orders, but only after you get a regulation haircut." I was sporting what was known as a "military-marginal" grooming standard. I was not the only one who was faced with that caveat since many of the other guys were Naval Academy graduates who were enjoying their first dose of freedom after being locked up for four years.

***

T-2A Adventures: On the way to jet training, I began to realize that in many ways I was now in the upper 50 percent of the student pilots. I thought, "Crap, now the competition will really get tough." The good news was that once we attained the jet pipeline, everyone seemed to settle into a less-competitive environment. It was more team-like, with everyone pulling for each other to succeed.

After checking into VT-7, and prior to completing ground school, I was hanging around the ready room when the Squadron Duty Officer (SDO) asked if I would like to jump in the backseat with an instructor. Of course, I would. Well, this was my first exposure to an oxygen mask. I had partied hard the night prior, and the flight involved a series of high-G maneuvers. It was a nasty first jet flight to say the least, resulting in the only time that I was ever airsick. *Come to think of it, I would probably not have made a good Naval Flight Officer (NFO).*

On my very first formation flight, which of course was dual with an instructor in the back, he would demonstrate proper formation, then he would give the controls to me, and I would attempt to stay there. Super colorful to say the least. This instructor was a good guy, but also a notorious screamer. In other words, his technique included berating me, and my entire family heritage. On the way back to the airfield he was supposed to demonstrate flying "formation" into the carrier break. Well, he didn't. I had the controls and he kept urging me to get closer. I could not help but see the instructor in the backseat of the lead airplane nervously watching us. That is when I glanced in the mirror and saw my instructor waving his hands in the air to show the now scared to death lead that the student is flying this aircraft. With all the screaming and berating, I was pretty sure that I was going to get bad grades. No. As I mentioned, although a screamer, he was a good instructor, and he gave me above average grades.

There was not a whole lot to do in Meridian, so Chris Visher and I used to head for Pensacola, or even New Orleans on the weekends. These trips were long and at times eventful. On one trip to New Orleans, we were on our way back to Meridian when we came across some of our fellow student pilots who had run into a wild pig.

They were driving a brand-new Corvette. On impact with the pig, the Corvette's mostly fiberglass body literally disintegrated. Fortunately, nobody but the pig was hurt (really, badly hurt). These guys had been partying hard so they were literally sitting in the ditch laughing about how one little pig could disintegrate a Corvette. I remember the Corvette, sitting there, with no fender and no hood, just a giant, very impressive engine that was still running. The passenger remarked, "Maybe the pig would not have done so much damage if we had not been going 125 mph when we hit him." I remember who these guys were. They both became successful fighter pilots, but I am leaving out names to protect the guilty.

***

Moscow, Mississippi: On one of the few weekends that we stayed in Meridian, I was hanging out with Bill Belden, a fellow north-country boy from Wisconsin and a good friend. It was Sunday, and we were grousing about not having anything to do. Suddenly, he came up with an idea. "Gil, let's go find Moscow, Mississippi." Moscow was one of the "initial points" that we passed over on the way back to the airfield. "OK, I have a map of Mississippi in my car, so let's go." We found this little town and decided we needed to toast our discovery with a beer.

We only found one little bar that was open, so we parked and went in. It was dark in there so we could not see much. We sidled up to the bar and ordered a couple of beers. The African-American bartender smiled and said, "You boys are from the Navy base, aren't you?" "Yes sir," I said. He asked, "You are also from up north, aren't you?" I said, "Yes sir, we sure are." He replied, "Boys, I would love to serve you a beer, but I don't think it is a good idea because this here is a colored bar." I said, "We don't care about that, because I bet the beer is just as cold as any other bar." He laughed, shook our hands and said, "Now for your own safety, you boys just go out the back door and get in your car. We have some folks in this town that might make trouble for both me and you if you don't."

That was Bill and my first real exposure to the extreme prejudice that existed in Meridian, Mississippi in 1967. Was that the end of it? No, we decided that we needed to leave our mark on this town, so we got in my car and drove it backwards from one end to the other. Not that big a deal because it was only a couple of blocks with no stop signs, but it made us feel good that we could lay claim to driving backwards through Moscow, Mississippi.

***

HOUMA, LOUISIANA: Myrden "Mert" Pellegrin was another friend of mine who was going through training at NAS Meridian. Mert was from a little town just south of New Orleans called Houma, Louisiana. I was always looking for a new adventure, and I knew that Mert often went home on weekends. I asked if I could go with him and he rather reluctantly obliged. He cautioned me on the drive down that this was a small, close-knit Cajun community. "There are a lot of pretty girls down here, but they all have big brothers." I could certainly take a hint, so I vowed to behave.

Mert had a good friend who was a couple of years younger than us. His name was Corky Fornoff. Corky's dad was Bill Fornoff, a very famous aerobatic pilot and air race winner. Bill had a beautiful WWII era (right at the end of WWII) F8F *Bearcat*. The *Bearcat* was a rather small fighter with a massive radial engine. It was one of if not the fastest propeller driven fighter ever made. Whenever Bill flew the *Bearcat*, he was always smoking a big cigar (or maybe he was just chewing it). This was his trademark. Bill also had a T-6 *Texan*, which was a WWII trainer, also with a radial engine, but not near as fast or maneuverable as the *Bearcat*.

While Bill was showing me around the *Bearcat*, Corky and Mert (in the backseat) took off in the T-6. Bill looks at me and said, "You just watch those guys. They will hide in that cloud over the field, and then when I take off, they will jump me and try to get a shot." Bill fired up the *Bearcat*, taxied to the end of the runway, and started a takeoff roll. As soon as he started the takeoff, out of the clouds came the T-6. Bill kept that *Bearcat* at about 10 feet above the runway as he retracted the gear and built-up speed. At the end of the runway, he pitched the nose straight up into the vertical, and the fight was on. Within two turns he was getting into a position where he was about to gun the T6 when they flew back into the clouds to save themselves.

It was to me one of the most awesome displays of airmanship that I would ever witness from both father and son. Bill was tragically killed during a mid-air collision at an air race, but Corky went on to become a very famous airshow performer. Mert went on to get his Wings of Gold, became an accomplished F-4 *Phantom II* pilot, and I think, eventually, a developmental test pilot. Houma, Louisiana will forever have a special place in my heart.

***

LAST FLIGHT IN VT-7: My graduation flight from what was called "intermediate jet training" was a four-plane formation flight with an instructor in a chase plane. He was a cool dude who allowed us to fly right over downtown Meridian on the way back to the field. I was one fired up student naval aviator as I headed back to Pensacola with orders to VT-4 for air-to-air gunnery and our first carrier landings.

***

BACK TO PENSACOLA: All the bachelors were assigned Bachelor Officer Quarters (BOQ) rooms at NAS Pensacola. Since we would only be there for three months or so, we were not given BAQ (funds to pay for a place off base). Although we kept our flight gear and most of our possessions in the BOQ, we also pooled our resources and rented a place on Pensacola Beach. Chris Visher, Jack Ewald and I shared a small place at Surf and Sand cottages on the bay side of Pensacola Beach. Our friends occupied all the other nearby cottages, so it was a very noisy, boisterous enclave, with parties pretty much every night. The flight schedule at VT-4 left lots of free time, which contributed to the party atmosphere.

One of my friends, Sergei Kowalcheck, moved into the cottage across the street from ours with his new bride, Patty. Patty had worked in Hugh Heffner's Playboy Club in New Orleans, so she was more than awesome to look at, and loads of fun. Somebody got hold of an old set of bedsprings and turned them into our special barbecue grill. We consumed many a burger/hot-dog/steak from that ugly, but effective grill.

***

AIR-TO-AIR GUNNERY: This was a wild summer to say the least. The T-2B *Buckeye* had two engines versus the single-engine T-2A, and much better performance. We were exposed to flying air-to-air gunnery for the first time. An instructor would tow a target banner, and we would set up a gunnery pattern around his flight path. We had 50-caliber gun pods with colored bullets so we could tell who got hits on the banner. It was great fun, but I was not good at it. I got messed up/confused in the pattern one day to the point where I was unsafe. This was my first and only "down" that I got in the training command. No big deal though, as I got a couple extra flights with a fellow who had been a fleet-fighter pilot, and with his help, I figured it out and finished the phase successfully. I must admit though that I never got a single hit on the banner (not that uncommon).

***

A LUCKY MEETING ON THE BEACH: Shortly after arriving back in Pensacola, I was walking the beach and passed by some extremely hot girls sun tanning on the sand. Struggling to potentially start a conversation, I asked if any of them had some suntan lotion that I could borrow. Being a typical Norwegian, my suntan consisted of a reddish burned coloring with some body part always peeling. The prettiest one stood up and said, "Oh my, you could really use some protection from the sun. And this is your lucky day because I just happen to have some extra lotion, and I think sunburned, sandy-headed boys are cute." "Thank you, Lord," I thought to myself. Thus began the relationship with Barbara Carroll, who would eventually become the love of my life and the mother of our three awesome children. *As it turned out Barbara was not your average beach bunny. She was a High School Math and Physics teacher.*

Barbara Carroll, a beautiful and brilliant southern lady.

FIELD CARRIER LANDING PRACTICE (FCLP): Finally, it was time to start field carrier landing practice. This phase is accomplished by painting a carrier deck onto a runway, setting up a Fresnel (pronounced fernel) lens for the pilots to use as a glide slope, and stationing a landing signal officer (LSO) next to the landing area to grade/critique the landings. It simulates the carrier environment pretty well, but the landing area does not move like the carrier deck will.

The LSO is a pilot who specializes in teaching the carrier-landing phase of naval aviation. In the fleet, he is responsible for the safety of the carrier landings, and he also assists the pilots who are landing by using a radio to talk them down, if required. Every landing on an aircraft carrier is assigned a numerical grade by the LSO. For training purposes, the LSO also makes comments on each landing using a shorthand language that is often politically incorrect. An example of such a grade follows: (-) HSCDLATOATIAR. This translates to "No Grade, High Start, Coming Down Like a Turd Off a Tall Indian at The Ramp."

The numerical grades may vary slightly between air wings, but generally break down like this: OK no comment (basically a perfect pass)–five points; OK with some comments–four points; (OK) fair pass (significant deviations from glide path and lineup corrected appropriately)–three points; (-) No Grade (significant deviations without proper corrections)–two points; (WO) Wave-Off (due to poor/dangerous pilot technique)–one point; (C) Cut Pass (awful pass that scared the shit out of everyone on the ship including the pilot)–zero points, or even minus one in some air wings.

A Cut Pass is guaranteed to make the replay of the day on the ship's TV. The results of these grades are posted in the squadron ready rooms via a chart called the greenie board. Many squadrons use the greenie board to form a Velcro landing ladder with pilot call signs attached in order of standing from first to last in landing grades. It is also a team sport in that all the squadrons within an air wing compete against one another to achieve the highest team grade. In other words, carrier-landing proficiency is extremely competitive within a group of already hyper-competitive people. Also, every landing is recorded by camera and shown on the ship's television system. And yes, they have replay just like professional sports. As we all know, team sports events or individual accomplishments often feature spectacular success. However, they might also replay colorful failures/bonehead mistakes. In the case of carrier landings, the more colorful (bad) landings are more fun to watch, so they are often featured on the television replays.

Once you are in the fleet and have a bad landing, you will probably suffer through your buddies watching that pass and laughing at you until somebody has one that is worse. As you may have guessed, the emphasis on carrier landing grades results in maximum effort toward perfection on every landing, which in-turn, promotes a safer environment.

The one LSO for our class that I remember was a combat veteran and colorful character by the name of "Super" Snyder. We conducted the FCLP landings at an outlying field, located just a few miles to the West of NAS Pensacola's Sherman Field. Bronson Field was much more remote than Sherman Field, therefore fewer complaints about noisy jets. This remote location offered us the opportunity to fly solo to and from Bronson Field. Super Snyder also flew an airplane out to Bronson, parked it, and then set up as the LSO. As I mentioned, he was a Vietnam combat veteran, a great LSO, and a colorful character who did not follow rules. In fact, he would put his flight gear on directly over his bathing suit (no flight suit) so that when he set up on the ground as the LSO, he could also work on his tan.

My logbook shows that I performed 84 FCLP landings in preparation to carrier qualify on the USS *Lexington* (CV 16). The FCLP landings were not just to prove that you could land accurately enough to catch a wire on the ship. They were also designed to make sure that you listened and responded to everything that the LSO said on the radio, and that you corrected deficiencies from one FCLP period to the next. The LSO was our savior (I don't remember if he used the name "God" or not, but he probably did), and as he often told us, "I am going to keep you from killing yourself at the boat."

On the transit back from my last FCLP flight, I was feeling pretty darn cocky about my aviation skills so as I got to the middle of Perdido Bay at about 1,500 feet above the water (where no one could observe my behavior), I decided to do an aileron roll. This did not go well. I did not pull the nose up enough prior to starting the roll, and the slow roll rate of the straight wing T-2B combined to result in the world's ugliest aerobatic maneuver and a recovery altitude of about 500 feet above the water. *It scared the crap out of me, but more importantly, it taught me a valuable lesson that kept me from ever doing anything that stupid again.*

***

AUGUST 6, 1968 — THE BOAT: We received one final briefing, which included the infamous "Crash Film" showing all sorts of awful mayhem from crashes aboard aircraft carriers. OK, here we go. We manned up as a four plane with an instructor pilot in the lead aircraft. Although the T-2B was a two-seat aircraft, no sane instructor was going to ride with a student for our first carrier landings. The lead's job was to get us to the carrier and to get us back to the beach after completion of our traps. Despite my extreme nervousness, I concentrated on the formation flying as we came into the break at the ship.

I was number four, so I had to work pretty darn hard to stay in position, and although I tried to sneak a peek, I really did not see the carrier until I was on the downwind leg. I don't remember all the numbers, but I worked very hard to get to the proper abeam position (in reality, I was totally brain locked and simply followed the guy in front of me). Our first two landings were touch and goes, versus actual traps. As Super Snyder said, "The touch and goes will get your bottom teeth to slam into your top teeth and that should finally engage your brain." He was right about that.

I honestly do not recall what my passes were like other than I managed to get four traps without any Bolters or Wave-Offs. As I was waiting in line to be launched off the catapult, one of the other aircraft catapulted and disappeared below the flight deck. The Air Boss immediately screamed over the radio "Aircraft off the cat pull your nose up, climb dammit!" He finally did, but the next guy in line would not move when the director tried to place him on the catapult. The Air Boss said, "Side number __, follow the director and get on the cat." Reply from student (I think it was Larry Dekker from Tupelo, Mississippi who had a Mississippi drawl), "No sir, I ain't going because that cat ain't working right." After a long pause during which I was certain the Air Boss was laughing so hard he couldn't talk, he said, "Son, there is nothing wrong with the catapult, you just have to pull the nose up after it gets you airborne."

The cat shot was exhilarating, as were all the daytime cat shots throughout my naval aviation career. After my 4th and last trap, the LSO said, "Congratulations Rud, you are now a qualified carrier aviator." I was absolutely elated. I cheered myself hoarse in my oxygen mask. They launched me off, I rendezvoused with the lead and my two buddies, and we were off to NAS Pensacola.

After landing in Pensacola, one of the LSOs flew in to debrief us. I don't recall what my grades were, just that they were good enough. I do recall that my buddy Chris Visher, my hippy roommate from AOCS, was the star with four OK passes. Next, we headed to the squadron administration office to pick up our orders. Mine were to report to VT-26 in Beeville, Texas the following Monday, which gave me about four days to travel 700 miles and be ready to start training in the F-9 *Cougar*.

This is about the time that I temporarily lost contact with Chris Visher. He did not follow us to pick up orders, but I did not think too much about it at the time. In fact, he was in the Commanding Officer's office, officially requesting to drop out of flight training.

As it turns out, he was doing this because he was a devout pacifist with no intention of ever killing anyone. He never shared this with me because frankly it was none of my business, but also because he knew I would try to talk him out of his decision. Chris ended up as a hangar deck officer on an aircraft carrier, and honorably served out his commitment to the Navy. He also proved to himself that he could handle flight training (in fact he was a superstar pilot). The last contact I had with Chris, I seem to remember that he was a lawyer, and I am sure a very successful one. A great guy that I will always consider a dear friend, and whom I greatly respect for having the courage to follow his beliefs.

**

MATURE DECISION-MAKING WAS CERTAINLY NOT MY STRENGTH AND WOULD NOT BE FOR QUITE SOME TIME: Larry Dekker and I headed over to the BOQ to finish packing, with the intention of departing on the first leg of the trip to Texas. Since all of us had already moved out of the beach cottage, the packing did not take long. "Say, wouldn't it be a good idea to celebrate our first carrier landings with a beer or two?" One problem: it was only 11:30 in the morning. Nothing on the base was open, but there was a dive bar a few blocks outside the gate that opened at noon. I said, "Let's go get a beer; I will drive."

We rolled the top down on the Tempest convertible and off we went to make the opening of this bar. Obviously, we were really pumped up. We had nothing to eat, had already successfully landed on a carrier, and all of this before noon. The place was a dump, but they had a pool table, and the beer was cold. All eight beers were really cold. Even though there were only two of us, with all the adrenalin we managed to get rowdy enough to be asked to leave the bar. "What? You can't kick us out, it isn't even 3:00 p.m. yet?" But they did.

While we were pounding down beers, a typical Pensacola thunderstorm had come through leaving a large puddle near my soaked convertible (yep, the top was down the whole time). Since we were going to be sitting in wet seats anyway, we decided to have a little water fight, splashing each other with muddy water from the puddle. This degenerated into a wrestling match that ended up covering both of us in red Florida mud.

Thank goodness this was pretty much before DUI was broadly enforced, so if you could still stand up and talk without slurring your words, you were good to go. We drove back to the base to get our stuff out of the BOQ and head for Texas. During the period of our ejection from the dive bar, one of the patrons had mentioned something about calling the cops. With this in mind, we decided that it would be a good idea to get across the state line into Alabama as soon as possible.

We were in what was at the time the newest BOQ. It was pretty nice with a swimming pool in between the two wings of rooms. The pool was very busy with lots of lady guests of the BOQ residents working on their tans. Larry and I looked at each other, decided that we were potentially already in big trouble, and took the shortest route to our rooms, which was a swim across the pool leaving a nasty red stain in the crystal-clear water. As I recall, once the screams died away, we got a standing ovation from the ladies that knew us. Now, it was *really* time to get across the state line.

One very important casualty of our hasty departure to Texas was a proper goodbye to my girlfriend, Barbara Carroll. As I mentioned before, I was incredibly immature and selfish. Not too far from the description of a typical naval aviator that a comedian came up with: *"A naval aviator is certainly capable of love and affection, it just doesn't involve anyone else."*

8. Earning the Wings of Gold

ADVANCED JET TRAINING COMMAND: There were a total of six advanced jet-training squadrons. Three were located at NAS Kingsville, Texas, and three were located at NAS Beeville, Texas. VT-21, 22 and 23 were in Kingsville, and VT-24, 25 and 26 were in Beeville. Kingsville was a good-sized college town, and it was located only about an hour's drive from Corpus Christi where there was a large Navy base and a beach community. Beeville, on the other hand was located smack dab in the middle of nowhere. It was at least two hours to Corpus Christi and about the same to San Antonio.

The only social life revolved around the officer's club on the base. The good news was that there was not much time for socializing anyway. It was now the fall of 1968; the Tet Offensive in South Vietnam and *Rolling Thunder* (the bombing campaign over North Vietnam) had taken a toll on naval aviation in both the carrier navy and the land-based Marine Corps. Replacements were badly needed so we literally flew seven days a week. I flew a total of 107 flights totaling 143 hours in less than four months. It was a grind, but except for the instrument (under the bag) flights, the flying was awesome.

All six squadrons were equipped with the two-seat TF-9J *Cougar*, and the single-seat AF-9J *Cougar*. The *Cougar* was a real fighter plane coming into the Navy inventory in 1956 to replace the F-9 *Panther* that was used in the Korean War. The *Panther* was straight wing, and the *Cougar* was swept wing with all the pluses and minuses of that aerodynamic design. It was fast and maneuverable at altitude, but under-powered, and unforgiving in the landing configuration. It also had a long takeoff roll, and the landing, especially in the single-seat version (mechanical one-to-one braking), required most of a standard Navy 8,000-foot-long runway.

Built by the Grumman Company, it was an especially sturdy design. So much so that one aircraft survived a crash that killed the student pilot. He got disoriented on a night takeoff and literally flew the aircraft into the ground. The impact killed the pilot, but the aircraft looked as though it might be flyable again (it wasn't). In another incident, a student pilot was practicing aerobatics on a solo flight when he collided with a bird. He felt and heard the impact, observed a dent in the wing, and returned to Chase Field. He reported the incident to the duty officer who at the same time was receiving a call from Randolph AFB in San Antonio. Apparently, a T-38 trainer had reported a mid-air collision with an F-9. The T-38 was so badly damaged that he barely made it back to base. As it turns out, that was the "bird strike" that dented the F-9. Sissy Air Force stuff needed to steer clear of our manly *Cougars*.

We did get some quality liberty in San Antonio. The 1968 World's Fair called "HemisFair" was on going from April 6 through October 6. My AOCS classmate and fellow student pilot, "Bug" Roach, was married to a girl who worked for Bell Telephone. She was employed at HemisFair, so he had the "gouge" (inside information) for having a good time at this spectacular event. Since the event was winding down, we decided to get a bunch of us together for a night of fun (debauchery). I can't recall all the folks involved, but some were from Kingsville, which was about a three-hour drive from San Antonio. Another one of my AOCS classmates, Jack Ewald, who was in a Kingsville squadron, had gotten into a bit of trouble for having a girlfriend living in the BOQ with him. He was put on restriction and ordered to stay in his BOQ room (sans girlfriend) for a week. Jack had a "crotch rocket" motorcycle so he blew off the restriction and headed for the big party at the HemisFair.

Unfortunately, on the way back to Kingsville, he lost control of the bike, and got banged up badly. Oh shit! Not to worry. Another student was following him in a car, picked up his battered body, and deposited him at the bottom of the steps leading to his BOQ room. He lay there moaning until one of the BOQ staff members found him. He said, "I tripped and fell down the steps." Are you shitting me? With the help of a friendly flight surgeon this ridiculous explanation was accepted and "Black Jack" continued on to be a shit-hot fighter pilot.

***

TRAGIC MISHAP AND LOSS OF A CLASSMATE: My class in VT-26 consisted of two Navy and two Marine student pilots. Don Roesh and I were the Navy guys. Mike Spivey and Dave Salisbury were the Marine students. Don was a Naval Academy graduate and matched me pretty well as being a bit on the immature/wild side. In other words, we were a lethal combination while on liberty. His father was a WWI fighter pilot and Don had awesome stories about his father's exploits. Oh, by the way, his father flew for the Germans.

All four of us quickly became close friends as we teamed up to study and prepare for the various training flights. Since the goal for naval aviators is to land on carriers, we practiced that type of landing on every flight. As we progressed through the syllabus, we all were scheduled for, and flew, similar flights. Near the end of the syllabus, all four of us were scheduled for two-plane solo night formation flights. We were flying the single-seat version with our respective instructors leading the flight in another aircraft. On completion of the area work, all four of us headed back to the field each flying formation on our respective instructors. The plan called for us to enter the carrier break, take proper interval on our instructors, and then perform a few night touch-and-go landings prior to making a full stop.

It was a very dark night, so I was concentrating hard on staying in position as my instructor led us to the "initial" and checked in with the tower. Just as he was about to check in, there was a frantic radio call from the tower, "Aircraft on the outboard runway, aircraft on the outboard runway, PULL UP, PULL UP!" I then observed a blinding flash in my periphery followed by the tower broadcasting; "Crash, crash on Runway 13R, Chase Field is closed!" My instructor told me to switch to squadron tactical, and we began a climb on a direct heading to NAS Kingsville, which was our briefed "emergency bingo" field. The awful part of this nightmare was that I was the last of the four of us to takeoff, so I knew that all my buddies were in that landing pattern ahead of me, and that now one of them was involved in this horrific crash.

I flew formation on my instructor's wing to Kingsville. I seem to remember performing a rather hairy night landing and then pulling into one of the Kingsville squadron's flight lines. I cannot recall if we spent the night or gassed up and returned to Beeville that same evening. In any case, it was not long before we learned that our good friend and classmate, 1st Lt Dave Salisbury had been killed in the crash.

I don't know what the final accident report stated, but we all suspected that he was the victim of some aircraft flight control issue, most probably related to the AF-9's hydraulic tail system. Dave was the only one in the class who was married, and his lovely young bride of just a few weeks, Bonnie, was now a 19-year-old widow.

Forty years after this tragic accident, I was attending an event in Washington, D.C. that included General Jim Amos, USMC. At the time I was representing the Boeing Company as a marketing executive. I knew the General from my time on active duty so as we shook hands he asked me, "Gil, have you ever met my wife?" I replied, "No sir, I don't believe that I have." His wife had her back to us at the time, so the General said, "Hey Bonnie, I would like you to meet a friend of mine, Gil Rud." She turned around, our eyes met, and she rushed over, hugged me and gave me a kiss. Then she declared in a very loud voice, "Gil Rud, it was forty years ago tonight!" The General, having a great sense of humor, said—also in a very loud voice, "God dammit Rud, you slept with my wife!" "No Sir, I am pretty sure I would remember that." Of course, she was referring to the night her first husband, my classmate, Dave Salisbury was killed, and the General knew that. However, the Boeing brass had no idea what was going on and could literally see millions of dollars of business with the Marine Corps being lost to my alleged affair with the General's wife. It really is a small world.

Although we were certainly used to accidents and even loss of life during our time in the training command, this was the first time that it really hit close to home. We were all devastated. The squadron gave the three of us a day off, and of course, we decided to honor our fallen brother with some hard-core drinking. We were still partying hard at 2:00 a.m., when we get a call from the squadron duty officer. I took the call and he said," Just calling to let you guys know that *Lexington's* carrier qualification schedule has moved up, so we have to start your Field Carrier Landing Practice tomorrow. "I assume you are talking about tomorrow afternoon?" "Nope, the reason I am calling at this absurd hour is that you are all scheduled to brief for your flights at 0530." "OK, got it. See you in three hours."

I looked at the two other inebriated pilots and told them what the SDO said. I then got this blank stare and then, "Holy shit! We have got to sober up, starting right now!" (There was no way that we would beg out of the flight, that just did not happen in 1968). Don Roesh and I immediately brewed some coffee and started to prepare some breakfast to initiate a semblance of sobriety. Mike Spivey, on the other hand, being a tough-ass Marine just grabbed another beer. As I recall, we were in my single-wide mobile home that I shared with Bill Beatty, who was in one of the other squadrons. There was enough room for all of us to eventually sack out for a couple of hours. Since my humble abode was only about 15 minutes from the base, Don and I awoke at 0500, and he headed for his BOQ room to get in his flight suit and make the 0530 brief. No Mike Spivey. I thought, "I guess he drove back to the base to get some sleep."

We accomplished our FCLP landings at an outlying field called Alice Orange Grove. It was about a twenty-minute flight from Chase Field. The only briefing requirement we had was to check in with the SDO to get our aircraft assignments. Don and I checked in at 0530 as required. We inquired as to Mike Spivey's whereabouts, but the SDO had not seen him. As we were walking down to maintenance to check the maintenance gripe sheets for our assigned aircraft, we heard this AF-9 start up and begin to taxi. I asked the maintenance chief what was going on and he said, "That is Mr. Spivey, he showed up here and said, "Give me an aircraft chief, I need to get in the cockpit and get some of that good old oxygen flowing." I thought he was acting a little odd, but you pilots are all a little crazy anyway, so I signed out an aircraft for him and sent him on his way."

On our way to pre-flight our assigned aircraft, we ran into Mike's plane captain. He was just a youngster, probably 18 or 19 years old. He said, "Mr. Rud, if I was a betting man, I would wager a whole lot of money that Mr. Spivey is shit-faced drunk. He did not even pre-flight his aircraft, he just got in, started it up and taxied to the runway." I quickly dismissed this observation and covered for him by saying, "No way is he drunk, he is just not a morning person."

Don and I did our pre-flight routine, manned up, and headed for Alice Orange Grove.

We completed our FCLPs and landed at Alice for an LSO de-brief and refueling of our birds. Spivey had not shown up, so we assumed he had some issue and had returned to Chase Field. About half-way through the LSO debrief, an AF-9 came roaring into the break, landed, taxied to parking and shut down. Out of the cockpit climbed a very white-faced, and now quite-sober Spivey. He told the LSO that he got a late start due to aircraft maintenance problems, and then he pulled Don and I aside to relate the real story.

"I was drunk and pissed about not getting an opportunity to properly mourn the loss of my friend. I thought that I'd better get an early start and suck up some of that oxygen to sober up before heading to Alice. I climbed to about 10,000 feet and headed in the general direction of Alice Orange Grove. That is the last thing I remember until I woke up at less than 1,000 feet with nothing but water everywhere. After the initial panic subsided, I correctly deduced that the only water this large had to be the Gulf of Mexico. I headed north, and after about 30 minutes of flight time, I finally spotted land. By now I was sober enough to identify this land as Texas. I successfully navigated my way here, and I will never drink another drop of alcohol again." That resolution lasted all the way to that same evening's happy hour, where of course his story began to achieve legendary status.

***

LIFE-CHANGING EVENT: Keeping in mind that we were in an era devoid of cell phones and computers, our main method of communication was via mailed letters. I had exchanged a couple of letters with my girlfriend in Pensacola. I really missed her, but we did not discuss anything too serious as we had only been together for a wild and crazy six weeks, and really did not know each other that well. Near the end of my flight training, as we were preparing for carrier qualifications, I got a call at the squadron from my cousin, Don Leland. Don was a Marine, a year or so behind me, going through early flight training in Pensacola. He had run into Barb, and gave me her phone number, stressing that she really needed to talk to me. "Great, I will give her a call."

I called her that same evening, and my life changed forever. She was pregnant with my child, who, by the way, would eventually become a better pilot than me. But, at this point, I was pretty much in shock. We agreed that Barb should resign from her teaching position and join me in Beeville as soon as possible. Once she arrived in Beeville, we would figure out where and how to proceed from that point. Barb had a new 1967 Ford Sedan, so transportation was not a problem. She arrived about a week after the phone call and moved in with me. Although we were not close friends, my roommate, Bill Beatty, was an incredibly caring southern gentleman. In fact, he seemed to be more understanding of the situation than I was.

Flight Surgeons often talk about mental distractions being primary causes of aircraft accidents. The cure for this is for the pilot to compartmentalize. This can best be defined as the ability to shut out all thoughts except those required for flying the aircraft. It certainly wasn't easy, but I managed to do it, and would rely on this skill to keep me alive through some mighty challenging situations in the coming years. The downside of compartmentalization is that to accomplish it, you must literally ignore problems; and especially people associated with those problems. This is not good for developing relationships.

Keeping in mind that this was 1968, the middle of an era that was known for acceptance of non-traditional behavior, it was not considered untoward to have children with a partner without being married. Although we were strongly attracted to one another, Barb and I barely knew each other. We discussed getting married but agreed that it would have little chance of success until we got to know each other better. I had orders to Navy Recruiting District Minneapolis for six weeks of what was called, "Feedback Recruiting." Barb had several close girlfriends who were in the process of moving to California. My destination after completion of the recruiting assignment was to the A-4 *Skyhawk* Replacement Air Group (RAG) in Lemoore, CA. We decided that Barb would join her friends in the Los Angeles area where there were plenty of jobs and support for her, while I continued to Minneapolis and eventually Lemoore. We made no commitments other than financial on my part. *The bottom line: Barb made a mature, thoughtful and selfless decision, while I continued my immature, self-centered journey as a naval aviator.*

***

Air-to-Ground Bombing: By this phase of advanced training in Beeville, I had pretty much decided that I wanted to fly Light Attack; either A-4 *Skyhawks*, or A-7 *Corsair IIs*. Most of these flights were solo with an instructor leader who would circle the manned target as a safety observer. The target was a ringed bull's-eye. The bombs were 25-pound practice bombs called Mark-76s. These bombs included a small charge that would create a white puff of smoke so that the observers in a tower at the range could call the accuracy of your hit. I absolutely loved this phase and excelled at it, earning my best overall grades as a student pilot.

During this phase, one of the other Beeville squadrons had what turned out to be a rather hilarious incident. The instructor leader was observing the students bombing. He had a passenger in the back seat, who was not a pilot. That passenger began to complain to the instructor that it was awfully hot in the back seat. This is a common reaction for a non-aviator who is beginning to feel airsick. Finally, he said to the pilot, "Is it normal to have fire under the floorboards?"

A few seconds later the aircraft exploded as they successfully ejected from the crippled machine. The instructor did not have time to call a "mayday," so the students just continued to bomb. They finished bombing, joined up, and headed back to Chase Field assuming that the instructor must have lost his radio and would get back on his own. The student now leading the flight did make a call to the range tower, "Just want you to know that there is a significant brush fire just west of the bombing range..." (this was the instructor's burning aircraft). The folks manning the range tower drove a truck over to check on the fire and found the instructor and his passenger in relatively good health, although the instructor was extremely upset that his students failed to recognize a crash site versus a brush fire!

***

CARRIER QUALIFICATION IN THE F-9 COUGAR: The USS *Lexington* (CV-16) was docked near Corpus Christi, Texas. Since there were more student pilots than aircraft available to fly out to the carrier, I was one of those chosen to walk aboard the ship. We all gathered in the ready room with the intention of "hot seating" into an aircraft after the fly-aboard pilot finished his qualifications. Hot seating meant that the aircraft would be tied down but stay running while we exchanged pilots. The aircraft was also refueled at that time so that we would have enough fuel to complete six carrier landings.

While gathered in the ready room, we watched the initial landings of our fellow student pilots. One fellow made his first trap, and then they parked him instead of shooting him back into the pattern. We were a bit confused about this until he walked into the ready room bleeding like a stuck pig from a huge cut over his nose.

He had forgotten to fasten his torso harness to the ejection seat so while completing his arrested landing; he was flung forward in the cockpit smacking his head into the instrument panel. As if we were not nervous enough, this event put the average blood pressure in that ready room to 300/250.

We finally got our chance to qualify, and although I do not recall each pass, my logbook shows two touch and goes, six traps and one Bolter. The good news was that those of us who hot seated got to fly the aircraft back to home base, while the other guys had to ride the ship into port.

Following Carrier Qualifications, we had a few more flights that were the most enjoyable flights in the training command. They were called road reconnaissance or more commonly "road recces" and were based on the experience of the Vietnam Veterans who had flown these types of missions, hunting down enemy trucks, etc. Once again, we students were flying solo in AF-9s with an instructor chase pilot.

The one flight that stands out to me was flown with a plowback (recently winged student assigned as an instructor as his first tour before being assigned to the fleet). His name was Ron "Running Bear" McKinney, and we would later be squadron mates in VA-215 on a wartime-cruise on USS *Oriskany*. Ron was one cool dude, and he was famous for allowing students to enjoy the flight, hunting down numerous farm vehicles, trucks and even buses for mock attacks. All of this was accomplished at extremely low altitude and high speed. We would locate a target, call it out, and then pop up to start a simulated low-level gun run on the unsuspecting vehicle.

I spotted a bus used to transport farm workers. It was on an old dirt road in the middle of south Texas somewhere. He was coming up a small hill, so I got down to about 100 feet above the ground and met him at the top of the hill at 400+ mph. He promptly left the road (fortunately there was no ditch) and bounced out into a field. According to our instructor, the driver got out of the bus and gave us the universal salute of his middle finger. Running Bear gave us all above average grades for that very fun flight.

I completed the advanced training command syllabus and finally earned my Wings of Gold. Normally this would be a rather impressive ceremony, however in 1968, it was more of a pin on the wings and be on your way to the fleet. The captain of the base was the officer that pinned on the wings. As he made his way down the line of students, he would ask each student where he was from. He asked the pilot standing next to me, Rick Maixner, where he was from. Rick was in a different squadron, and I had not met him up to this point. Rick replied, "Sir, I am from New England, North Dakota." The captain lookeds a bit taken aback and said, "I have never met anyone from North Dakota." I then interrupted with a smile, "Well captain, today is your lucky day." He looked at me with a huge frown and said, "What do you mean by that?" "Captain, I am from Portland, North Dakota." He was truly amazed.

This was the beginning of a life-long friendship between Rick and me. He was recruited out of Minneapolis and had orders to feedback recruiting and then on to Lemoore to fly A-7 *Corsair IIs*. Eventually we would become ship-mates on a Vietnam Combat Cruise embarked in *Oriskany*. He was a member of the VA-155 *Silver Foxes,* and I was in the VA-215 *Barn Owls.*

9. Feedback Recruiting

NAVAL RECRUITING DISTRICT MINNEAPOLIS: Barbara had already left for Southern California, so immediately following the winging ceremony; I was off to Minneapolis. It was quite a shock to go from south Texas to Minnesota in December. It was also a shock to my Pontiac Tempest, costing me a water pump replacement somewhere in Nebraska. Following a quick stop in Fargo to see my parents and sister, I checked into Navy Recruiting District Minneapolis. It was located on the military side of Wold Chamberlain Field, which was the main airport for the twin cities, and is also now known as Minneapolis Saint Paul International Airport. The Navy Reserve was stationed there flying P-2V *Neptune* Anti-Submarine Warfare aircraft.

The Recruiting District office used an area of the reserve base where they not only recruited from, but also gave aviation physicals, and swore-in new recruits. My boss was a Navy helicopter pilot, Commander "Rock" Rowell (I don't recall his real first name because everyone referred to him by his call sign). He was a no-shit warrior, with hundreds of combat missions in Navy Helicopter Attack (Light)-HAL-3. He had a photo on his desk of a crashed *Huey* in a rice paddy with him standing on a levee firing a Browning Automatic Rifle (BAR) at the bad guys.

Rock did have a T-34B, but due to the awful winter weather and some maintenance issues, it was not available for us to fly. So, we just kind of hung around the office, and talked up the program with new recruits. In other words, I was bored stiff, which usually leads to trouble.

Rock had a very attractive young lady that served as a receptionist. She was also responsible for giving newly sworn-in Aviation Officer Candidates an overview of the process they would go through to earn their Wings of Gold.

She had pictures of the training command aircraft on the wall behind her desk. She had them arranged with the T-34 first, branching off to the jet, prop and helicopter pipelines. I noticed that she had memorized her presentation, simply pointing over her shoulder at the various aircraft. Of course, she was such a knockout that the average newly sworn in applicants were mostly focused on her anyway.

My bored and slightly evil mind began to develop a plan. While she was at lunch, I swapped out all the aircraft photos for playboy centerfolds. When she returned, she did not even glance at the photos. In a few minutes, a couple of excited and newly sworn in Aviation Officer Candidates showed up for her presentation. They desperately tried not to laugh as she began pointing over her shoulder to the photos, initially suppressing giggles that soon turned into outright laughter. She was astounded at their reaction to her well-intentioned briefing. Then she turned around, saw the centerfolds, and began to scream, "Oh my God! Oh my God!"

I had retreated to a small office that I shared with Rick Maixner when my boss, Rock Rowell, bellowed at the top of his lungs, "Ruuuud! In my office now!" As I entered, he threw a set of car keys across the room at me. "Get your ass on the road right now! Take the Chief with you and do not come back except to check out and head for the fleet!" *Well, I thought, so far I have certainly impressed my first boss!*

The good news was that the road was exactly where I belonged. Our first stop was the University of Minnesota, Duluth (UMD). Our goal was to administer the Academic Qualification Test/Flight Aptitude Rating (AQT/FAR) qualifying examination to as many students as possible. If they couldn't pass this test, then we were wasting our time trying to recruit them.

The first step in this process is to get the attention of a potential recruit. We did this by setting up a display in the student union. We were decked out in our dress blue uniforms, so it was easy to gain the attention of the students. This was January 1969, so things were a bit testy on all the campuses. Shortly after we set up our booth, the Students for a Democratic Society (SDS), a rather radical anti-war group, set up right beside us. No big deal, and they were really pretty nice kids (even radical anti-war folks from Minnesota tend to be polite). After we had been talking to potential recruits for a couple of hours, a student showed up in a Vet's Club jacket. He took me aside and suggested that the Chief and I go get a cup of coffee. Then he winked and said, "Don't worry, we will watch your stuff."

We headed into the coffee shop, ordered and sat down to drink our coffee. Suddenly, many of the students got up from their tables and headed toward the area where our booth was. The Vet's Club boys were in the process of destroying the SDS booth and physically removing the folks who were manning it, depositing them unceremoniously into a snow bank outside the student union.

My other memory of UMD is much more pleasant. One of the sororities set up a booth near us. It was a kissing booth. Yep! The pledges were selling kisses to make some money, and I would have been happy to help them fulfil their requirement for earning active status in the sorority. They did not seem to be getting much business. (Believe me, I was sorely tempted to contribute a few dollars, but the Chief convinced me that it would be un-officer like). One of the girls did ask me if I had any advice that would improve their business.

Oh no! My slightly demented mind came into play, and I said, "You know, back when I was in college, the most successful kissing booths named their various kisses and charged accordingly." She was cute, but very blonde and unbelievably naive. "Really?" She said, "Yes. I recommend you start with just one." "What should we call it?" "I recommend you call it 'Around the World.'" Business rapidly picked up until one of the older sorority girls came by the booth. It wasn't long until she looked over at me with her hands on her hips, giving me a rather stern adult to child stare. *It would not be the last time that I got one of those looks.*

We did exceedingly well at UMD, successfully testing several promising candidates. Our next stop was Winona State University, which is right on the Wisconsin border. It was a great party school, and we added a few tested recruits to our list.

Now it was time to head for the recruiting gold mine: the mini twin cities of Fargo, North Dakota and Moorhead, Minnesota. With three colleges to work, we stayed here the rest of our road trip. At North Dakota State, rather than set up a booth in the student union, I would work the fraternities, even conducting the tests at their facilities. I also worked with the ROTC folks to offer another potential path to becoming a pilot. *Bottom line: A huge number of tests, which was our mandate when Rock sent us on the road.*

10. Introduction to the Scooter

A NEW RIDE AND MY FIRST TRIP TO THE WEST COAST: I checked out of NRD Minneapolis and headed to Fargo for a few days with my family prior to beginning the trip to California. Since my Tempest was paid for, I decided to check out a potential new ride that was worthy of this self-described, shit-hot naval aviator. I found it in the showroom of the Oldsmobile dealership in Fargo. It was the ultimate muscle car, a 1969 Oldsmobile W31 Cutlass. It was red with white racing stripes on the hood. It had a 350 cubic-inch, 325-horsepower engine with a 4-barrel carburetor that included direct ram air induction with two scoops located underneath the front bumper. I don't recall the exact gear ratio, but it was built for 0–60 awesomeness rather than top-end speed. The darn thing looked like it was speeding when it was still parked. I added an after-market tachometer that strapped onto the steering wheel to give it even more testosterone. I was now ready for the big trip.

It was February in North Dakota so the weather was unpredictable to say the least. I got all my belongings into the W-31 and started west on Interstate Highway 94. I got just past Jamestown (about 100 miles into a 2,500-mile trip) when I ran into a nasty snowstorm. It was a typical blizzard with very low visibility due to blowing and drifting snow. I got behind an 18-wheeler that gave me some protection and served as a bit of a snowplow.

After a few miles of slow speed progress, he finally put on his flashers and stopped. We sat there for a few minutes until I decided to walk up and get into his cab. He was a typical over-the-road driver. He had his CB radio tuned in and was busy talking to another truck that was just a few miles west of us. That fellow told us that it seemed to be getting better and that since it was moving from west to east, we should be able to proceed in an hour or so. It finally cleared enough so that we could start moving. Before I got out of the cab, the driver mentioned that he was going to try make it over the Missouri River bridge to Mandan, North Dakota because they were on Mountain Standard Time versus Central, so the bars would still be open. "Sounds like a plan to me."

I followed him to Mandan. We pulled into this country western bar that was indeed open and hopping. Having no real plan at this point, I decided to have a beer and then look for a motel. Now of course, Mr. Cool here was adorned in my leather flight jacket as I entered the bar. As I mentioned, it was a very active place. Above all the racket, I hear this very familiar voice, "Oh my God, Gil Rud, is it really you?" It was Karen Fosse, a former girlfriend from college, and as I mentioned earlier in the book, a partner in a misadventure or two. It was a great reunion as we caught up on all our mutual friends, and she was kind enough to provide a roof over my head for the night.

The next day found me back on the road with the intention of eventually picking up Interstate Highway 80 (I-80), which would take me to California (keep in mind that I had never been west of North Dakota before, and this was long before GPS). Prior to reaching the interstate, I got off the highway to grab a bite to eat in a small town in Wyoming. I was sitting at the only stoplight in town when a police car came up behind and signaled me to pull up to the curb. He got out of his vehicle, took a long look at my car (I had Texas license plates because there was no state tax, and having been stationed there, I could claim residency). The W-31 did not like to idle so I had shut it off. He asked me for my license and registration, looked in the back seat, and saw my flight gear and helmet. "Are you some kind of race car driver with this wild-ass car and all that stuff in the back? If you are, I want you to get this here vehicle the hell out of my town before some kid sees it and wants to race. Then I will have to pick up his bloody body and go tell his mother. Now start this thing up so I can determine if it is even street legal."

I started it up and of course it barely idled and sounded like what it was—a virtual rocket ship. He told me to shut it off, then said, "What the hell are you doing way up here anyway?" I told him I was a Navy pilot, that my flight gear was in the back, and that I would not race any kids in his town. His demeanor immediately shifted, and he said, "I gather that you will be on your way to Vietnam pretty soon. Thank you for serving our country, and if I had the time, I would buy you lunch. Now get this thing back on the highway, and out of my jurisdiction."

I thoroughly enjoyed the trip through Salt Lake City and finally into Nevada. Surprise—no speed limit. I certainly abused that privilege, probably taking about 10,000 miles off my tires at a steady 100 mph trip to Reno, where I stopped to get some gas. The attendant complimented me on my muscle car, and then offered some advice, "You better get yourself a set of tire chains for the ride over the Sierras to California." "What? No way, I am from North Dakota, and just navigated a big blizzard up there. I don't need chains to navigate a few inches of snow in the Sierras."

He replied, "Donner Pass currently has 15 feet of snow, and they will not let you over the pass without a set of chains on." "Holy shit! Did you say 15 feet of snow? OK, give me a set of those chains." The trip through Donner Pass was a real eye-opener for this North Dakota farm boy. I had seen plenty of snow, but nothing like what accumulates in this area of the world. But the temperatures were still sissy-nice compared to what we endured in North Dakota.

***

ARRIVAL AT NAS LEMOORE, CA: The San Joaquin Valley is often cloudy/foggy during the winter, but on this day, as I drove down the west side of the Sierras on I-80 toward Sacramento, I was greeted with the most beautiful view imaginable. The visibility was probably 50+ miles and all I could see were these beautiful green fields. Wow! Paradise is what it looked like to me, and I thought to myself, I am going to love it here.

Of course, it wasn't quite as enticing when I drove through the gates of NAS Lemoore, but the flat farm country with all sorts of machinery working the fields right up to the gate made me feel at home. I checked into the Bachelor Officer Quarters (BOQ), and then headed down to the operations area of the base, which is a good 5 miles from the rest of the base. With a 100,000-acre easement controlled by the Navy, Lemoore was both the newest and the largest master jet base in the Navy. The easement was necessary to keep the public from building too close and then complaining about jet noise.

Unlike the Air Force Strategic Air Command (SAC) bases, the Navy leased the farmland included in the easement back to the farmers so that they farmed right up to the edges of the massive 13,000-foot runways. Since the base was only a few years old (opened in 1961) everything was relatively new, including the hangars.

VA-125, my destination, was the first hangar on the left (west side of the field) as one drove through the underpass that supported a taxiway connecting the two sides of the field. Wow! The parking lot was full, and the ramp was absolutely jammed with A-4 *Skyhawks*, affectionately called "*Scooters.*" I am not sure of the exact number, but I seem to remember it as being close to 100 of them, just in VA-125. I was not sure what to expect, so I was pleasantly surprised by the attitude of both the instructor pilots and the administrative staff who checked me in. *It was obvious that although once again a student, this was no longer the training command.*

I headed for the Officers' Club and soon ran into several of my buddies from the training command. It became clear that staying in the BOQ was not the best option, especially since we were authorized BAQ (extra money) to find a place in town. Lemoore was the closest town, about 13 miles from the base with a population of about 10,000. Hanford, a town about twice the size of Lemoore, was another option, but was located 23 miles from the base. I ended up teaming with Doug "Mini" Hiatt to rent a small, two-bedroom apartment in Lemoore. Doug, an Oregon State NROTC graduate, had just checked into VA-122, the A-7 *Corsair II* Replacement Air Group (RAG), so our timing was perfect.

These were all one-story furnished duplexes. The good news was that the complex had a swimming pool, which was perfect for conducting happy hours and cookouts. Our neighbors were mostly single folks or young married couples. Our next-door neighbor happened to be employed by the local Union 76 station as a mechanic. He was infatuated by my W-31 Cutlass and offered to drive it to the Union 76 gas station and perform a tune-up at cost.

Hmm, I should have seen this coming. I gave him the keys in the morning and caught a ride to work with Doug. When I got home that afternoon, the mechanic pulls up in my muscle car, which sounded even louder than normal. He was super excited and began to relate what a great day it had been. "We tuned-up your car, and then thought we'd better test it out against a GTO Judge, the beast that everyone in town considers to be the #1 muscle car in Lemoore. We lost the first run, but here is the great news. We cut off your exhaust pipes, turned the ends on them right and left so now they are much shorter and a little louder. The exhaust now comes out just behind your doors versus out the back. And we went back out and hammered that GTO Judge. You are now officially the fastest 0–60 street vehicle in Lemoore, and we had so much fun that we are not even going to charge you for the exhaust pipe modifications." From that point on, it was not unusual to be stopped at a light, and have some kid roll up, asking for a race.

***

SURVIVAL TRAINING: Next on the agenda was mandatory survival training, which was required for all potentially Vietnam-bound aircrew. The initial training took place at NAS North Island, located on Coronado Island near San Diego. Of course, we Lemoore folks were super excited to get an opportunity for a road-trip to San Diego. The folks who were flying S-2 *Trackers*, E-1s/E-2s, or helicopters on the west coast were stationed at North Island. The fighter guys flying F-8s and F-4s were stationed at Miramar, the Marine fighter pilots were stationed at MCAS El Toro about an hour north of Miramar, and there were also Marine Ground Force bases plus Navy Black Shoe bases (surface warfare), and even a submarine base. The good news was that most of these military facilities had awesome officer's clubs.

To spread their business, they would each pick a night to have happy hour, a band and invite lots of ladies. As I recall, it was Miramar Wednesday, Marine Corps Recruiting Depot (MCRD) on Thursday, North Island I-BAR and Officers' Club on Friday, and a special treat was the North Island Down Winds Bar near the beach on Sundays. Whew! It could be hard on the body keeping up with all these opportunities, and still fly airplanes.

The survival training was intense. It involved a simulated Prisoner of War (POW) camp that was pretty darn realistic. All the aircrew were mixed, including some very young, enlisted folks that were a bit more impressionable than some of us "more-worldly" pilots. We were all trucked up into the mountains near Warner Springs, California, which is about 60 miles northeast of San Diego. We were offloaded from the trucks as though we had just survived a crash in enemy territory.

We were given a destination on a crude map and given two hours to reach that destination. The problem was that a large contingent of "bad guys" was hunting us down during that two-hour period. I decided that there was no way to make it to the destination without getting caught, so I found a good hiding spot and took a nap. At the end of two hours, they sounded a siren to indicate that this portion of the exercise was over, and those of us who had not yet been captured, came out of hiding.

That is when the shit really hit the fan! I found out after the completion of the training that they concentrated on those of us who basically blew-off the exercise to get some rest. The treatment we received was realistic. It included lots of physical and verbal abuse, ending in confinement. The confinement was to a box or crate-like device that was not all that uncomfortable unless you were claustrophobic. It was warm so every few minutes a guard would knock on the box to check on our health. You were ordered to respond to this knock with your name and serial number. During our briefing prior to this entire POW experience, we were encouraged to provide reasonable resistance to our captors.

I decided that reasonable resistance could be accomplished by not answering when the guard knocked on my box. This action forced him to open the box, pull me out, knock me around a bit and stick me back in. I did this a couple of times, feeling quite proud of myself. The third time I did it, instead of pulling me out of the box, they grabbed one of my neighbors, who was a young P-3 *Orion* crewman. He was probably 19 years old and scared to death of this whole situation. He pleaded with me to answer when they knocked so that he would not get beaten up. Wow! This tactic by our captors was very successful, as my resistance ended immediately, and I learned a great lesson about teamwork in any POW situation.

After several hours of enduring interrogations, including a form of water boarding, we were sent to a more central area of the camp for the night. Oh, by the way, before the exercise began, we were told that if we somehow escaped from the compound, we would be presented with a peanut butter and jelly sandwich. The idea of escape became an obsession with me.

I observed the watchtowers around the prison compound and decided that the one closest to the camp latrine was not manned. Being a farm kid, and having fenced miles of cattle pastures, I was very familiar with barbed wire. As I recall, the fencing around the compound consisted of several layers of barbed wire. The latrine building provided a screen of sorts from the normal guard patrols, so I began my escape. I carefully crawled under several layers of barbed wire, until I was out of the camp.

Elated, I stood up turned around and was looking down the barrel of an AK-47 (or whatever they were using to simulate one). It had taken me at least an hour to accomplish the escape. At this point the guard got this big grin on his face, pointed at the barbed wire, and said, "Stupid Yankee Air Pirate, you got out, now you will go back in the same way." I spent most of the rest of the night carefully working my way BACK INTO the prison compound. Embarrassing to say the least. *At the conclusion of the training, I was informed that due to my stubborn attitude and actions, I would not have survived being a prisoner of war. Believe me, I took this training seriously, and felt that I was now much better prepared to survive honorably in a POW situation.*

***

BACK TO FLYING IN VA-125 RAG: I was extremely excited and more than a little nervous as I prepared for my first flight in the TA-4F on March 25, 1969. As I mentioned when I covered checking in to the squadron, there were a massive number of A-4s on the flight line. Everyone had a primary aircraft assigned for a particular flight and a backup if that one went down. In other words, if scheduled, you were going to go flying. There was no doubt that the Navy was in a hurry to get us to the fleet. As I review my logbook today, I find that I flew 100 flights totaling 156 hours of flight time in just over four months.

During one of my initial two-seat training flights, the instructor demonstrated asymmetric leading edge slat deployment. He did this by slowing the aircraft to just above stall speed and then initiating a turn that would deploy only one of the two slats, resulting in a very uncomfortable aerodynamic reaction from the aircraft that required careful control manipulation to recover safely. After demonstrating the situation and recovery, he gave control of the aircraft to me.

I was a bit timid in my set-up for the situation, so he said, "Go ahead and pull that nose around sissy!" I responded with a bit more exuberance than required and that slat not only deployed asymmetrically, but also caused the aircraft to depart from controlled flight (stall) with a nasty adverse yaw (the nose slid across the sky in the opposite direction of my turn). I initially let go of the controls to see what would happen. The nose pitched straight down, and the aircraft then began a tight right turn. The airspeed was not increasing so I correctly deduced that we were in a spin. I applied anti-spin controls and the aircraft recovered after a couple of turns. "Aha!" I thought, "What a clever way for him to introduce me to a spin!" I thought it rather odd that he did not comment other than to say, "Nice recovery."

We completed the flight, and I was in the locker area changing out of my flight gear when I overheard my instructor talking to the Squadron Duty Officer (SDO) in the ready room. He said, "Holy shit! I just encountered my first spin in an A-4. I was about to take the aircraft from the student when he calmly made the correct control inputs and recovered the aircraft. The kid has nerves of steel." The SDO said, "He doesn't have nerves of steel, he is just too stupid to be scared." He was right of course. I just assumed that the instructor had set me up to teach me how to get out of a spin. *The good news: I never got into another one.*

***

April 16, 1969: I returned from a flight (we were flying twice a day, six days a week), to get the news that I was the father of a beautiful baby daughter, Valerie Erin Rud. She was born in Orange, California where Barbara was living and working as a substitute teacher and nanny for a Marine Lieutenant Colonel fighter pilot, and his family. I was both elated and terrified. I am a 24-year-old, immature, irresponsible, self-centered, egotistic Navy pilot, and now a father. Due to the flight schedule and distance to Orange from Lemoore, it was a couple of weeks before I had a chance to see my precious baby.

Since Barb and I still had a lot to work through to get to know one another well enough to make a marriage work, we agreed that she and Val would stay in Southern California, where she was now substitute teaching, until I knew where fleet orders would take me. Thank goodness that one of us was strong enough (it sure as hell was not me) to make this arrangement work.

***

Fallon Deployment and a Spectacular Midair Crash: During this period of our lives, both Barb and I had our hands completely full. She was single-handedly managing the countless duties of being a new mother, and I was utterly immersed in one of the most intense, demanding, and time-compressed aviation training periods of a "nugget" (first tour of sea duty) aviator's career. To get by, we both had to compartmentalize the things going on in our crazy lives until we could come up for air long enough to make thoughtful decisions about our future.

The syllabus in the RAG did not let up. It consisted of familiarization (FAM), which include formation and of course landings, all of which were performed using the Fresnel lens in preparation for landing on the aircraft carrier. We then transitioned to weapons. This included a two-week deployment to NAS Fallon, Nevada. This was the wildest two weeks of my naval aviation adventure, up to that point at least. We flew two to three flights every day, and they were all really fun flights, dropping practice bombs from 30 or 45 degree dives, and from nuclear delivery profiles.

And in the evenings, we continued to charge hard. Remember, all the instructors were Vietnam veterans, glad to be alive, survivors of *Operation Rolling Thunder* (bombing of North Vietnam) and not wanting to waste any opportunity for fun. Our final graduation flight was a live ordnance exercise with real 500-pound bombs. The targets were derelict trucks/buses/tanks, etc. There were six of us scheduled early Saturday morning in two flights of four aircraft each (the lead aircraft in each flight was flown by an instructor).

The problem is that we had all spent Friday night in Reno, which is about 80 miles from Fallon. We did not get home until probably two or three in the morning, so it was a sorry looking bunch that showed up for the brief. One of the instructors, who had been with us in Reno called the SDO and supposedly said, "Operator, trace this call, for I do not know where I am." As it turns out, he was in a trailer house, out in the desert with some awesome gal (according to him).

The first flight took off, and the rest of us started up and got ready to taxi. That is when we noticed a large/dark/ugly/scary smoke cloud a few miles south of the field. What had happened was that during the rendezvous of the first flight after takeoff, #3 ran into #2. I will leave off names to protect the guilty here, but #3 had not even been to bed before the brief, so he was basically "shit-faced." As he recalled, "I knew something was wrong when my flight controls stopped working, and then I noticed that there was fire pretty much everywhere. At that point, I ejected." Meanwhile, #2 was on the bottom of the collision as both aircraft headed for the desert.

He could not eject because the two aircraft were stuck together. Luckily, when #3 ejected the aircraft separated from each other and #2 was able to eject, getting only a couple of swings in his parachute before hitting the ground. As luck would have it, the #3 pilot (this was his last flight in the Navy) landed right next to Highway 50. He wasn't injured, so he walked up to the highway and hitchhiked back to the base. Unfortunately, the person he got a ride with was a local radio newsman, so the whole world soon learned of the accident before the Officer in Charge of the detachment could even inform the squadron commanding officer. Not good. And the Search-and-Rescue (SAR) effort recovered #2 close to the crash site, but no #3.

Assumption: he must have died in the crash. No, #4 swore he saw two chutes, so they started hunting for #3. They eventually found him in his BOQ room drinking wine and "chilling out." He was also known to partake of a toke or two, which more than likely showed up in his post-accident blood test. Remember, these were the days of the draft, so not all of us were straight-laced lifers. My guess is that #3 is hanging out on a Southern California Beach with his surfboard, right now.

***

PRACTICE BOMB HUNTING: Most of the bombing ranges were closed on Sundays, so we decided to drive out to the raked target on B-16, which served as the nuclear delivery practice target. Nuclear deliveries were challenging and fun to perform. I know, you considered practicing delivering nuclear weapons fun? Of course, we did. It was the methods of delivery that made it fun, and we all realized that in a "real" scenario, we would more than likely be on a one-way mission anyway.

There really were three methods. The first was called laydown. This involved a 450 knot run-in to the target at 100 feet of altitude. YAHOO! Talk about flat hatting. The practice bomb was retarded, so that "theoretically" you could get far enough away before it went off. The second method was called over-the-shoulder. This involved that same "YAHOO" run-in to the target. Now, however, you passed over the target and then started a 4G "half Cuban eight" maneuver, releasing the bomb at the top of the maneuver, when you were, in fact, upside down; therefore, the term "over the shoulder." After releasing the bomb, you finished out the backside, and hauled ass back out the run-in line at 100 feet, and as fast as you could get that "Scooter" to go.

The third method was called a "65-degree loft." This was the most practical and survivable of the deliveries because it involved tossing the bomb, and then finishing out a "half Cuban eight" with much more distance from the target than the other methods. Of course, considering the destructive potential of a real nuclear weapon, you did not have to be very accurate.

The bull's-eye was a large garbage dumpster. During our bomb hunting adventure, just as we got to the dumpster, we heard some jet noise. "What the hell?" It was an A-6 *Intruder* roaring down the run-in line at 100 feet and 450 knots. So, we took off on a dead run to get the hell out of there. Except for one fellow, who just got into the target dumpster. His philosophy was that with the type of delivery the A-6 was using (loft maneuver), there was no way they would get close to the bulls-eye. He was right. By the way, the practice bombs made for perfect backyard ashtrays.

***

LOW LEVEL NAVIGATION: A key part of the nuclear delivery option was the ability to get to a target undetected. The best way to accomplish this was to stay below the level of radar detection. How low was that? Really, as low as you could safely fly an A-4 at 360 to 420 knots, which by experience was determined to be 200 feet above the ground. This was easy to do on level ground or over water. However, it was much more of a challenge to do this over mountainous terrain. Our graduation flight from this phase of training involved hauling a "shape" that, although completely fake, closely resembled the weight and characteristics of a real weapon. It was so heavy that to get airborne before the wheels on the little Scooter disintegrated on the takeoff roll, we took off at first light in the morning, before it got too warm.

In preparation for this flight, we put together detailed maps, and then cut them up into strip charts that we could flip through as we progressed along the route. The A-4 *Super Charlie* that I flew on that mission had absolutely no navigation system other than a TACAN (Tactical Air Navigation System), which was useless once we got down to 200 feet. My instructor flew his own aircraft and followed me along the route as a safety observer and source of instruction in case I got hopelessly lost.

The route began at high altitude over the Pacific Ocean to simulate operating from an aircraft carrier. We then descended at the Oregon coast and commenced the route. Since we were so low and fast, you really needed to memorize the route versus relying on the strip map. Although struggling to be an average light attack pilot, I was good at this phase of training. Now, not only did we need to stay on the route and find the target, but we also had to do so within a finite time frame. I managed to do all these things, dropped the "shape" on B-16 within the time constraints, and then began the final phase of the flight, which was a high altitude return to Lemoore.

The instructor had not said anything on the radio for some time, but that was perfectly normal, as he had briefed that unless he needed to talk, he would not. This added a touch of realism to the event. As I began my climb to head back to Lemoore, he said, "I am low on fuel, and I am going to land at Fallon to gas up." He then added, "You have done a great job on the flight, so if you have enough gas, go ahead and continue on to Lemoore." I checked my gas and replied, "I have enough to do that safely, see you in Lemoore."

In those days, the controlled airspace from Oakland Center did not begin until you reached 24,000 feet or FL240. With that in mind we had planned to go straight to Lemoore at something less than FL240. As I climbed, I noticed for the first time, a most ominous thunderstorm directly on my course to Lemoore. No problem, I am sure I can get over it. Shit! As I got closer, I realized there was no way around or over this monster.

So, for the first of many times that I used old Chuck Kundert's (my C-150 Private License instructor) advice on common sense/ask forgiveness instead of permission; I shut off my transponder "IFF" (a device that tells the controller what my altitude is) and climbed that A-4 *Super Charlie* with its altitude-loving, J-65, W-30 engine to FL 370. That got me over the thunderstorm, and the follow-on idle descent, got me back into Lemoore with plenty of gas.

Now, it just so happened that one of my classmates and his instructor were just a few minutes behind me. They ran into the same situation, played by the rules, turned around and landed at NAS Fallon. They saw my instructor getting fuel and asked where I was. He told them that I had optioned to head for Lemoore. They then related having to turn back due to the giant thunderstorm. He then became concerned and called back to Lemoore to see if I had made it OK. The SDO replied, "Rud? Yep, he is taxiing in as we speak."

Relieved that I had made it, he asked the SDO to have the maintenance chief check my fuel level. They called him back and responded that I had 1,200 pounds left in the tank, which is more than anyone expected. I hung around until my instructor got back from Fallon. He gave me good grades, and an above average in headwork for doing whatever it took to get home safely. He said, "I figured out your fuel usage and decided that you had to take that baby way up into controlled airspace to get over that thunderstorm. Screw Oakland Center, I would have done the same thing." *As I mentioned when I first checked into the RAG, this is no longer the training command.*

***

INITIAL NIGHT AIRCRAFT CARRIER QUALIFICATIONS: We spent most of July preparing for night carrier landings. I cannot recall our exact class makeup, but I do remember that it was very large, consisting of 16 total students. Fourteen of us were nuggets, getting our first exposure to night carrier landings, and two of the pilots were experienced folks, who were coming off shore duty. To be perfectly honest, I was not good at this. Of course, neither were the other 13 nuggets.

Our FCLP grades were awful, and the LSOs regularly told us that unless we improved, they would not take us to the boat. The practice landings were conducted on Runway 32L at Lemoore. Although the runway was a massive 13,000 feet long, the portion that was lighted at night was only the size of a carrier deck. Since there was nothing out there but farm fields, it was dark as hell, and closely simulated an actual night carrier landing. When you set up for your final practice landing, they would turn on all the runway lights, so you could safely make a full stop landing.

***

WHAT IT LOOKS LIKE JUST BEFORE YOU DIE: During the last night of practice prior to heading for the boat, I was working my ass off to fly the ball well enough so that they would take me to the boat. I felt good about my performance, as I lined up for my final pass, which was to be a full stop. I called the ball and rolled out on final with it centered. The ball began to go a little low, so I added some power. Nothing happened. The LSO called, "Power," so I pushed the throttle all the way forward, with no engine response. The LSO next called for more power and finally a "Wave it off!" The engine did not respond at all.

Thankfully, I had the presence of mind to lower the nose and safely land way short of the simulated carrier deck, but still on the runway. So, the last thing I saw was the meatball going low, turning red, and disappearing off the bottom of the lens. In other words, I saw what I presume many pilots attempting to land on real carriers saw just prior to dying.

The LSO did not say much after I landed other than, "Geez kid, did your engine fail?" "Yes sir, it sure did." I was able to taxi it to the flight line and shut it down. I described what happened to the maintenance chief, who immediately and correctly diagnosed a fuel control failure. When I got to the ready room, I must have been pale as a ghost, because the other students all asked if I was feeling OK. I replied, "Hell no, I am not feeling OK, since I just witnessed what it must be like to die on a night carrier landing." The LSO (I think it was "Bonnie" Baker) debriefed us all and complimented me on safely landing the jet. Then he said, *"Rud, how about describing to the rest of these guys what it looks like just before you hit the ramp and die."*

***

USS *Bennington* (CVA 20): As I mentioned earlier, I do not remember much of the training command carrier landing evolution. However, I distinctly remember my landings on *Bennington*. She was an *Essex*-Class Carrier that served in WWII as a straight deck followed by conversion to angled deck in 1952. Unlike her sister ship, USS *Lexington* (CVA 16), *Bennington* had hydraulic catapults. Prior to attempting night landings, we each needed to get two touch and goes, plus six-day landings.

Like most naval aviators, my biggest concern was not to hit the ramp, followed very closely by not wanting to miss all the wires (bolter). My first trap was awful! I settled at the ramp and got a taxi one wire. The LSO chewed me out over the radio prior to my next attempt. This resulted in a super-over correction and two Bolters in a row. Finally, I settled down and got some decent landings. Because of my marginal performance, I got a total of 15 day traps versus the normal 10. Then it was time to take on the challenge of a night carrier landing.

Unlike landing ashore at night, where you have buildings, vehicles, and all sorts of clues that define the earth from the sky, at sea, you have none of those. In other words, it is the most vertigo-inducing environment that exists in aviation. You can be the greatest bomber or fighter pilot in the world, but if you cannot fly well on instruments, you will not survive night carrier landings. *Neil Armstrong summed it up best when a reporter asked him how hard it was to land on the moon. He said, "It was easier than landing on a carrier at night."*

***

The Night Catapult Shot: Awful experience, especially with hydraulic catapults. Steam catapults are bad enough, but at least they build up the stroke and assorted violence during the 200 or so feet of deck travel. The hydraulic cat just gives you the whole nasty force right at the start. Think about going from zero to 150 miles-per-hour in 2.5 seconds.

As we were lining up behind the catapult the ship began a turn into the wind. During this evolution, I had to keep pressure on the brakes, so as not to slide over the side. While holding the brakes, my leg began to twitch up and down quite violently. I took my fist and slammed it onto the quadriceps on the top of my leg between the knee and hip to stop the twitch. Then the other one started to twitch. The bottom line: when I finally got home, I had huge black and blue marks on both quadriceps muscles.

Once you are on the catapult, you are given a signal by flashlight to run up to full power, and tension is taken on the bridle that hooks you up to the catapult. You cycle all of the flight controls, make one final check of all your instruments, then you concentrate on the attitude gyro, and you turn on your external lights. That is a signal to the catapult officer that you are ready for launch. You place your head back against the ejection seat headrest, grip the throttle hold device so that your hand doesn't inadvertently slide back during the launch, and you place your right hand on the control stick so that you are ready to fly once airborne. Oh, by the way, you cannot see a damn thing, including instruments, during the catapult launch. It is just one shaky-blurred mess.

At the release point you are only 60 feet above the water, you regain vision, and you get the nose up into a climbing attitude followed by raising the landing gear, and eventually the flaps, if you are going to marshal (a holding pattern that the aircraft are sent to prior to commencing an approach to the carrier). Usually during carrier qualification, we just stayed in the landing configuration leaving the gear and flaps extended. During this whole evolution you must pay especially close attention to your airspeed/angle of attack so that you do not enter an aerodynamic stall. Geez! I am beginning to sweat just writing about this.

All of us hope for a nice moonlit night, but it is not going to happen. Instead, we have a dark night with an 800-foot overcast that is solid up to 1,500 feet. The night pattern calls for us to climb to 1,200 feet, then turn downwind and follow the directions of the radar controller on the ship. Depending on other aircraft in the pattern, you can normally expect to be turned toward the final bearing at a distance downwind that will allow you to intercept the glideslope and begin a descent at three nautical miles.

During all this time, you have not looked outside, concentrating only on flying instruments. The controller on the ship will continue to give you heading and glideslope information until you reach a point 3/4 of a mile (about 18 seconds) from landing. At that point he will transfer control to the LSO who will give you corrections as necessary all the way to touchdown. In theory, you do not look outside the cockpit until you get to 3/4 of a mile. That will help to keep you from getting vertigo.

But alas, I am still a rather undisciplined nugget, so I sneak several peaks prior to 3/4 of a mile, which of course results in massive vertigo. In fact, it looked to me like the ship was standing on end and although wings level, I felt like I was in a 30-degree angle of bank. I believe that the best way to describe this situation is to share the radio conversation as I recall it.

Radar controller, "*Skyhawk* 510 you are approaching 3/4 of a mile, you are slightly above glide path and slightly left of centerline, call the ball." I transition my scan from inside the cockpit to visually acquire the Fresnel lens meatball and I say, "510 *Skyhawk* ball, 3.0 (3,000 pounds of gas), pilot Rud." The LSO responds, "Roger ball, you are a little high, a little right for your line up, put some power back on, POWER!" and I land safely, although far from perfectly.

Somehow, I managed to get all six of my night landings accomplished, albeit with a colorful display of ineptitude. After my sixth trap, the yellow shirt (aircraft director) parked me and began to refuel my aircraft. He gave me a full internal load, and then the Air Boss transmitted, "510, you are a qual. Your signal is bingo NAS Lemoore, bearing 340 at 210 NM." "Shit hot, I made it!" I thought as I headed for the catapult. By now, I was just a little more comfortable, which was a good thing because a new adventure was about to begin.

As I climbed above the overcast and turned toward the heading for Lemoore, everything except the engine quit. My rear end immediately sucked up the seat cushion (description of typical pilot reaction to an aircraft emergency). I suspected an electrical failure, so I deployed the emergency generator. That gave me back some lights in the cockpit and primary instruments, but my radios did not work at all.

OK, time for farm boy common sense. It was a long way to Lemoore, but I had plenty of gas and it was 2 a.m., so there was very little other traffic to contend with. The weather over the beach was good, so I flew toward Los Angeles, which was easy to find. I then picked up Highway 99, which I knew would take me over the Grapevine to Bakersfield. From there I continued to follow Highway 99 until it intersected with Highway 198 at Visalia. I made a left turn and followed Highway 198 to Lemoore, landing on 32R without a radio, which at that time of night was no problem. I taxied to parking, shut down and checked in with the SDO, who quickly sent a message to the ship announcing my safe return to Lemoore. In 48 hours, I had gotten 15 day and six night landings. *I was exhausted, elated, and looking forward to assignment to a fleet squadron.*

11. Rocket 21 in the VA-216 Black Diamonds

VA-216 Black Diamond A-4 Super Echo landing on USS Forrestal *(CVA-59).*

MOST OF THE WEST COAST Light Attack squadrons were stationed at NAS Lemoore, but there were three exceptions. VA-152, VSF-1 and VA-216 were all stationed at NAS Alameda, California. Alameda is located right across the bay from downtown San Francisco with a whole lot more opportunities for liberty than existed in Lemoore.

VA-216 was flying the A-4 *Super Echo.* They were equipped with the latest electronic warfare gear, which was in the distinctive hump on the top of the aircraft. The squadron call sign was "Diamondback," and the hump had diamonds painted on it. These were probably the coolest, most testosterone-laden birds in the Light Attack Navy, and I was elated to get an opportunity to fly them.

I was the last nugget (first cruise) pilot to join the squadron, and the most junior in rank of the 21 pilots assigned. Therefore, I became Rocket 21, the SLJO (shitty little jobs officer). These duties included Mess Treasurer, Movie Officer, and any other job that was considered beneath the dignity of a more experienced pilot. As is typical of a single-seat squadron, I also had a primary (real) job in the Maintenance Department as the Airframes and Parachute Riggers Branch Officer. The squadron had recently completed a combat deployment on USS *Coral Sea* (CVA 43). It was a tough one with the loss of pilots either killed or captured. The Navy leadership decided to give the squadron a break and assigned us to *Carrier Air Wing 17* for a Mediterranean Sea deployment on *Forrestal*.

The ready room (all the officers in the squadron) was an amazing mix of nuggets like me, second-cruise junior officers who were combat veterans and the department heads, all of whom were veterans of 200 or more combat missions. The Commanding Officer was Jim Busey IV (he would eventually become a four-star admiral). He was an awesome leader in every respect. He walked the talk, and he was a terrific pilot. The Executive Officer (XO) was quite different. Unfortunately, shortly after I got to the squadron, it was time for Commander Busey to move on, so the XO became the commanding officer. Let's just say that in just a few weeks, I was exposed to two very different styles of leadership, both of which I learned valuable lessons from.

The nuggets included Mike Gellman who was a classmate of mine in AOCS. My favorite liberty mate was John Davison. He was a U.S. Naval Academy graduate, often grousing about finishing near the bottom of his class (the bottom man gets a dollar from all the rest of the graduates). Don't make the mistake of misinterpreting his low standing in the class. It was not due to academics (he was super smart) but rather due to some raucous behavior that resulted in the maximum number of demerits, etc. Much like my training command running mate, Don Roesh, John was an absolute wild man on liberty. The two of us made a toxic pair, but we had a blast.

Although we had broken up as a couple before I entered the Navy, my old college girlfriend, Beth McLaughlin, was teaching school in the Bay Area.

We were still good friends, so I invited her to one of our squadron social events. She met John, and they hit it off from the start, eventually getting married with me as the best man.

At this time, my personal life had deteriorated a bit with lots of bad decisions made under the influence of booze and fellow hell-raising bachelors. Barb and I had spent very little time together, so I had not really participated in Valerie's first few months of childhood. As I mentioned earlier, Barb and I had made no real commitment to one another yet, so I was living a rather loose and frankly, crazy life.

I shared an apartment with a couple of other single guys from VA-152. It was located just a couple of miles from the base, so it was very convenient. On Friday afternoons, all the squadron bachelors (I think there were eight of us) would pool our meager monitory resources, hire a limo, and head over to San Francisco International Airport to gather up as many flight attendants as possible, and transport them to the officers' club at NAS Alameda for happy hour. In those days, security was not an issue at airports so we would be in uniform and go right to the various airline gates in the terminal, sometimes even using signs to attract potential party attendees. I know, pretty demented, but although not overtly worried about survival, we certainly lived every day like it might be our last.

Shortly after arriving at the squadron, we deployed to NAS Fallon, Nevada for a weapons detachment. We performed the same sort of maneuvers as we had done in the RAG with the addition of some more sophisticated weapons. One of these was the Bull Pup missile. This was a crazy weapon in that once you fired the missile; you then had to visually control it using a little stick that was located on the left-hand console near the throttle. There were two sizes, 1,000-pound and 2,000-pound versions. The only weapons station on the *Scooter* that could handle the big version was on the centerline (directly below the cockpit). The good news was that our target was B20—just a big rock out in the middle of nowhere.

I rolled in and released the 2,000-pound beast several miles from the target. It initially dropped off the weapons station, and then it lit-off, literally pushing my aircraft away and causing a significant negative-G load. I watched this monster head in the general direction of the target, reached down to start controlling the missile and the entire control box came out of the console. I now literally have the aircraft control stick in between my knees so I am flying with my legs while I try to control the missile with the control box and the little control stick in my lap. The control box included a bunch of electrical wires that I was afraid were going to electrocute me or even worse, potentially eradicate the family jewels.

Obviously, this did not go well. The missile ended up pitching straight up into the air, aerodynamically stalled, headed for the ground, and exploded a good mile short of the target. After we landed, the CO chewed my ass out about screwing up the missile shoot. In one of several insubordinate acts with this particular leader, I literally grabbed him by the arm and told him to climb up into the goddamn cockpit to see why I screwed up the missile shoot. He did, but as was his style, at least with me, he just shrugged and walked off.

One big difference between the RAG and the fleet was the competitive attitude of the pilots. The ready room we were using at Fallon overlooked the end of Runway 31. It was considered quite manly to complete a low transition on takeoff. This involved raising the landing gear, while staying very low over the runway, letting the speed build up, then raising the flaps while still staying very low for the entire length of the 14,000-foot-long runway. When performed properly, this action would allow you to transition from takeoff to a very impressive high-performance climb.

To encourage competition, we literally drew grease pencil lines on the window overlooking the runway. The lowest line defined the "shit hot zone." The next one up was the "average zone," and anything above that was considered the "sissy zone." Oh, I almost forgot, the safety officer insisted that we draw a line even with the ground, which he labeled "tied the record zone."

***

GROWING UP; BETTER LATE THAN NEVER: As we came closer to deployment, my 16-year-old mind trapped in a 24-year-old body finally began to mature. I realized that I had the opportunity to be the luckiest man in the world. I was the father of a beautiful little girl whose mother was totally out of my league in so many ways. She was smarter than I was, much better looking, and just a superior person in every respect. Barb had a very good friend who was married to one of the VA-152 pilots. We arranged for her and Valerie to come up for a weekend. They would be staying in the same complex as the one I lived in, and it would give me some time with my daughter.

I had been working up the courage to propose to Barb, but I was deathly afraid that she would answer in the negative and might even laugh at me because I was such a jerk. On looking back, I guess I pretty much stayed in the dumb-shit mode, as my proposal was something like this: "I love you babe, let's jump in the car, drive to Lake Tahoe and get married." I can't remember exactly what she said, but it certainly was not an enthusiastic "yes." The important thing is that she did not say no. Later, I got a bit emotional while holding Valerie, so I credit that wonderful child with helping make up her mother's mind to give me a chance.

Barb's friend offered to babysit Valerie. Barb borrowed a dress from another neighbor in the complex, Donna Hezlep, and we were off to Lake Tahoe. We had about a three-hour drive to get to know each other a bit better, and to make some plans for the future.

Unfortunately, it was after midnight when we got to the Nevada side of Lake Tahoe and all the wedding chapels were closed. No problem, we just pressed on to Carson City, found a 24-hour chapel, hired some dude hanging out at the chapel to be a witness, filled out the paperwork, said, "I do," and we were married. It was October 12, 1969, and to that point, the best day of my life. Our honeymoon was spending a wonderful night in Lake Tahoe, and then it was back to Alameda as Mr. and Mrs. Gil Rud. I know, what are the odds of this marriage ever working? Well, it certainly did, for a wonderful 36 years until my love finally lost a courageous battle with breast cancer. *I really was the luckiest man in the world!*

My roommates were deployed on workups so Barbara and Valerie stayed with me in Alameda. My squadron mates had never met Barb, and they really did not know me that well, so when I announced that I was happily married and introduced her at the next social function; there were some shocked bachelor bubbas. "Rud, you ugly Norwegian, how in the heck did you ever catch this awesome lady?"

***

No-notice Deployment: A few weeks after we were married, the phone rang at about midnight. Barb answered and passed it to me. "This is the SDO, you need to get your butt in to the squadron right now. We are conducting an emergency deployment on *Forrestal,* and you are scheduled to depart at 0500." "Holy shit! Is this for real?" "It sure is, and you are wasting time talking to me, pack a bag that will fit in the hell hole of the *Scooter* and get out here now!" *"Welcome to the life of a Navy wife, Barb."*

We did deploy the entire squadron within a 12-hour period. This involved getting 12 out of 14 aircraft all the way across the country. To accomplish this feat, our maintenance folks begged, borrowed and even stole whatever they needed to get these birds airborne. We stopped for refueling at Tinker AFB in Oklahoma, and then proceeded to NAS Oceana, Virginia. Since we had no maintenance personnel, a few of the junior pilots, including me, became plane captains. This included performing both hot and cold refueling in the refueling pits.

Totally illegal and unsafe, but we did what was needed to accomplish the deployment. I remember the Carrier Air Wing Commander, Cal Swanson saying how amazed he was when he saw the first *Diamondback Scooters* land aboard *Forrestal.* As it turned out, there was some sort of crisis involving Cuba that had precipitated this whole evolution. We stayed on board and flew work-up-type missions for a couple of weeks before heading back to Alameda.

***

HARVEY WALLBANGER DEBACLE: As we got closer to the full-cruise deployment, Barb and I attended a squadron party at one of the married pilot's homes, which was a good distance from Alameda. It was a Harvey Wallbanger/hippy costume-themed event. We had a blast, but I managed to consume one too many wallbangers, so Barbara drove back to our apartment.

At some point, she realized that she was lost, woke me up, and told me to ask for directions. We were stopped at a stoplight in San Leandro, so I got out to ask a pedestrian for directions. At that point, the light changed, and Barbara had to drive around the block before coming back to pick me up. Well, it turned out that a normal "around-the-block" option was not available, and she got lost, so I was basically left on the street corner in my hippy garb, which included flared pants and a string of beads around my neck. The pants were those weird hip-hugger types with no pockets, so I had no wallet, and no money.

This was long before cell phones, so we had no way of solving this issue easily. My dear wife managed to find her way back to our apartment sans her dumb-ass husband. Meanwhile, I managed to find a bar that was open, and talked the bar tender into hanging out there until I could figure out a way home. He finally closed about 2 a.m., but by this time figured out that although stupid, I really was a decent fellow, so he let me sleep in a booth until morning. By the way, I was the only white man in this part of town.

The next morning, I decided that I would try and get on a bus. I attempted this move, but without any money, I was rebuffed by the driver. Fortunately, getting on behind me was an older lady who felt sorry for this ridiculous looking white boy and offered to pay for my trip. When I finally got back, I was looking forward to a big hug and maybe even a few tears of joy. Instead, I was greeted with the hands on the hips, arms crossed, grown-up-to-child look of disapproval. During my absence, Barb had contacted the police and hospitals fully expecting that they would find my body in some alley. It was not my finest hour, but I was still married and after a few days on the couch, and multiple apologies, all was well again.

***

DEPLOYING ON *FORRESTAL:* Just before leaving on deployment, the squadron was informed that we would be decommissioned upon our return. This situation left lots of questions as to what was next. So much so, that we decided that Barbara and Valerie would return to the LA area until I received follow-on orders.

I made out an allotment that sent most of my paycheck to them. As I recall, I had a total of $18 in my pocket when I walked aboard *Forrestal*. While walking aboard, John Davison and I were tasked with getting the allowed liquor on board to be used at the various port calls during the cruise. Of course, in addition to the allowed liquor, we also sneaked a private supply for illegal use while at sea. We were struggling with a Navy metal cruise box full of beer when the ship's Catholic chaplain offered to give us a hand. "Goodness gracious boys, what do you have in here that is so heavy?" Without hesitation John replied, "Father, this cruise box is full of Budweiser beer." He gave us a quizzical look, and then burst out laughing at the top of his lungs. "Oh, you west coast boys have such a good sense of humor." He was the perfect cover for successfully getting this precious cargo past the Officer of the Deck.

***

As the SLJO, I was assigned to an eight-man bunkroom. My roommates included my good friend and fellow VA-216 pilot, Terry Freedman, plus the two most-junior Radar Intercept Officers (RIOs) in VF-211, the *Red Rippers*, Dave Bardal and Rick Zwiebel. The four of us were the only permanent residents in the bunkroom. It was located on the 0-3 level (just below the flight deck) directly across from the barbershop, and pretty much isolated from other officer berthing. In other words, it was the perfect place to host parties at sea.

John "Lightning" Davison and me engaging in some illegal activity in the 8-man bunkroom on USS Forrestal *(CVA 59).*

***

BOAT OFFICER INCIDENT: We were a west coast squadron now assigned to an east coast ship and carrier air wing. East coast rules were quite a bit stricter, and basically followed peacetime protocol, which included an in-port watch bill that required all officers to remain aboard the ship every fourth day. Of course, our combat veterans would have nothing to do with this rule. We got around it by doubling-up on required watches when we did stay aboard. In other words, if it was my day to stay aboard the ship, I would stand a couple of four-hour integrity watches (keeping an eye on the aircraft in the hangar or on the flight deck), followed by a Boat Officer watch that night.

Boat Officer watches were required because the ship never actually tied up to a pier. We always anchored out using liberty boats to transport the crew to/from the respective port. Even though we non-Academy pilots knew absolutely nothing about boats, we were line officers, and therefore in command of the individual liberty boat that we were assigned to. I loved this watch, especially the late-night one where we were picking up all the inebriated sailors. It was a three-man crew, which included the coxswain (normally a first or second-class petty officer boatswain mate [BM] who drove the boat), a very junior enlisted engineman (EM) to keep the engine going and to assist with the lines, and the boat officer. On the late night/early morning runs, we would be completely empty on the way from the carrier to the port.

My first and most memorable watch was from an anchorage off Barcelona, Spain. The weather was marginal with low visibility. The channel was well marked though, so as a crew, we were comfortable heading in to pick up the last load of sailors. So much so that the coxswain asked if I would like to drive the boat. "Hell yes, I would love to." As I was heading up the channel, I noticed two lights in addition to the normal channel markers. I asked the coxswain about the red and green lights that were well above us, and rapidly getting closer. "Holy shit!" It was a huge tanker coming out of the harbor. I gave that liberty boat full right rudder, and fortunately that move together with the bow wave of this massive vessel saved our ass, but just barely. *Now had we hesitated a minute or so, my Navy career and more than likely my life would have ended right then (you are probably noticing a pattern of potentially career-ending events occurring quite regularly, and this would not be the last).*

Normally, the last few sailors that we picked up were well spent from all the fun that they experienced ashore. However, alcohol affects people differently, so occasionally we would get a nasty drunk who wanted to fight or was just plain annoying to everyone. The way we handled these situations was to simply stop the liberty boat, let it bob around in the almost always-rough seas (which would make the passengers seasick), and then I would announce, "We are not going anywhere until you (the offending sailor) knock it off." If he still did not quiet down, his fellow shipmates would step in and make sure that he did.

***

GROUNDED: We were obviously not at war in the Mediterranean Sea, so we were not getting shot at. I would estimate that we averaged three weeks at sea between each port visit. During these at sea periods, we flew a very challenging schedule including lots of night operations. When I first checked into the squadron, I asked if I could begin training to become an LSO. I did this mostly because I wanted desperately to get better at landing aboard the carrier. Geoff Dillon and Al Chesterman were already filling those positions, so I was not allowed to officially train to be one. I did, however, spend a good deal of time hanging out on the LSO platform in the hopes that I would improve my below-average landing grades.

Also, I volunteered for and flew several extra night events including several "Recovery Tanker" flights. Nobody liked to fly these because with the added weight of the refueling buddy store, your maximum landing weight only allowed for two landing attempts at the boat before reaching "bingo" fuel (a predetermined fuel state requiring the pilot to either tank or head for an airfield ashore). This was especially critical on the last launch of the night since you would be the last aircraft to trap, and there weren't any other tankers to refuel from. These situations resulted in several one-wire traps since in this case, a wire behind you was considered a wire wasted, even by the LSOs.

Because these volunteered-for flights were often scheduled and recorded only by the schedule writer (my buddy John Davison), an interesting leadership lesson emerged late in the cruise. Our skipper was about to become the first pilot in the air wing to achieve centurion status (100 landings) for the cruise. Always a self-promoter, he told John to make sure that his achievement was announced via the 1MC (a communication system that is designed to be heard throughout the entire ship), and that the appropriate cake-cutting ceremony would be arranged.

For some reason John was angry at the skipper, so he looked him in the eye and said, "I would do that except that you are not the first pilot to reach centurion status." The skipper was incredulous, "What do you mean I am not the first pilot to get 100 landings for the cruise? Did one of the other squadron commanders beat me to it?" John smiled and said, "No sir, Gil Rud has 107 traps." The skipper went ballistic, "How in the hell did he get ahead of me in total number of traps?" John replied, "He got ahead of everyone because he volunteered for all the last-event night spare tankers that ended up launching." Instead of being impressed by my diligent efforts at self-improvement, he said, "Gil Rud is immediately grounded, and will not fly again until I have more traps than he does." *This was certainly a leadership lesson that I would never emulate over the next 25 years.*

***

NEW YEAR'S EVE SHORE PATROL: One of the watches that we could volunteer for involved heading up a shore patrol team. This was normally only available to more-senior officers, but I decided it would be a great adventure, so I applied, and was assigned to a team responsible for Gulfe Juan, France. The ship was anchored off Cannes, France, which was an awesome liberty port, albeit expensive and therefore not all that popular with the sailors. Gulfe Juan, was located a few miles east of Cannes and was much more affordable, with numerous eating and drinking establishments that were welcoming to our crew. Since prostitution was legal in France, it was also the center of this thriving business. In fact, our shore patrol headquarters was located in the most infamous hotel in town.

As the only officer on the team, my role was mostly to stay at our headquarters and handle the administrative duties as well as liaison with the local gendarmes. It was New Year's Eve, so the action and the number of sailors ashore were much more than a normal night. To make it more interesting, the weather was very marginal for liberty boats to operate safely. At around 11 p.m., I got a phone call from the commander in charge of the entire shore patrol effort. He said, "I have some bad news. The ship has just canceled all the liberty boats until tomorrow morning. This means that we will have to find lodging of some sort for all the sailors currently ashore. Good luck." And he hung up.

I was fortunate to have a couple of Senior Chief Petty Officers on my team who had experienced situations like this before. Since Gulfe Juan was a relatively small community, there was no way to get lodging for all the sailors. One of the Senior Chiefs said, "Mr. Rud, I recommend that we head out to all of the bars right now and see if we can get them to stay open until tomorrow morning. We will need you to talk to the owners."

While we were seeking permission for the bars to stay open, we were also informing the sailors regarding cancellation of boating. The operation was going smoothly until we arrived at one of the bars where an altercation was in process between two drunken sailors. By this time, my patience was wearing thin, and my infamous Norwegian temper was beginning to take over.

I walked up to the offending pair and said, "Knock it off right now!" One of them immediately backed down, but the other one had a knife in his hand and said, "Screw you!" I was wearing a heavy Navy Pea Coat which I felt would absorb most of a knife attack, so I advanced on him and said, "You better put that knife away now, or I will shove it up your ass!" I can be imposing when I was pissed, so he looked at me, put the knife away and sat down. The team arrested him and we took him back to our headquarters.

The Senior Chiefs both took me aside with the following advice: "Sir, don't ever do something like that again. If he had assaulted you with that knife, he would be spending the rest of his life in Leavenworth." As it turned out, he was an ethnic Filipino assigned as a Mess Specialist and a very good sailor, at least when he was sober. I apologized to the Senior Chiefs for my actions and told them to take the kid back to the bar and settle him in for the night. *There will be many more instances where I sought out and adhered to the advice of the heart and soul of the leadership of the Navy, the Chief Petty Officers.* By the way, word of my decision to give this young sailor another chance soon got back to the leadership of the Mess Specialists. This resulted in special treatment for all the members of the 8-man bunkroom. We could party hearty without worrying about being turned in by anyone working in the officers' mess.

***

ROTA, SPAIN: As part of my attempt to get better at flying, and to save spending money, I often volunteered for various detachments that would sometimes be offered as alternatives to participating in a port visit. One of these involved sending a division of A-4s to Rota, Spain while the ship was in port in Naples, Italy. I don't remember the operational purpose of this boondoggle, but I do remember attending the absolute world's worst bullfight. The Matador was simply unable to kill the bull, to the point where he was getting whistled and booed openly by the crowd, who cheered for the bull.

***

SIGONELLA, SICILY: The most unique and spectacular detachment was to Sigonella, Sicily from January 10–14, 1970. There were four of us picked out to be on this detachment. Geoff Dillon was the senior person in charge of the mission, and he was only a second-cruise lieutenant. The rest of us were all nuggets. As it turned out, there was a reason for not having any senior leadership involved.

Once we arrived in Sigonella, we were sent to a secret area on the base to get a briefing on our mission. It turned out to be the most fun we had flying on that whole cruise. Our mission included finding a Russian fleet that had recently entered the Mediterranean Sea through the Strait of Gibraltar. The intelligence we were provided indicated that they were anchored off the northern coast of Algeria. Once we located them, we were to specifically seek out the newest Russian helicopter carrier, *Moscova*. We were then given free rein to basically aggravate them into painting us with their radar and electronic equipment.

Aha! This is why they sent a bunch of junior officers. If it became an international incident, there would not be anyone senior enough involved to get in any real trouble. We decided (over pizza and many beers) that we would act like a bunch of dumb-shit American aviators, who were lost and about to accidently land on a Russian aircraft carrier. When we found them, we did a fly-by in diamond. Next, we entered the carrier break in echelon with our arresting hooks down. We then formed a landing pattern with our landing gear down, and each of us made a low pass on the ship as though we were going to land.

Every ship in their anchorage started painting us with their respective radar and electronics, which was the purpose of the mission. I remember a huge *Hormone* Helicopter on the flight deck that was "turning" when I waved off and flew by. We may not have been flying combat missions, but we sure had some fun on cold-war missions like this one. The only downside was that we were sworn to secrecy and could not share our adventure with the rest of the squadron.

***

CLOSE CALL: On January 24, 1970, I was scheduled for a flight with the carrier air wing commander (CAG). Cal Swanson was a fighter pilot, but like most CAGs he flew all the aircraft in his air wing. I recall that our mission was air-to-air tactics against the F-4 *Phantom IIs* that day, which included carrying the newest AIM-9 *Sidewinder* missiles in the fleet.

We briefed to rendezvous overhead the ship at 17,000 feet. The CAG launched well ahead of me, so after the cat shot, I proceeded at the usual low altitude of about 500 feet to stay below the aircraft that were now coming into break to land. At 7 miles, I pitched up to arc at 10 miles, as I climbed toward the rendezvous. I saw the CAG across the circle and started my rendezvous, but noticed that I was not able to gain on him, even with the throttle wide open.

I called CAG on the radio and said, "*Diamondback* lead, this is *Diamondback* two, I may have an engine issue and I am turning to a heading of 090 in the direction of Araxos, Greece. I am currently at your seven o'clock; recommend that I take the lead and you join on me." CAG quickly turned toward my position and joined up to look me over. I was now down to about 80 percent power, (the J52 P-8 engine at full throttle should be showing about 92 percent) so I definitely had an issue. I pulled out the emergency procedures for airborne engine issues and followed the recommendation to bypass the fuel control by selecting manual fuel. It did not help, and I was now starting to lose altitude.

The CAG called back to the Air Boss on the ship and advised him of the situation and my intention to head for our emergency bingo field in Araxzos, which was about 100 NM to the East. Apparently, our squadron representative in the tower decided to inform our skipper of the situation. It wasn't long before he came up on the radio. He did not ask me to explain the situation, he simply ordered me to return to the ship and land with the recovery in progress.

I responded that I did not have enough engine performance to safely recover aboard the carrier. He reiterated his demand (it was not a suggestion) that I return to the carrier. That is when the CAG stepped in, "The pilot knows best what the performance of his aircraft is capable of, and I can tell you that at full power, he has already lost several thousand feet of altitude. We are continuing on a heading of 090 and are now approximately 50 NM from Araxzos. I will let the ship know when he is safely on deck."

As if things were not bad enough, I pulled the throttle back a bit to see if it would respond to my inputs. It did, but when I tried to advance it back to full throttle, I started getting compressor stalls, so I tightened my harness in preparation for a possible ejection.

Thankfully, it was a beautiful day, the CAG made all the radio calls to Araxzos and I concentrated on keeping the crippled *Scooter* flying. By the time I lined up for a long straight-in approach to the runway at Araxzos, I had lost

14,000 feet of altitude and was at 3,000 feet. I held the landing gear and flaps until I knew I had the runway made. Once safely on deck, I was instructed to pull off at the end of the runway to wait for security.

They arrived in a few minutes along with two fire trucks. I noticed that the security folks were heavily armed with rifles and machine guns. I must have been quite a sight as it was winter, so in addition to my flight gear, I was clad in an all-black wet suit (normally it would be a dry suit, but our air wing was experimenting with this new wet suit). They asked me to remain in the aircraft until the base commander arrived. I thought that was a bit weird, especially when I saw this bright-red Jaguar convertible roaring down the runway at what had to have been 100+ mph. As the Jaguar pulled up, this very dapper Greek colonel jumped out, ran up to my cockpit, saluted me and said, "Welcome to Araxos my friend. My men will make sure that your jet and especially that missile you are carrying, are secure." At that point, just as I exited the aircraft, they placed a blanket over the *Sidewinder*.

Since CAG was on his way back to the ship with all the information needed to get started on a plan to get me—and hopefully my aircraft—back to the ship, I relaxed a bit to ponder my situation. The colonel was a very emotional person. He gave me a giant hug, directed me to his Jaguar and said, "Now it is time to celebrate you cheating death, and our good fortune to host a valued ally." Off we went in the Jaguar. The colonel had this awesome white aviator's scarf that he threw over his shoulder, so it flapped in the wind as we roared back down the runway to his office.

A couple of hours later, the ship's Carrier On-board Delivery (COD) aircraft arrived with some mechanics from the squadron. They did a quick inspection and confirmed that the engine was toast. Apparently, I had ingested a foreign object (FOD) during the catapult shot, causing major damage to the compressor blades in the engine. They were amazed that it held together long enough for me to land safely. Bottom line—this A-4 was not going anywhere until it got a new engine.

We scheduled a COD pickup for all three of us for the next morning, then it was off to celebrate my survival. Of course, I had no change of clothes, so the colonel said, "No problem my friend, I have pants, shirt and even a dinner jacket that you can borrow." "Thank you, sir, but I am not sure they will fit." The colonel was maybe 5 feet 7 inches tall, but he did have a big enough waist and upper body to match my clothing needs. So, although the pants looked more like pedal pushers, with a good deal of lower leg exposed, once I put my flight boots on (no way were my size-13 feet going to fit in his shoes), we were ready to party. It was my first exposure to the famous Greek liquor, Ouzo, and to a Greek custom that involved tossing candy at each other after the meal. At least I think that is a custom, or maybe the colonel just made it up in my honor for the evening.

The next morning, we made sure that the aircraft was in a secure hangar. The COD brought some ordnance people to download the *Sidewinder* so we could return that to the ship and we all headed back to *Forrestal*. Upon my return, I debriefed my CO and the CAG on what the mechanics had found. Since we were going to be coming back to this area of operations later in the cruise, they put together a plan to get a new engine for the aircraft. Obviously, it was lost to our use for a significant amount of time. I learned another valuable lesson in leadership here as my CO still felt that I should have brought it back to the ship and told me so. I just gritted my teeth, looked him in the eye and said, "I may not be a high IQ nuclear physicist like you, but fortunately I am a North Dakota farm boy with enough common sense to ignore your order and elect to stay alive!" *No wonder my fitness report from this man drew outright laughter many years later when I was before a promotion board that selected me early to the rank of Captain (0-6). I ended up ranked eighth of eight junior officers, two slots below the seventh ranked guy. For some reason, my first fleet CO did not like me.*

***

ANOTHER CLOSE CALL: I don't recall the exact date on this one, but it involved a night flight around the middle of the cruise during some very challenging weather conditions. We had great airplanes, but they had one glaring weakness: it was the ARC-27 radio, which was very unreliable. I was flying wing on a newly minted section leader, Jim Frazier. He was a good pilot, but not very experienced. Near the end of the flight, I lost the ability to transmit on the radio. I could still receive in a degraded mode, but only when I was close to the person or aircraft transmitting.

No problem, Jim will just bring me down on his wing and drop me off at 3/4 of a mile for a night landing. We set up to do this using a flashlight as a signaling device for lowering the hook, landing gear and flaps. Keeping in mind that it is difficult enough on a dark and nasty night to just fly your own airplane, imagine Jim flying with me on his wing and him having to look at me to give these signals. The final signal for me to detach and fly my own pass is a flashing of his external lights; followed by him adding power and climbing away while I transition to the meatball.

He gave the signal. I shifted my scan from his aircraft to the meatball. No ball on the lens and of course I could not hear the LSO. I felt high so I reduced some power, which immediately resulted in "wave off" lights. As I added full power and started a climb, the ball showed up on the bottom of the lens blinking red, steady red and finally amber. That scared the shit out of me, but I did not have time to be afraid since I was low on gas, needed to find Jim, join up and try again.

I finally did accomplish the rendezvous and we shot another approach. This time he dropped me off in good shape, but I managed to screw it up on my own by adding too much power in close, which resulted in a Bolter. Now I was in trouble because I was too low on fuel to try another pass. I knew that the recovery tanker, an A-3 *Skywarrior*, was circling overhead at 5,000 feet. I climbed to 5,000 feet, saw his green rotating beacon (everyone else has a red rotating beacon) and commenced a rendezvous. This is done using all his external lights to define him as the target and affect a safe and efficient join-up.

Of course, since I did not have a radio, he did not know that I was attempting to join-up for fuel. By this time, I had a nasty case of vertigo, got way too far acute in the join-up and literally ended up passing under the tanker inverted. I got my shit together and finally joined on his left wing just in time to hear him transmit (I was now close enough to hear his transmission) to the ship: "We just saw that NORDO (No radio) *Diamondback* A-4 go underneath us, inverted. We think he went into the water!"

By this time, I had joined so close that he could feel me on his wing. He looked over and saw me, and put his hose out. Since the rest of the recovery was already complete, he signaled me to top off, which I did. I then remained on his wing for another few minutes until one of the more-senior *Diamondback* pilots that had just launched, joined up and put me on his wing for another 1.5 hours of night formation culminating in my third section approach, from which I finally managed to make a decent landing. *This was one of those nights where I was 24 years old when I launched, and I felt like I was 70 when I finally landed.*

***

DÉBUTANTE BALL IN CANNES: Newly married and a father, I had very little money to accomplish much more than sight-seeing around the local ports-of-call and having a few drinks with the boys in the Squadron Admin, ashore. In other words, I was looking for funded methods of liberty. At one of our Cannes port calls, the CAG Office sent out a memo looking for volunteers to escort young ladies (we are talking early teens here) for a débutante ball. Apparently, it was considered tradition to have uniformed military officers as escorts.

What the heck, it sounds better than standing a watch on the ship, and food/drinks are all paid-for, so I signed up. Like most opportunities for socializing with the locals, this event was an absolute blast. We were only 25 years removed from WWII, so many older French people considered us heroes and liberators. I saw an older gentleman in a general's uniform, walked up, introduced myself and started a conversation. As it turned out, he was a veteran of WWII and a wonderful character with amazing stories.

He related the following regarding his one experience in an aircraft early in the war, "I was an artillery observer. I was riding in the open front cockpit of an older-model bi-wing aircraft with the pilot flying from the back seat. We were directly over the impregnable Maginot Line, and I was watching the German soldiers as they impregnated our impregnable Maginot Line." (He said all of this with an awesome French accent and a grin that made you want to giggle at his humor, but of course did not, out of respect). "Suddenly, the pilot pokes me in the back and frantically points behind us. Right there, just a short distance from our tail, is the dreaded German Me109. I look back into my cockpit, and when I turn around, the pilot of my aircraft is gone! He has bailed out of the aircraft, so I very quickly do the same. So, I get both my first flight and my first parachute jump in the same plane on the same day. After that experience, I have no desire to ever fly again."

***

RELAXING ABOARD SHIP: As previously mentioned, although it was illegal to consume alcoholic beverages aboard ship, we did it anyway. And the 8-man bunkroom was big enough and remote enough to be the favorite spot for both VA-216 and VF-11. One evening, my roommates and I decided that we would punk some more-senior folks in VA-216 who had given us a hard time. To keep the passageways clear, all the stateroom doors opened into the stateroom. We got a roll of wrapping paper, cut it to the size of the door, painted a giant middle finger on it and taped it to the outside of the door opening. We then got some duct tape, cut a piece the size of my size-13 flight boot, filled the void with shaving cream, put another piece over the top and cut just a pin-sized hole in the toe. We slid the toe of this duct tape bomb under the door and I stomped on it. There were four department heads playing cards in LCDR Jack McCauley's stateroom when the bomb exploded. Of course, they rushed to the door only to find this giant middle finger that they had to bust their way through to catch us. They never did!

Another 8-man event that comes to mind involved some midshipmen who moved in for a couple of weeks during their summer cruise. These two young guys arrived while my roommate, Terry Freedman, and I both were flying. Since we were senior to Rick and Dave, we shared a small, semi-private alcove in the far-end of the bunkroom with the other six bunks in the front.

It was an awful night, dark as hell with a pitching and rolling deck that made the already-difficult night carrier landing evolution downright dangerous. Terry landed first with a right to left drift that exceeded the limits of the arresting gear to the point where it was necessary to remove the #1 wire. I was next and did almost exactly the same thing, only because the #1 wire was already removed, I damaged the #2 wire. As we entered the ready room, the SDO gets a call on the communication box. "This is the captain of the ship. You tell those two nuggets, Freedman and Rud, that they owe me $900 each for those cross-deck pendants!" Of course, he was just pulling our chains, but we did not know that at the time, and in my mind I began to form a plan to accomplish this via a monthly payment.

Stressed, and very much in need of solace of some sort, we headed for the bunkroom. As we entered the bunkroom, drenched in sweat, more than likely pale as ghosts and still shaky from our experience, we were greeted for the first time by these two fresh-faced middies. They jumped to attention and saluted us as we headed for our little alcove. I replied, "At ease boys, we will introduce ourselves in a minute." Terry beat me to the safe, spun the combination and pulled out a bottle of Chivas Regal Scotch. We each took a big swallow right out of the bottle, and then offered a shot to the stunned middies. "Welcome to naval aviation boys!" They were great kids, but after having this experience, I ran into one of them a couple of years later and he was a Navy SEAL. *I guess he figured he had a better chance of survival in that duty.*

***

Souda Bay, Crete: We were operating in the area between Greece and Crete when an opportunity for another adventure came up. Since my buddy John Davison was the schedule writer, he managed to set up himself, and me as his wingman, for a trip into Souda Bay, Crete. Our mission was to fly in, spend the night, and then fly a mission simulating bad guys attacking the carrier.

Not only were John and I a bad combination on liberty, but we also tended to feel as though we were the most shit-hot A-4 pilots in the fleet. We briefed a fan break for Souda Bay. There were some hills as I recall, and he was going to enter the break at as fast a speed as possible, so he emphasized the need to pull up to 6-Gs to stay safely within the confines of the field. We entered the break at close to 500 knots, completed an awesome fan break, and landed uneventfully. After landing, we were invited into the Base Commander's office. He complemented us on our very impressive carrier break and then said, "Please do not do that again, because now my pilots will attempt to do the same and more than likely they will kill themselves."

Next, it was time for some liberty. Since we had no knowledge of the area, we hired a taxi just outside the gate to the base. We made a deal for a fairly good sum of drachma (Greek currency) that he would transport us to a good restaurant, and a couple of bars before returning us to the base. We set off on this adventure and all went well until the end of the evening, when he returned us to the base. At this point, he demanded more money. A rather raucous argument ensued.

Finally, John walked up to the guard at the gate, who was maybe 18 years old and said, "May I borrow your rifle for a few minutes?" The incredulous guard handed John the rifle. John then walked over to the taxi driver, pointed the rifle at him and said, "We have paid you what we owe you and what we agreed to. Now get the hell out of here before I get angry!" Needless to say, the taxi was gone in a flash. John handed the rifle back to the now-laughing guard and said, "Thank you." The guard then replied in broken English, "That man is a real asshole. He will no longer cheat riders."

***

END OF CRUISE AND NEXT ORDERS: Near the end of the cruise, I had a discussion with LCDR Tom McGraw who was a great officer and mentor for all the JO's. During that discussion, he asked if I would like to change my designator from 1315 (active-duty reserve) to 1310 (regular line officer). He explained that as a 1315 I would be subject to release from active duty as soon as the war wound down. As a 1310, however, I would need to request a release via a letter of resignation. "Since you are married with a child, and the airlines are not hiring right now, I would strongly suggest that you change to 1310. The good news is that right now they are approving all requests." (If they had scrutinized my performance at all they would never have approved the request). I decided to make this move, which, at the time, meant little to me, but turned out to be the only thing that saved me from getting a pink sheet (release from active duty) when the war ended.

12. Learning to Fly the SLUF

ABOUT TWO WEEKS PRIOR to the end of the cruise, we all got our next set of orders. I was elated to be assigned back to the A-7/A-B *Corsair II* (often referred to as the *SLUF – Short Little Ugly Fellow)* RAG, which was now VA-125 in Lemoore. The orders varied greatly as folks received assignments to A-4s, A-6s and RA-5Cs. John Davison, Dan Howe and Mike O'Brien, all top ball fliers were assigned to RA-5C's since they were especially challenging to land on a ship. Al Chesterman, Mike Farmer and Bill Harrison got A-4s Lemoore. Terry Freedman got A-7Es Lemoore. Mike Gellman and Jim Frazier both got assigned to A-6 *Intruders*. As soon as I got my orders, I sent a letter to Barbara with the necessary paperwork for her to execute a move to Lemoore. Our first choice was base housing, but at the time, there was a waiting list, so she got an apartment in town. Our address was 110B Quandt Place, Lemoore, California. It was located near the Meadow Lane grade school and our neighbors were two colorful bachelor pilots, Dave Jackson and Dave Ashworth.

Shortly after settling in to our first home as a family, our names came up for a house on the base. Since I knew that within a few months, I would be deploying to Vietnam, we jumped at the opportunity and moved-in to 157 Crusader in base housing. Our immediate neighbors were Bart and Sue Creed and the Browns. Although the housing was not great, the support system and the convenience/security afforded for deployed pilots, was perfect.

I checked back into VA-125 and joined a colorful RAG class learning to fly the A-7A/B. I can't remember all of them, but most ended up going from the RAG to squadrons on the east coast based at Cecil Field in Jacksonville. Mike Johnson, Howard Sussman, Whitey Droesel and Mike Tkach all ended-up on the east coast with the rest of us staying in Lemoore. Since I was a second cruise/experienced guy, I did pretty well in the RAG this time. On my second flight (there were no-two seat aircraft to train in, so instructors chased students with each in his own aircraft), Pete Grubaugh was my instructor. We briefed the flight, during which he said, "You are an experienced guy Gil so just fly the aircraft like I am not even there."

Barb and Valerie, hanging out at the NAS Lemoore Officers' Club swimming pool, which was walking distance from our house on base.

We lined up on the runway where we had briefed a 10-second interval for takeoff. I started my takeoff roll, lifted off, raised my landing gear and flaps, when I heard over the radio, "Shit!" followed by the distinct sound of an emergency beeper signifying an ejection. I rolled my aircraft up on the right wing just in time to see this massive explosion as Pete's aircraft cartwheeled along the ground, spreading flames and parts along its path. I also saw Pete in a parachute, appearing as though he would land safely away from the crash. At that time, another instructor in a T-28 *Trojan* announced that he had command of the Search and Rescue (SAR), and that I should just head out to the practice area and burn-down some fuel before landing.

Wow, I thought. I am all by myself out here on only my second flight. I decided to make the most of it working in an area just west of NAS Lemoore within sight of the field. I did all sorts of aerobatics and was just having a hell of a good time when I get a call on "guard," "NJ-505, call VA-125 base radio, now!" I check in with the base and the SDO said, "Where the hell are you?" I answered, "In the operating area burning down fuel." He said, "Dump the fuel down to landing weight and get your ass back here now." Unlike my previous experiences, I was not in trouble for the decision to play around in the work area by myself. The CO, CDR "Frenchy" Leblanc, just wanted me back to participate in the submission of the accident report, which turned out to be a total engine failure just after liftoff.

Much like my previous experience in the A-4 RAG, VA-125 had a slew of aircraft for our use, so if we were scheduled for a flight, we completed it. I flew 85 flights for a total of 125 hours in just four months. My flight performance improved except for the carrier-landing phase. I distinctly remember one of my heroes, Phil Gay, who was a decorated combat veteran and our LSO, taking me aside after a less-than-stellar FCLP session. "You know Gil, I have come to the conclusion that you are just not a very good pilot."

Wow! That comment really struck home. I decided that the only way that I could overcome this deficiency was to become an LSO. I started spending as much time as possible with the RAG LSOs watching passes and writing comments in the grade book for them. I realized that I was getting a late start not having been an LSO on my first cruise, but I was determined to do whatever was within my power to get better.

***

LOSS OF A FRIEND: In early December 1970, my friend and former VA-216 squadron mate Geoff Dillon, now an instructor pilot with VA-122, disappeared on a flight from El Centro to Lemoore. He was ferrying an aircraft with a cracked windscreen. Because of the windscreen issue, he was flying VFR below 10,000 feet with no flight plan. The next few days were spent searching for him along what was considered to be the most likely flight path. When he was not located by air search, we started a ground search. Acting on a tip that someone thought they saw an aircraft go down near the Tejon Ranch property in the mountains near the Grapevine, several of his friends, including me, drove to the Tejon Ranch and started a walking search. Although a very somber process with no positive results (he was not found until the next summer when the snow melted near a ski resort known as China Peak), the search did provide an opportunity to quell the feeling of hopelessness one experiences from this type of loss.

On the morning of the ground search, I stopped by to pick up Harlan Gerardy. Harlan was a west Kansas cowboy and I believe, a former rodeo rider. He was a friend of Geoff's, a fellow LSO and now one of my instructors in VA-125. When I got to Harlan's place, he was just getting up after a late night of playing Pinochle (for big money) at the infamous "Red Barn" located on the north side of Lemoore. Nursing a hangover, he pulled out a bag of Rice Krispies, grabbed a container of milk, looked into the bowl, changed his mind, grabbed a bottle of Chivas Regal Scotch, poured it on the Rice Krispies and said, "Snap, crackle and pop now, you little bastards." *We were not wanting for characters in the Light Attack community, but they all felt the sadness of losing a friend, especially under these circumstances.*

13. Back to the Fleet with the VA-215 Barn Owls

UPON COMPLETION OF THE A-7A/B RAG, I was assigned to the VA-215 *Barn Owls*. The *Barn Owls* were in *Carrier Air Wing 19* aboard USS *Oriskany* (CVA 34) and were already in the middle of workups to deploy to Vietnam. I checked in with the Commanding Officer, Paul Phillips, and the Executive Officer, Jim Crummer. Both men were great pilots and leaders. Their leadership was reflected in the overall high morale and pride throughout the entire organization.

I felt extremely fortunate to be assigned to the squadron. I immediately faced a challenge, however, since just like my A-4 squadron, the LSO slots were already filled. Jack "Gator" Paschall was the squadron LSO and Terry "T.L." Hightower was the designated trainee. I literally begged skipper Phillips to let me become a second designated trainee. "I will take whatever job that you assign to me and I will work my ass off to accomplish that task. The LSO work will be secondary, but I believe that I will be really good at it." The skipper looked me in the eye and said, "Were you ever debriefed by your previous CO on your VA-216 fitness report?" Oh shit, I thought, here it comes. "No sir", I replied. "I didn't think so, but I consider that water under the bridge anyway. You have an opportunity for a fresh start here as a *Barn Owl*, and I can use your experience having already completed a cruise in A-4s. You are going to be my Line Division Officer and my second LSO trainee. Welcome aboard LT Rud!"

***

LIFESTYLE CHANGE: I was absolutely elated. Having settled into married life and my role as a father, I had now progressed just a little with the maturity issues, at least on the ground. I would describe myself at this point as a 21-year-old pilot trapped in a 26-year-old body. Speaking of bodies, I had certainly been riding mine hard. I was smoking way too many unfiltered cigarettes, including Camels and Lucky Strikes. When I had a cold I would buy a menthol brand, usually Salems. I would then bite off the filter and smoke them. The good news was that I had a furnace for a digestive system, so I had not gained any weight, and even though a heavy smoker, I also liked to run so that had kept me relatively fit.

Al Chesterman and his wife Joanie, VA-216 squadron mates and friends, came over to our house on the base for dinner. Al was an avid runner, and I think a reformed smoker. In any case, after a couple of Scotch and waters, he challenged me to a race. We decided that a 440 yard dash would be the right distance (you must be shitting me). Of course, we did not have a track or anything like that, but there was a row of trees behind our house that we estimated to be about 220 yards away. I was blessed with decent sprinting speed, so I got to the trees first. As I turned around to head back to the house, I met a smiling and confident Al, who had barely broken a sweat. It wasn't long before he sprinted past me laughing the whole time. When I finally finished (I was still running, but it was more of a shuffle), I immediately began throwing up. This earned me one more adult-to-child look, and probably a night on the couch from my bemused wife.

I credit that incident with initiating a huge change in my life. Although it was one of the most difficult things I have ever done, I stopped smoking, which I am sure saved me from myriad health issues in later life. Since to stop smoking is so challenging, I needed to find a replacement for the habit. It was running. I always enjoyed running, but now I really got into it. I found that I could do it without any excuses. In other words, I did not need a partner, a gym or special gear other than a good pair of running shoes. I never ran with a transistor radio earpiece or anything like that. I just ran with my own thoughts, and later in my career I was able to do problem solving, and even compose speeches while running.

Now I realize that you might have questions about relapsing back into smoking? I promise at some point I will be sharing a true success, but in this case, following my first combat mission on June 16, 1971, I relapsed. Imagine returning from your initial baptism of someone shooting at you to find that the other three guys in the division are all lighting up smokes in the ready room. I thought, shit, if I live long enough to die of lung cancer, I will be a lucky guy.

And so, I was back to smoking, but only until we got back from Vietnam, and then I really did quit for good.

***

WORKUPS ON USS *ORISKANY* **(CVA 34):** *Oriskany* was a 27C-class veteran of the Korean War. She was a great little carrier, but she was relatively small compared to *Forrestal* (CV 59), which I had completed my first cruise on. The distance from the ramp to the number one wire was only 98 feet versus about 140 feet on the bigger boats. The bigger decks used a 3.5-degree glide slope, while we used 4-degrees. A smaller target requires a steeper approach. A steeper approach also means a higher rate of descent, which requires the pilot to be farther back on the power; so overall, we LSOs had far-less tolerance for error. The F-8J fighters and the RF-8 photo-reconnaissance birds were especially difficult to bring aboard the carrier safely. If not flown correctly, they could easily get behind the power curve, which would require quick recognition by the LSO and an earlyWave Off to avoid a potential ramp strike (a crash into the back of the ship).

With several new pilots added to the air wing during the turnaround between cruises, the workups were always colorful, especially at night. I absolutely loved the challenge of being an LSO trainee. Our CAG LSOs were "Moon" Vance and "Butch" Tolbert. Moon was an F-8 pilot and Butch was an A-7 guy. Both were great teachers with the patience to let us trainees make a few errors without over-reacting. The LSO team leaders were my contemporaries in seniority, who had completed the previous cruise as trainees. During the early part of the workups, the team leaders, who included my AOCS classmate "Bug" Roach, my fellow North Dakotan, Rick Maixner, plus Jack Paschall, Chris Guluily and Russ York did most of the night waving, allowing we trainees to handle day operations. I felt some pressure to excel, since I was only going to have this one cruise to attain the qualifications necessary to continue in a teaching billet ashore.

The good news was that waving aircraft aboard ship came very naturally to me. I was quick to discern situations that required corrections and made timely calls to the pilots. I also began to improve my own flying performance around the ship to the point where I was maybe just a tad above average.

One great workup story from this time period involved the unusual make-up of our air wing. Although very different in performance and mission,

system was not as sophisticated as the newer A-7E, but it did consider enough parameters to allow for deadly accuracy in comparison to a manual system. Because of this capability, the FACs favored the A-7 over the Air Force F-4s, who did not bomb that well, and were always short on fuel/playtime. "Playtime" being defined as the amount of time you could stay in a target area. For troops in contact, the Marines were definitely preferred, especially if they were working with a Marine FAC.

***

NORWEGIAN ONE AND NORWEGIAN TWO: We did not always fly with the same flight leader or wingmen. However, I did get several opportunities to fly wing on a combat veteran and shit-hot squadron mate, Gus Gudmunson. Since we were both of Norwegian heritage, if things were a bit slow (several flights cued up with a single FAC), Gus would check in with the FAC using a very heavy Norwegian accent, "Dis is Norveegian One, I got Norveegian Two wit me, we are a couple of Vikings wit a bunch of bombs and bullets, and ve vant to kill some bad guys!" That approach would usually get us right to the front of the line. Yes, we were getting shot at, but a sense of humor is always appreciated, especially by a FAC, who literally spends his day trolling for trouble.

Gus had loads of combat time, having a couple of previous cruises flying the A-1 *Skyraider* prior to transitioning to A-7s. He was also an avid midget car racer, a very popular sport in the Lemoore/Hanford area. In fact, he often won both the sprint and main events. About six months before we deployed, Gus was attempting to pass another racer when he rolled his right front wheel over the top of his opponent's left rear wheel. Not good! The result was a rather horrifying crash involving multiple end-over-end rolls that he was lucky to live through. He was alive, but he did literally break his neck. He was in a full body cast for a month or so and had just gotten clearance to resume flying prior to our deployment. As part of his reinstatement, the flight surgeon warned him that if he ever had to eject, he'd better be sure to use the face curtain versus the handle on the lower part of the ejection seat. This would help to ensure better body position to survive an ejection.

***

NORWEGIAN ONE EJECTS: About the middle of the cruise, I was in-bound on the start of a mission as Gus and Alex Whitmore were returning to the ship from the completion of a combat mission. At this point both of our flights were monitoring our squadron tactical frequency. Just as we "checked-in," I overheard conversation between Gus and Alex. From Gus, "Alex, check my tailpipe to see if there is a fire." Alex's response, "Gus, there is no fire, no smoke, no fuel vapor, absolutely nothing. She is dead as a doornail." Gus said, "Well at least I am over the water. I am going to trim this baby up and slow down before I eject." Alex replied, "OK, I have plenty of gas and will follow you down so that I can vector the SAR helicopter for a quick pick-up."

Gus slowed down to a safe ejection speed, trimmed the engineless aircraft for a glide, followed the flight surgeon's advice about using the upper face curtain (to keep from injuring his recently broken neck) and ejected. As it turned out, he had trimmed the aircraft so well that it hit the water in a perfect landing attitude and even floated for a couple of minutes (unheard of in an A-7) before sinking. Gus came through this situation in great shape and was back in the air in a couple of days.

***

FIRST PORT CALL — SUBIC BAY, THE PHILIPPINES: Our normal combat schedule consisted of roughly 30 days on the line in the Tonkin Gulf followed by five to seven days in port. After our first 30-day line period, we pulled into Subic Bay, in the Philippines for some much-deserved liberty. The base was divided into a surface ship area called Subic Bay, and an aviation side called Cubi Point. The carrier pier was on the airfield side right next to a standard 8,000 x 200-foot runway along with excellent repair facilities for the air wing's aircraft. The base was perfectly set up to handle a full complement of a battle groups officers and sailors.

There were three officers' clubs. One was a very formal Surface Warfare club, and one a less formal country western-style club called the Chuck Wagon. Both clubs were located on the Subic side of the base. The Cubi Point Officers' Club was just up the hill from the carrier pier, and that was the preferred destination for the air wing pilots. It was an awesome club that absolutely rocked when an air wing was in port. Drinks were dirt cheap, and of course happy hour offered them for something like 10 cents a drink. There were very few rules, especially during happy hour. One very important, albeit officially unwritten, rule; was that the Cubi Point Club was for aviation folks.

As soon as *Oriskany* tied up to the pier, we all headed directly for the Cubi Officers' Club. Being our first day in port, happy hour was basically in effect for the entire day. I was sitting at a table with a bunch of squadron mates. It was not unusual for a few beers to be spilled and rather carelessly tossed/slid down the table. As I joined in some sort of toast to us all still being alive, I noticed cubes of ice being thrown, not tossed, in our direction. I took a closer look just in time to see this fellow, who I did not recognize, wind up like a baseball pitcher and fire an ice cube at our table.

My Norwegian temper immediately took control. The thrower was at the bar, which was above our table. I jumped on top of our table, stepped from that to the shuffleboard table, and then over the bar rail to confront the thrower. He was in the process of winding up to throw another cube when I arrived at full speed. I grabbed his throwing arm and threw an elbow into his neck that sent him off his stool and pinned him to the bar. I was about to punch him when a more-senior aviator suggested that I might get in trouble if I hit him, so I did the next thing that came to mind.

I took a big drag on my Camel cigarette and then shoved it up his nose. He was screaming like a stuck pig as I picked him up by his belt (creating a bit of a wedgie in the process) and threw his ass out of the club. Everyone cheered, and I returned to my table to continue celebrating surviving our first line period. No one knew who this guy was, and he was not wearing wings, so I did not give it another thought.

The next morning, I was awakened by a loud knock on our 4-man stateroom door. Since we were all nursing some nasty hangovers, nobody jumped up to answer it until we heard, "Open the door, this is the skipper, and I need to talk to Gil now!" I jumped up and opened the door to find the skipper standing there in his dress whites. "Yes sir! What can I do for you?" He said, "Get into your dress whites and meet me in the Ready Room in 15 minutes." "Yes sir, but may I ask why?" He replied, "You are going before Captain Haak (*Oriskany* CO) for a disciplinary hearing called "Captain's Mast" in 30 minutes. Now get dressed."

He did not elaborate so in my hung-over haze, I tried to recall what I could have possibly done to end up at Captain's Mast. I arrived in the ready room and grabbed a quick coffee just as the skipper entered. "Let's get going up to the bridge; we do not want to be late." I followed him, darn near passing out as we climbed the ladders up the seven decks to the bridge. As soon as we got to the bridge, the skipper asked permission for Commander Phillips and LT Rud to enter the bridge. All of this was very formal and extremely intimidating. As we entered the bridge, I at least had the presence of mind to salute the Captain and I said, "LT Rud, reporting as ordered sir!"

At this point, I noticed another LT in his whites with a huge bandage on his nose. Oh shit, I thought, now I know what this is about. Captain Haak than said, "LT Rud, LT (I can't remember his name, but he was a black-shoe ship's company officer) says that you assaulted him in the Cubi Point Officers' Club yesterday. Is that true?" "Well, no sir, I was going to assault him, but I decided that I might get in trouble if I did that, so I just shoved my cigarette up his nose."

Captain Haak turned away from me to blow his nose or something (I found out later that it was to suppress a giggle). Then he asks me why I did that, and I related the ice cube-throwing incident in detail. He then asked the alleged victim of the alleged assault if my description of the incident was accurate. "Yes sir, but everyone was throwing ice." I then further described the difference between tossing a few cubes out of a cup versus firing them like a baseball with obvious harmful intent.

The captain took a few minutes to contemplate the situation, and then he turned to us both and said, "LT Rud, I don't ever want to hear of any more boisterous behavior like this from you. I dismiss the charges of assault, but to teach you a lesson I recommend that Commander Phillips put you in hack (take away your liberty) for the next 12 hours." Since it was only 0830, I would be ready to roll the same evening. I was dismissed and returned to the ready room. When Skipper Phillips got back from the bridge, he related that the captain told the victim that he better never set foot in the Cubi Point Club again. Captain Haak then stated, "While you were pushing papers in your office for the last 30 days, these pilots were out there flying combat missions and getting shot-at. Your officer's club is on the other side of the base." *Whew! I managed to survive another potential career-ender.*

***

COLLATERAL DUTY CHALLENGES: As the Line Division Officer I was given a great opportunity to make a difference. The Line Division is responsible for the security, movement, fueling and scheduled maintenance of all the squadron aircraft. Each aircraft is assigned a plane captain. The plane captain does not actually taxi the aircraft, he does however sit in the cockpit and applies the brakes, as required, when his aircraft is being towed by either flight deck or hangar deck personnel. It is the largest maintenance division in the squadron. It is also populated by the most junior sailors, who learn about the aircraft as a plane captain before "striking" for a more specific role in the maintenance department.

I was lucky to have a great Chief Petty Officer, ADJC Morgan, acting as the den mother of this group that probably averaged about 19 years old. Keeping in mind that this was not an all-volunteer force, we had a wide range of draftees in the department. Most were great sailors, but of course it was Vietnam in 1971, so we also had some malcontents, who tended to bitch and whine about the job, which, aboard ship, was extremely dangerous. Considering that an A-7 or F-8 engine intake could ingest a 200-pound sailor from up to 15 feet away (this would almost always be fatal), and the exhaust could easily blow a person down or even over the side into the water, there was a good reason for the 200+ sailors assigned to flight deck operations to average 19.5 years of age. *This was not an environment for old guys.*

One of my plane captains was an older draftee. He had completed a master's degree in math at MIT prior to being drafted. Of course, he was not a happy camper. He was, however, very charismatic and had influence on the younger guys. Unfortunately, at the beginning of the cruise, that influence was not good. The chief and I even discussed getting rid of him by sending him to some "paperwork" type of job. Before doing this, I decided to see if I could turn him around a bit.

I asked what he intended to do with his math background. "Well sir, I intend to become a college professor." Since we had several kids that did not even have a high school diploma, I gave him a challenge. I said, "If you really want to become a professor, how about seeing if you can tutor your squadron mates in the division toward achieving General Educational Development (GEDs for High School equivalency) prior to the end of the cruise. Personally, I don't think you are capable of doing it." Bingo! He puffed his chest out and said, "LT, if you get me the material, and the use of a classroom, I will have every one of these boys a GED by the end of the cruise." He did just that, and I think got a Letter of Commendation in the process. *Leadership Lesson—You must assume that there is potential in everyone, figure out what motivates them, and let them loose to succeed.*

The other part of the Line Division consisted of the Trouble Shooters. Unlike the inexperienced youngsters serving as plane captains, the Trouble Shooters were highly qualified sailors picked from various rates to correct discrepancies that occurred during live flight operations. This means that they had to know a little bit about everything, as they made the last second checks on the aircraft while it was on the catapult. It was often their decisions as well as their miraculous ability to fix a last-minute discrepancy that determined the success or failure of a critical combat mission. These men were the most-respected individuals in the maintenance department. With that in mind, I feel it is appropriate to name the five that I remember from those hectic days on the flight deck. Hats off to: "Billy" Daniels, "Hud" Hudson, "Mex" Cardwell, "Whit" Whitacre and "Pete" Von Essen.

***

REST AND RECREATION (R&R) IN JAPAN: At the midpoint of the cruise, *Oriskany* got the usual break in action that was to involve a port call in Singapore followed by another in Yokosuka, Japan. We were elated to hear that there was going to be an opportunity for ship and air wing wives to meet us in Yokosuka. I jumped at the opportunity, wrote a letter to Barbara, and encouraged her to get a spot on the charter flight so that we could spend some time together in Japan. She did and managed to get a seat on the aircraft. As part of the preparation to travel overseas, the wives needed to get some required shots just like we sailors did. Unfortunately, during this process, Barb reacted to one of the vaccinations and got sick enough to require hospitalization. Fortunately, she recovered in time to still make the airlift.

At this stage in our lives, we pretty much lived paycheck to paycheck. With this in mind, and desperately wanting to have enough money to show my lady a great time in Japan, I volunteered to skip the Singapore port call to fly test flights on six A-7s that were undergoing maintenance in the Philippines. Each of the A-7 squadrons had two aircraft that needed test flights. My task was to get all of them into an up status so that they could be flown aboard *Oriskany* as the ship passed by the Philippines on the way to Japan. I had about five days to accomplish this, so I was one very busy pilot. I flew at least two flights every day and stayed in the BOQ, so I was well on my way to a goal of having spending money for my reunion with Barbara in Japan.

As *Oriskany's* transit from Singapore to Japan got close enough to Cubi Point, A-7 pilots were flown by carrier on-board delivery (COD, C-1 *Trader*) to pick up their respective aircraft that were now in an up status. Of course, since I was already there, I was scheduled to fly one of the two *Barn Owls* birds back aboard ship. Unfortunately, there was one *Barn Owls* aircraft that, despite several test flights, I was having a heck of a time getting in good enough shape to fly to the carrier. And there was one more fly in the soup—a major typhoon was bearing down on Cubi Point. Finally, the maintenance folks gave up on getting that aircraft ready and we put it in a hangar to ride out the storm.

Shit! As the weather rapidly deteriorated, I desperately searched for a way to get both me and the maintenance crew from the air wing the hell out of there before we would all end up stuck in the P.I. while the ship headed for Japan. I went over to Base Operations to see if anything was going to Japan. "Well LT Rud, your timing is great. An EC-121 *Super Constellation* just landed. They are going to refuel and head for Atsugi, Japan." I approached the plane commander, and he was more than willing to stuff the five of us into this giant machine. We all got aboard, and he took off in heavy rain and high winds, but we were out of there!

When we landed at Naval Air Facility Atsugi, Japan, which is just a few miles from Yokosuka where *Oriskany* was going to tie up, the Japanese customs officials asked me for a shot card. Luckily it was raining and nasty, so he was not paying close attention. I had a shot card, but the troops did not so I showed mine to him, passed it to the next guy in line, he to the next, etc. until we had gone through all five of us with the same shot card. *Maybe not totally ethical, but sometimes you do what needs to be done and accept the risk.*

We got a ride to Yokosuka and arrived just about the time *Oriskany* was due to tie up. But it wasn't there. I asked someone on the dock where the carrier was and he said, "Oh she had to alter course and delay her transit because of the typhoon." "What about the contracted airliner that is bringing all of the wives?" I asked. He did not know but directed me to the operations office where they hooked me up with a liaison officer responsible for coordinating the arrival of the wives. He informed me that *Oriskany* would not arrive until the following morning, and that he was heading into Tokyo to meet the wives flight. "Would you like to accompany me?" "Heck yes, my wife is on that flight, let's get going."

The liaison officer was all decked-out in his nice-white uniform, while I attempted to be super low key in civilian clothes. He and I made a deal. I would stay out of sight until I spotted Barbara. At that point, I would whisk her off to the nearest hotel, while the liaison officer dealt with what we rightly assumed would be some very disappointed wives expecting to be met by their husbands. I know, a more mature, career-oriented officer would have stepped up to take charge of this situation and ensure all 260 wives would be delivered safely to their respective spouses when the ship pulled in. Well, I was not that officer. In fact, I was still in the immature, self-centered, and now in love-dominated state of a 26-year-old Navy pilot that had not seen his beautiful wife in four months, so I quickly rationalized that I was only responsible for one of these women—mine!

When I left on cruise, I weighed 190 pounds. After the completion of three combat line periods, manning aircraft on a flight deck where the temperature often was over 100 degrees, I now weighed 165 pounds. Having spent countless hours on the LSO platform in the blazing sun, I was also burned (a sort of reddish tan) to a crisp. When I saw my drop dead gorgeous 5-foot 10-inch, exceptionally well-proportioned Barbara enter the terminal, I literally hid behind a support pillar until she walked by. I grabbed her in a wonderful hug, and she literally jumped back with a "who the hell are you" expression. My hope of sneaking her off to a hotel without being noticed was quickly quashed when she screamed, "Oh my God, Gil!" At that point we entered a loving embrace that lasted quite some time.

When we finally separated, I was looking into the eyes of the Captain and the CAGs wives, who had their arms, folded in front of them and were giving me the familiar adult-to-child, look of disapproval. "Where are the rest of the husbands?" I related the typhoon delay information and introduced them to the liaison officer who was awesome. He already had secured hotel rooms and transportation for the wives to meet the carrier when it pulled in the next morning. The captain's wife was incredulous, "Well, if they all got delayed how did you make it here?" I responded, "There is no typhoon that is ever going to keep me from my lover!" The junior officer wives all cheered at that point as I whisked Barbara off to the Sanyo Hotel.

We had a wonderful time in Japan. When we rode on the trains, Barbara got a bit uncomfortable when the usual "packing" occurred. This involved a conductor literally pushing folks in the door just before departure. Since the typical Japanese man was quite a bit shorter than my wife, they would end up with their faces at chest level, which they seemed to thoroughly enjoy. Not being the least bit shy about her physical assets, Barb was a great sport throughout this experience.

One evening we were attending a dinner/show in downtown Tokyo. We ended up sharing a table with a Japanese couple. The waitress asked us for a drink order in broken English. She then turned to the Japanese couple and asked them in Japanese. The guy turned to me and said, "Hey buddy, will you tell her that I am a third-generation Japanese-American, and don't understand a word she is saying." We had a wonderful evening with these folks who were visiting Japan for the first time just like us.

Eventually we made our way down to Yokosuka, and I checked-in with the squadron. I had previously corresponded with them by phone explaining the circumstances of my predicament with the aircraft in the P.I. and they had sent John Snow, who was married to a Filipino lady and had family in the Subic Bay area down to cover that mission. Still, the skipper was not happy with my decision to leave the airplane, and head for Japan. He played around with me a bit by threatening to place me in hack for the remainder of the Japan in port period. If he had been serious (he was not), I would have ended up AWOL. *In the end, I managed to sidestep another rock on the road to a successful career.*

Atsugi had an excellent repair capability. A few of the air wing's F-8 fighters had flown off the ship prior to it docking to have some modification installed. When we left Japan, those aircraft launched from Atsugi to land back aboard ship. During the takeoff phase from Atsugi, one of the F-8 pilots (I will leave off the name to protect the guilty) lost sight of the formation he was joining on. He had inadvertently entered a cloud, got vertigo, and came out of the cloud 90-degrees nose-down in full afterburner, just a few hundred feet above downtown Yokohama, Japan.

Yokohama, like most of the Tokyo area, was densely populated. The pilot realized that he was about to die, and promptly ejected from his perfectly functioning fighter. Thank God the aircraft went into a small community garden area in full afterburner. It was literally vertical, so it just dug a huge hole in the ground and by some miracle no one was even injured. A Japanese civilian had filmed the potential disaster, which clearly showed the fighter on fire as it plunged to earth. That of course was the afterburner, not an aircraft fire. The Japanese media jumped on the heroic actions of this brave fighter pilot who expertly guided his crippled fighter into this small garden prior to ejecting. This fellow stayed in Japan for a week or so basking in the accolades of the grateful Japanese citizens. When he finally got back to the ship, which had stopped for a short visit to Hong Kong, he was promptly placed in hack.

***

BACK TO THE LINE: We soon returned to combat operations in the Tonkin Gulf. Most of our cruise was during the monsoon season, which involved nasty weather over land and relatively good weather in the gulf. The heavy rains and resulting muddy conditions slowed the normal movement of supplies on the Ho Chi Minh Trail and tended to quiet the enemy anti-aircraft artillery (AAA) fire. The reason for this was that if they shot at us, and we spotted where the guns were, it was difficult for them to move before the next strike got them.

As we were returning from a mission involving a FAC working in Laos, we were asked if we had any 20 mm rounds still on board. We had not fired our guns, so I replied in the affirmative. I was the flight leader of a section of *Barn Owls* (I do not recall for sure who was my wingman, but I think it was Mike Prince, a first-tour nugget). We were asked to contact a FAC who was flying an OV-10 *Bronco*. He, in turn, was in contact with a little O-1 *Birddog* Cessna that was in the process of helping a "Bright Light" rescue team. These are small special-forces units that are often inserted into extremely dangerous and hostile areas to locate and rescue downed pilots or even POWs.

While in the process of carrying out their mission, this Bright Light team had run into a large enemy force, and they were now seeking to disengage from this very bad situation. The weather was awful, and the area they were in was very rugged (south of Tchepone as I recall). I had my wingman hold at altitude while I worked my way in between thunder bumpers until I got sight of the OV-10 who then put my eyes on the Cessna. He told me that the Cessna was just to the south of the bad guys, and that I should do a strafing run just to the north of the Cessna. He also warned that the clouds on either side of my target were filled with rocks (it was a valley between two mountains).

I kept my eyes on the Cessna, armed my gun and rolled in. I was hauling ass at 500+ knots as I squeezed the trigger on my 20 mm gun. "Bang!" I got off one round before the gun jammed. By this time, I was just a few feet above the jungle, so I hauled back on the stick and climbed through the clouds (fortunately none had rocks in them), back to altitude. "Shit! Sorry my gun jammed." The FAC replied, "well your gun may have jammed, but you got so close to the ground that you scared the crap out of the bad guys, and they have broken contact."

When I got back to the ship, I got together with the Ordnance Chief and a few of the more senior enlisted folks in the Weapons department. I first complemented them on the wonderful job that they were doing, and then I related the aforementioned event, in detail. To my knowledge, we never had another gun jam on that cruise.

***

Aircraft and Pilot Losses: We did not lose any air wing aircraft or pilots to hostile fire on this cruise. We did, however, lose several pilots to aircraft accidents, and had others that successfully ejected from crippled aircraft. One of these accidents involved the failure of the nose gear on an A-7 during a catapult shot. Tom Frank, the XO of VA-153 and a great guy, was lost when the nose gear literally disintegrated, dropping his nose to the flight deck, which put him out of the ejection envelope, and cost him his life. The CAG initiated an inspection of all the other A-7s following this incident, which ended up grounding all of us for a significant amount of that line period. We also lost LT Jim Borst during a practice mining exercise before we even started our first line period. CDR Metzler from VF-194, LT John Painter and LTJG Ray Debalsio from the A-3 squadron were also lost to accidents.

To be honest, our greatest fear during these combat missions was having an engine failure. To put this in perspective, nearly half of the 21 pilots in the *Barn Owls* had, at some point in their career (not necessarily in VA-215), ejected from a crippled aircraft. Such was the nature of the single-seat/single-engine Light Attack community.

***

Scary Night Emergency Divert to Da Nang: The trees in Laos were often triple canopy thickness and could grow as tall as 300 feet above the ground. We were all trained on how to rappel and carried the appropriate survival gear to hopefully get from the trees to the ground. That being said, we knew that the chances of rescue from most of the areas around the Ho Chi Min Trail were fraught with risk.

I was on a night mission in which we joined up with an Air Force F-4 pathfinder aircraft. We were to fly his wing while he used what I believe was a Loran-based navigation system to bomb a known target. This was an awful mission involving night formation flying on a dissimilar aircraft. On this mission only, we also dropped our bombs from straight and level flight, which was much riskier than a dive bomb run, which afforded us more separation from potential bomb-to-bomb collision or early detonation.

Shortly after joining on the F-4, I was working my ass off as the outside man of a section of A-7s (there were a total of four A-7s, two on each wing of the F-4). Of course, I was not looking directly into my cockpit, when out of my periphery I saw a momentary flashing master caution light. "Shit! I hope this is something minor," I thought to myself. Pretty soon it came on again, and it stayed on. I moved out a bit from the formation so that I could see what it was.

"Engine hot." At this point, training kicked in. Instead of panicking, I immediately turned away from the formation, reduced my throttle and began a turn toward the Tonkin Gulf. My flight leader soon followed (you never left a single-seat aircraft alone in an emergency). The engine over-temperature followed the movement of my throttle, which pretty much verified that it was indeed running too hot. Fortunately, the fire warning light stayed off. I decided that bypassing Da Nang on the way to the ship was not a good idea so that is where I headed.

There was an emergency bomb jettison area just off the coast of Da Nang and I had eight 500-pound bombs to get rid of. I did that on the way into Da Nang and requested an arrested landing on arrival. By this time, I had relaxed a bit because I knew that if the engine quit, I would be ejecting over friendly forces. As I began my approach, I saw several lightning flashes near the field so I thought that weather might now be a factor as well. I landed into the short-field arresting gear and was immediately surrounded by fire trucks and emergency vehicles. As I exited the aircraft, I noticed some debris on the runway. I inquired as to the source and the response was, "Oh don't worry about that, just a lucky hit from one of those mortar shells the Gooks are harassing us with." That is when I realized that the flashes were not from lightning.

They towed my aircraft to a protected revetment and sent me to the area where our A-3 crews were billeted. After a couple of stiff drinks, I headed for my assigned rack and fell asleep. The next morning, I woke up with a desperate need for a shower. I asked where it was, stumbled into the community shower and begin to enjoy the experience. That is until I spotted what appeared to be a black pajama-clad Viet Cong (VC) covering his mouth and appearing to be giggling. What a way to go, shot by a VC, naked in a damn shower.

On further inspection, I realized that my black pajama clad adversary was really an 80-year-old grandmother who worked as a cleaning lady. The big brave naval aviator then quickly decided to see how I could get back to the ship. The COD (C-1 *Trader* carrier on-board delivery aircraft) soon showed up with some mechanics to check out the aircraft and give me a ride back to the ship. As it turned out, the aircraft problem involved a faulty thermal couple that gave very real indications of engine failure.

***

Last Combat Sortie — Mu Gia Pass: On November 21, 1971, I was scheduled for the very last combat flight for *Carrier Air Wing 19* for that cruise. It was a night strike to Mu Gia Pass, which was the most northerly of the three main passes through the mountains from North Vietnam to Laos. It was a nasty, dark night over the Tonkin Gulf. Bill "Duke" Mahew was the flight lead and I was the wingman. We were teamed up with four other A-7s, two from VA-153 and two from VA-155.

You try not to contemplate it being the last flight, but as I manned up the aircraft, I could not help but think, "I can't believe how dark it is, and why me?" A day cat shot can best be described as a thrill ride and just plain fun. A night cat shot is an evil, nasty, out-of-control experience, fraught with danger and really shitty odds of survival if anything goes even slightly wrong.

I taxied onto the catapult shuttle, got the run-up signal from the Cat Officer, pushed the throttle full forward, grabbed the throttle grip to make sure it did not retard during the catapult stroke, took one last look at all the gauges, placed my hand on the control stick, my head back on the headrest, flicked the switch on the throttle that turned on my external lights (signals to the cat officer that I am ready to go) and focused on the attitude gyro, but out of my periphery watched the Cat Officer's wand touch the deck and BOOM! Down the deck I went from 0 to 150 mph in 2.5 seconds. As I have previously mentioned, you cannot see shit during that time frame. Just a blur, and then suddenly you are airborne a scant 60 feet above the sea snake-filled waters of the Tonkin Gulf. Check attitude gyro, angle of attack gauge and vertical speed indicator, gear up, check airspeed, flaps up, now climb like a homesick angel into the overcast (as if it was not already dark enough).

I flew instruments intercepting a pre-briefed TACAN (Tactical Air Navigation System that provides bearing and distance from the ship) radial and broke through the overcast. I found Duke, rendezvoused, and we headed for the corridor just below the DMZ (designated area separating North and South Vietnam) on the first portion of our journey to the target. Although all six A-7s would eventually rendezvous with an Air Force F-4 Fast FAC in the target area, we transited as three separate sections (thank God). I swear the bad guys knew that it was our last mission before heading for home, because they shot at us at Tchepone and Ban Karai, on the way to Mu Gia. Now don't get me wrong here, no hero stuff. We were well above most of the AAA being shot at us; it is just that at night it gets your attention. During the transit, we passed over a C-130 gunship, *Puff the Magic Dragon*, working over a spot on the Ho Chi Minh trail with his Gatling guns (very impressive). It was a long flight, but we finally arrived in the target area.

As per the brief, each section of A-7s held at different altitudes as we commenced check in with the FAC. This is what we referred to as "Indian Country." We were right on the border with North Vietnam (NVN) going after their main supply route. Therefore, the only lights we had on were the flight leader's formation lights. We were armed with eight 500-pound bombs, each. However, four of the eight, or every other one that was dropped, was a type of mine that was designed to close the trail for an unpredictable amount of time, up to a total of 33 hours before exploding.

The FAC had already marked the target, which was a road through a mountain pass. He had done this with a marking device called a log. It sits on the ground and glows so that you have something to aim for (remember, it is a nasty, dark night). The FAC reminded us that due to the number of guns and even potential SAMs from the NVN side of the border, that it would be a single run for each A-7 from east to west. "No lights boys, as they might hear us, but they don't see us yet." Duke and I are the last section to be called in. As luck would have it, one of his bombs gets a direct hit on the target log. Now, there was absolutely nothing for me to aim at. The FAC said, "Bulls-eye, *Barn Owl* One. *Barn Owl* Two, standby and I will drop another marker." There had already been some AAA firing at the sound of the first five guys but thank goodness no radar-controlled stuff.

FAC said, "Unfortunately, I am out of logs, but I do have flares, so I am going to drop those." The FAC then called, "Flares away!" I was now almost to my roll-in point for an east to west run starting at about 16,000 feet. Standard procedure for a flare drop is to release them at an altitude low enough so that the bottom of my bomb run will still be above the flares. The good news was that the flares were right over the pass, placed perfectly to light up the target. The FAC said, "*Barn Owl* Two, cleared in hot east to west, pull off west!" I rolled into a nice steep dive of about 50 degrees, and about halfway down the run, "Daylight!"

It took just a second or so for my brain to realize that I had gone from invisible to one juicy target. I pressed and held the pickle down as I began pulling what turned out to be 8.6 G's in an aircraft designed for a maximum of 7 G's. The bombs soon came off accurately with the aid of the CP 741 bomb computer. The FAC said, "AAA, jink!" Thank goodness I could not see all the AAA being shot at me as I jinked (aggressively maneuvered the aircraft to spoil the aim of the AAA) my way west into the welcome darkness. As it turns out, the FAC had dropped the flares much higher than normal. To them of course, it was no big deal, a bit of entertainment and a chance to locate where some of the guns were for the next day's sorties. For me, not having been seriously shot at very often, it was a bit unsettling to say the least—especially on my last combat sortie of the cruise.

I finally got rendezvoused with Duke and we headed back to the carrier, now late, due to my misadventures. The weather at the boat was best described as shitty, forecast to become shittier (we had now gone from monsoon bad weather over land to bad weather over the gulf). We were the last section to arrive and as they split us up to enter marshal, I ended up tail-end-Charlie of that night's strike group. Normally, the last aircraft to land would be the recovery tanker, but that night the tanker was an A-3 who would head for MCAS Da Nang once all the aircraft were safely aboard. Also, since this was the end of our line period and the winds were favorable, the ship was already on a course for Subic Bay.

What did this mean? It meant that the captain of the ship and the Air Boss, plus the entire air wing and ship's company, were waiting for me to land so we could all head for home. As I began my Case III instrument flight rules (IFR) approach I heard pilots calling "Clara" (this means that they do not see the Fresnel lens) at 3/4 of a mile with the controlling LSO, Jerry "Biter" Arbiter talking them down to what was probably around 300 feet where they finally called the ball. At this point, my blood pressure was probably 300/295 with a heart rate of 200 or so. I worked my ass off on instruments in the goo (my logbook says 1.8 actual instrument time), got to 3/4 of a mile and nothing. I called, "Clara!" Which in this case meant that I didn't see the meatball or the ship for that matter.

Biter calmly replied, "You are a little high, keep it coming down." I finally broke out of the clouds saw the ship and landing area but no meatball. I called again, "Clara!" and Biter replied, "You are still a little high just keep it coming." In close, still no ball. Biter said, "POWER!" I added full power, transitioned my scan to the centerline of the deck and prepared for a Bolter, when suddenly, I got pitched forward into my straps and I was on the deck, somehow having snagged the number four wire. The Air Boss transmitted, "Welcome back, now gets those lights off on my deck." It was a shitty pass resulting in a "No Grade," but in this case, any one you can walk away from is a good one. During the maintenance debrief, I wrote-up the aircraft for the potential overstress of the airframe. On inspection, they found several airframe rivets popped, which was a hell of a lot better than patching bullet holes.

***

LAST LSO PARTY: Since all three of us *Barn Owls* LSOs lived in the same four-man stateroom, our abode was chosen by the CAG LSO, Moon Vance, as the best choice to host the last LSO party. It was also an opportunity to get rid of all the illicit alcohol before entering Pearl Harbor where there would be a customs inspection before we headed for home. At one point, we probably had 15 LSOs in this one little space, sharing mostly true stories about the cruise. We finally finished up all the booze (at the end somebody was mixing Vodka and mouthwash) and everyone left. Well, not everyone. One of the more junior trainees had passed out on the floor so we covered him up and let him sleep it off.

After the party, we had carefully placed all the empty bottles in a large parachute bag, which we intended to toss over the side later that night, while still at sea. We misjudged the timing a bit, however; when we woke the kid on the floor up, gave him the parachute bag and instructed him to toss it over the side from the hangar bay on his way back to his room. He did just as instructed. Unfortunately, it was now mid-morning and the ship was making the turn at Hospital Point in the middle of Pearl Harbor. The ship's First LT, responsible for all sorts of pre-docking procedures, just looked at the parachute bag thrower, shook his head in resignation and told the kid to get his ass to his stateroom. *This sort of behavior would end a career in the post-Vietnam era, but fortunately was tolerated in 1971.*

Clowning around with a "sissy" flare pistol prior to a combat mission. I carried a .357 Magnum in addition to this peashooter for survival purposes.

***

HOMECOMING: Following a lengthy transit, on the morning of December 17, 1971, I manned up for the fly-off to NAS Lemoore. In many ways, the fly-off was a dangerous and challenging evolution. We briefed and flew all three A-7 squadrons in a massive 36-plane flyover of the cities of Lemoore and Hanford. We did this using basically a diamond of diamonds formation consisting of nine separate diamonds joined into one massive formation. Since no other air wing had more than two A-7 squadrons, we set the record for number of aircraft in a flyover of our two hometowns. It was a thrill to be a part of this evolution, but we were all very anxious to land and see our loved ones.

It is difficult to adequately describe the emotions one feels as we taxied to parking, all the while attempting to pick out our loved ones in the crowd. What a shock it was to see how much my little girl had grown in seven months. Many children were reluctant to approach a father they had not seen for so long. Not my Valerie. At just over two and a half, she was already a precocious, super-active bundle of energy. Barbara was at least to me, the most beautiful woman on earth, and it was a wonderful feeling to be home just in time for the Holidays. *I knew that I was a very lucky man.*

Barb, myself, Mary and T.L. Hightower celebrating the return from the 1971 Combat Cruise. T.L. was my roommate on cruise and next-door neighbor in Lemoore.

15. The Ultimate Bagger as a Flight Instructor

I HAD DONE WELL ENOUGH as an LSO trainee to achieve Squadron LSO designation. This was a huge factor in getting orders to the A-7E RAG in Lemoore as a flight instructor and training LSO. It was my number one choice for orders, allowing for family stability and the opportunity to fly the newest and greatest Light Attack aircraft, the A-7E. At this point in time, I may not have had a whole lot of combat experience, but I certainly had the most patches on my flight jacket. I had been in the fleet for just 2.5 years, however I had already served in two different fleet squadrons, attached to two different carrier air wings, on two different aircraft carriers, on two different coasts, flying two different types of aircraft. I had also achieved centurion status (100 carrier arrested landings) on two different aircraft carriers. I was just a junior LT and I was running out of space for patches on my jacket.

***

LEARNING TO FLY THE A-7E: The A-7E *Corsair II* was state-of-the-art in every respect. It had an additional 3,000 pounds of engine thrust over the Bravo, and a much more sophisticated navigation computer, ordnance delivery system and a ground-stabilized Heads-Up Display (HUD) that was good enough to use for landing aboard the carrier. Prior to becoming an instructor, I spent several weeks immersed in learning these new systems. Since this was my third type fleet aircraft transition, I jumped into the challenge with an open mind and an intense desire to make use of all the new gadgets and capabilities (this was not the case with all transition pilots as some liked to stick with the old way of doing things).

1972 was a wild year for the fleet, as the bombing of North Vietnam resumed. The North Vietnamese had used the bombing pause to upgrade their air defense systems so that they were absolutely deadly. The best description is that the environment over Hanoi and Haiphong in December of 1972 was comparable to that of Berlin for the Eighth Air Force in 1944. This resulted in lots of aircraft damaged and lost. The replacement aircraft mostly came from our inventory at VA-122.

We got so low on A-7Es at one point that those of us still current in the A-7A/B chased A-7E students on familiarization flights with A-7Bs. Since the two cockpits were not even close in design, I would cut out and paste a photo of the A-7E computer over the top of the old one in the A-7B. That way I could answer questions that the student might have regarding the use of the new equipment.

***

LIFE AS AN INSTRUCTOR IN THE RAG: It was crazy busy and included great flying opportunities. My collateral duty in addition to being a flight instructor and LSO was to work in operations as a schedule writer. It is one thing to write a flight schedule for a fleet squadron with 20 or so pilots and 12 aircraft, but something totally different to do the same task in a squadron with 70 aircraft, 50 instructors, probably close to 100 students, with several phases of training occurring simultaneously. Since this was before computers, we did all of this manually via a giant grease board that took up a whole wall of our office.

I worked for the Operations Officer, Jack Calvert. He was a combat veteran and a great guy to work for. He did, however, possess a very deep and authoritative voice, which he used occasionally to scare the crap out of students who failed to read the flight schedule properly. Although I do not recall their exact job titles, Frank "Tank" Bledsoe and Jerry Palmer were in my chain of command and were both great to work for.

The good news was that for a newly anointed instructor pilot (IP), I was in a very powerful position. I figured out very quickly who the "baggers" were and who the "swans" were. Baggers will generally fly any flight at any time day or night. Swans on the other hand, would give you every excuse in the world why they couldn't fly. One of the favorite excuses was from folks who were working on their advanced degree or maybe a real estate-license or some other form of education. To make sure that the baggers got some good deals to go along with the late night and bad weather sorties they often flew, I would place a gold star next to their name on the grease board. For the more notorious swans, I placed a nasty black lump of coal-looking mark.

One day a fairly senior swan asked what the black mark by his name meant. I readily shared what it meant, and since he never volunteered for anything, he sure as shit was not going to get any good deals. This did not go over very well with the squadron leadership, so I was forced into a more clandestine method of discriminating against the swans. Not to worry, I still managed to get the good deals for the pilots that deserved them. As a schedule writer, I was also the first to know about opportunities to get qualified in as many phases of the training syllabus as possible. It was not long before I was able to fly all the different types of training flights. *Yes, I was the consummate bagger.*

***

Getting Qualified in the T-28 *Trojan:* VA-122 had eight T-28 *Trojans.* We used the T-28s mostly to serve as Forward Air Control aircraft and as safety monitors for low-pullouts on night dive-bombing runs. We also used them to ferry both people and parts to Fallon, El Centro and Yuma for weapons detachments, as well as occasional runs to Travis AFB and/or NAS Alameda to pick up folks traveling from overseas, usually on emergency leave.

There were only a few of us that were checked out in the T-28, but all of us had one thing in common: we absolutely loved to fly. And we loved the challenge of piloting this avionics-free, 1425-horsepower radial engine (it could produce 52 inches of manifold pressure and sounded like a WWII fighter), throwback-beast of a machine. Unlike the training command orange and white paint job, ours were painted in a combat color scheme much like the A-7s. Realizing that this was as close as I was going to get to fly a WWII type aircraft, I jumped on the opportunity to become one of the few T-28 drivers in the squadron. The good news was that the T-28 hours did not count against your total allotment of instructor hours allowed per month, so it was also a baggers dream.

When we were faced with getting a qualification in a short amount of time, we often did so via a cross-country flight. I sought out just such an opportunity and ended up talking a fellow IP (Instructor pilot), "Muddy" Rivers, into a cross-country to North Dakota. It was one of the dumbest stunts we ever pulled. We got an extremely late start, which resulted in us stopping for gas at Hill AFB near Salt Lake City. We then flew to Ellsworth AFB in Rapid City, South Dakota. These two flights totaled 5.2 hours of nighttime with three hours flown in the clouds and logged as actual instrument time, including the approach into Ellsworth.

It was early April, so the mountains were still covered in snow, and it was cold as hell. Icing? Shit no, it was "too cold" to get icing. We obviously had some good tailwinds, because our cruising speed was only 180 knots and it was a 900-mile trip. I distinctly remember checking in with Denver Center some place over Wyoming. The Denver controller asked us to repeat what kind of aircraft we were in. When I said a Navy T-28 *Trojan*, he replied, "Are you nuts?" We pressed on to Fargo, North Dakota, then Muddy went on to Nebraska, all without a break other than to refuel. *By the way, my dad was more impressed by me flying this aircraft than any of the jets I would later bring to Fargo.*

***

First Carrier Qualification Detachment: Shortly after getting qualified in the A-7E, I was assigned, as one of three LSOs to take a class through FCLPs, then to an aircraft carrier for their first exposure to night landings. Technically, I did not yet have a Training-LSO designation, so I was basically an Instructor Under Training (IUT). Pete Nichols and Muddy Rivers were the other two LSOs. They were great guys and were excellent at teaching the students how to stay alive around the boat at night, so I learned a great deal in a short amount of time. Both Pete and Muddy had cruised on larger-decked aircraft carriers so when we were told that our boat was going to be *Lexington* (CV-16), they were glad to have my experience from working that class of ship.

We walked aboard *Lexington* on a Sunday evening while it was docked at the carrier pier in Pensacola. We went to check out the LSO platform, and both Pete and Muddy immediately noticed that there was no Pilot Landing Aid Television (PLAT) monitor. Instead, where the monitor would normally be on a larger-deck carrier there was just some Plexiglas to record side numbers and pilot names.

At first, they were a bit taken aback, until I pointed out that the aircraft will land so close to us that you will not need a television to ensure proper line up. We all agreed that since I had the experience with the little deck and the 4-degree glide slope, that I would wave the first few passes (Pete or Muddy backing me up), until they became accustomed to the environment. We did the same thing for the initial night landings. This situation worked perfectly for me to quickly get my official Training LSO designation.

I do recall that prior to our kids getting their first night landings, my AOCS buddy and fellow *Oriskany* LSO, "Bug" Roach, was working some F-8 *Crusader* nuggets. Pete, Muddy and I arrived on the platform just as "Bug" waved off an F-8 that had settled at the ramp. He must have given a POWER, POWER, WAVE-OFF call because the kid lit his afterburner and roared by so close that we could feel the heat. I turned to them and said, "Welcome to night operations on a 27C."

***

FLYING A T-28 TO FALLON IN A SNOWSTORM: Some of the greatest flying we did both in the A-7E and the T-28 involved the weapons detachments to NAS Fallon, Nevada. These detachments were normally two weeks in duration and involved an intense flight schedule that required at least two and sometimes three flights per instructor every day. In the winter, the weather over the Sierra Nevada's could be very nasty. The normal transit route from Lemoore to Fallon was a direct flight over the highest part of the Sierra Nevada range with mountaintops more than 14,000 feet. This was usually not a problem with the A-7s but could pose a real challenge to the T-28. We did have a turbo charger and oxygen masks, but icing and performance issues during bad weather often required us to find an alternate route of flight.

I was scheduled to fly a T-28 to Fallon with an enlisted engine mechanic riding in the back seat. These guys were essential to keeping these old birds flight worthy, and all of them loved to fly (even though they were not pilots, we gave them lots of stick time from the backseat). I looked at the weather and decided that there was no way that we could fly direct to Fallon. Instead, we flew to Bakersfield, followed Highway 58 east through the pass, and then picked up Highway 395 going north in the Owens Valley toward Mammoth Lakes. As we got close to Bishop in the north end of the Owens Valley, the ceiling associated with the storm begin to drop and it became obvious we would not be able to continue that route to Fallon.

I turned around and began to head for NAS China Lake where I intended to refuel. As we made the turn, we entered an incredible updraft. I literally pulled the throttle to idle and we still were experiencing a thousand foot per minute rate of climb. Luckily, there was very little turbulence, so we just let the aircraft climb for a few minutes as we enjoyed the surreal feel/view of this deadly, but beautiful, force of Mother Nature.

We made an uneventful gas stop in China Lake, topped off and then followed Death Valley until we picked up Highway 95 that runs from Las Vegas to Fallon. The weather was much better along this route, until we got close to Fallon. As you may have figured out from our IFR (I follow roads) flight plan, so actually a Visual Flight Rules (VFR) flight plan, we needed to stay clear of the clouds. As we continued to follow the highway, I began to realize that the ceiling all around us had lowered to the point where we could not safely turn around. As it continued to get lower, and we entered the last mountain pass prior to Fallon, I lowered the landing gear with the intention of landing on the highway, if necessary.

We ended up with a couple of hundred feet to spare and eventually landed at Fallon. Nobody else transiting to this detachment made it that day, including the jets. After arriving at Fallon and checking in with the SDO, who had driven up a couple of days earlier, he told me to call the CO of VA-122. I called him and said, "Good afternoon, Skipper. LT Rud here. What can I do for you?" He said, "LT Rud, where are you calling from?" "Sir, I am calling from the SDO's desk in Fallon," I replied. "How in the hell did you get from Lemoore to Fallon in this awful snowstorm?" "No problem, Skipper, we flew over to Las Vegas and came in from the east." "Forget it Rud, I don't want to know how you got your crazy ass to Fallon in a snowstorm in a goddamn T-28." He then promptly hung up the phone. *Hmm, I wonder if this is going to be a plus or a minus in my next fitness report?*

***

TULE FOG: Winter in the San Joaquin Valley brought with it a weather condition known as Tule Fog. It often reduced ceiling and visibility to near zero, even though the top of this cloud deck was usually just a few hundred feet. Although, it could last all day, normally, it was the worst early in the morning. It would often then lift to VFR conditions during the middle of the day, then drop back down near sunset. As one might imagine, this weather phenomenon created lots of issues and some mostly true stories about surviving the challenge.

The A-7E, with a HUD and super navigation system, was reasonably easy to land safely if the ceiling/visibility was 200 feet and 1/2 mile. The visibility was also measured in Runway Visual Range (RVR), with a minimum requirement of 2,400 feet. The primary takeoff runway at Lemoore was runway 32R, and the primary landing runway was 32L. Since the thresholds for these runways were almost a mile apart, weather conditions, especially Tule fog, could be quite different, at the two different locations.

As I mentioned earlier when defining T-28 missions, we would often fly to Travis AFB to pick up sailors that had arrived from the Tonkin Gulf, normally in an emergency leave status. During one of these runs, which occurred in December of 1972, I picked up a sailor at Travis AFB. The weather was OK on the flight up to Travis but closed-in on our way back. When I checked in with Lemoore approach control, they vectored me to Stratford, which was the holding fix for Runway 32R. The controller passed along that the weather was currently below minimums for landing anybody.

I had plenty of gas to make our alternate, NAS China Lake, so I decided to hang out at the "holding fix" to see if the weather improved. As we were holding, the controller announced that they had reached 2,400 RVR (measured at the approach end of the left runway) and he cleared an A-7 to start his approach. He then gave me, the only aircraft holding for the right runway, a frequency change to another controller who then cleared me for an approach to the right runway. Now, in those days when two aircraft were shooting simultaneous approaches to both runways at the same time, the jets going into 32L were initially assigned to begin their approach from 1,500 feet while I was assigned 2,500 feet to start the ground-controlled approach (GCA). Since the T-28 instrumentation was antiquated and power/speed changes resulted in a real challenge trying to maintain heading, I set my approach speed and power a bit higher than normal and commenced the GCA.

The terrain was flat as a board with no obstructions, which was good news. I instructed the sailor in the back to look straight down for the ground. As we approached the minimum descent altitude, I could see nothing but fog. The sailor suddenly shouted, "Sir, I see the ground." I said, "OK, let's take it a little lower. Aha, there it is—pavement." I transitioned to a landing flare only to discover that it was the taxiway, not the runway. No problem, I just made a little adjustment to the right, added some power and set her down on the runway.

At this point the tower called out on guard, "The field is closed with zero visibility and zero ceiling." No shit! We sat on the runway and I had to call the tower to get a "follow me" truck to come out and lead us to parking. I showed up in the ready room, and the skipper invited me into his office to discuss me landing the T-28 while an entire flight of our A-7s had to divert to China Lake. My explanation, "Well Skipper, the weather on 32R was a lot better than 32L. His reply, "Bullshit! But thanks for getting the kid home, now get your crazy ass out of my office." *Hmm, I wonder if this one will be a plus or minus on that fitness report?*

***

Cleared in Hot: Certainly, one of the most challenging and fun missions we had in the T-28 was acting as a Forward Air Controller (FAC). We did this to teach the students how to perform Close Air Support (CAS) missions. We carried 2.75" White Phosphorus (Willy Pete) rockets that we used to mark targets in the B-17 impact area at Fallon. The students on this flight were carrying live MK-82, 500-pound bombs for the first time so it was a very exciting experience for them. On this particular sortie, I had four students in the target area awaiting my marking of a target.

I found an undamaged truck, rolled in using a grease pencil spot on the windshield as a gun site, and fired my rocket right into the window of the truck. As I pulled off target to clear in the first bomber, I noticed that one of the students had already started his bomb run. "Holy shit!" I rolled the T-28 nearly inverted over the next ridgeline, then rolled level in that depression just in time to see and feel the blast of the 500-pound bomb explosion. A wheel of the blasted truck came flying over the hill nearly hitting my aircraft. "OK boys, I know you are anxious to kill something, but please do not make a bomb run until I clear you in goddammit!" *By the way, I gave that student a below average in headwork and an above average in bombing accuracy.*

***

T-28 Dual Crew Pranks: Tom "TMac" McClelland was in my back seat on a flight from Fallon to Lemoore at night, in the winter. At just about the half-way point across the snow-covered Sierras, I mentioned that this was probably the worst possible place to have an aircraft problem, since we were not within gliding distance of a survivable situation. At that point, the engine rpm began to fluctuate. I swear that my rear end literally sucked up the seat cushion, while in the back TMac began to giggle like a little kid. I thought the engine was going to quit, and I had a nut-case in the back seat! In fact, every time the rpm fluctuated, he giggled. As it turned out, he was pushing the primer button, which in-turn caused the rpm to fluctuate. This was very funny to him, not so much to me.

Now it was my turn to pull the prank. I was flying with a fellow instructor and awesome guy by the name of Gary Simpkins. I was in the front and he was in the back. He was known for his love of Hostess Twinkies, and sure enough, I saw in the mirror that he was unwrapping a package of them. He put them up on the top of his instrument panel as he prepared his feast. I asked if he could spare one for his buddy. He replied that I should have brought my own Twinkies. I then proceeded to push the nose over into a little negative-G situation, which launched the Twinkies into the front cockpit. The look on his face was so pitiful, that after some negotiation, which of course involved beer, I gave him back his Twinkies.

***

TROJAN CARRIER QUALIFICATION BOONDOGGLE: While serving as an LSO in the RAG, Don Simmons and I decided that it would be a great idea to get carrier-qualified in the T-28. Our logic was that we could then fly on and off the ship to properly debrief/brief folks that were on the beach between CQ periods. Of course, our real intent was to bag as many traps as possible. Well, we did some field carrier landing practice while this proposal was sent up the chain to the Commander Naval Air Forces, Pacific (COMNAVAIRPAC). According to a friend of ours that was on the staff, when it came up at a meeting with the chief of staff, he inquired as to who was proposing this initiative. When he heard the names Rud and Simmons, he burst out laughing and threw the proposal into the wastebasket (shit-can). *Sometimes a reputation can limit potential success.*

***

T-28 300-KNOT BREAK AT NAS NORTH ISLAND: Even though we did not get to take the *Trojan* to the ship, we still used it to fly parts/people into North Island for further movement to whatever carrier we were using for CQ. During one of these events, I was flying wing on one of the other RAG instructors who at one time had been an S-2 pilot. He decided that the S-2 guys would really appreciate

189

a T-28, high-speed, fan break. We looked up the "redline" airspeed for the *Trojan* (I believe it was 310 knots). To achieve this speed, we descended out of about 10,000 feet and hauled ass into the break. He said we were going 300 knots; I have no idea, because I was attempting to stay as close as possible in formation. The good news was that folks on the ground described the sound like WWII P-51s. The bad news was that we were almost in Tijuana on the downwind before we got slow enough to lower the landing gear. *It is not always easy to be cool.*

***

NIGHT HAMMERHEAD STALL: Juan Garcia was a fellow instructor at VA-122. He had more T-28 time than most of us and was fearless about taking it to the edge of the flight envelope. We were on a weapons detachment to Fallon, NV. He and I were assigned to serve as safety observers for A-7 students who were about to conduct their first night dive-bombing flight. Since target fixation and disorientation were a hazard, we would circle the target at what was a safe pullout altitude for the student. If we did not see him start a pull, we would transmit, "Start your pull now!"

Juan was in the front cockpit flying the aircraft and I was in the back with the responsibility to monitor the bombing pattern. The target was B-16, which served as both a conventional bomb target and as a practice nuclear option. As such, it had a plowed run-in line that ran several miles into the target. Because of the difference in speed, we took off ahead of the jets. As I previously mentioned, this was at night! Juan asked, "Hey Gil, have you ever done a hammerhead stall in the T-28?" I said, "No." He replied, "Well tonight is your lucky night." He proceeded down the run-in line at a respectable 100 feet or so (there was a moon). He flew over the "bulls-eye" and pulled that T-28 straight up in the air until it stalled at a couple of thousand feet, ruddered it over for a simulated practice bomb run and exited the target at 100 feet. Then he said, "Now do you want to try it?" I said, "Maybe after the sun comes up..."

***

AIR-TO-AIR TACTICS PHASE INSTRUCTOR: About a year after arriving at VA-122, I changed jobs from being a schedule writer to a member of the Tactics Phase department. Although the A-7 mission was "light attack," which consisted mostly of air-to-ground weapons delivery, we also carried AIM-9 *Sidewinders* and a very effective Gatling gun capable of firing up to 6,000 rounds per minute. The gun was a viciously effective air-to-ground weapon but could also be used for self-protection against enemy fighters. We did not teach students to go looking for enemy fighters, however, we did teach them how to fly the A-7 to the absolute limits of its aerodynamic flight envelope, so that they could defend themselves if attacked by an enemy fighter.

Over the first few years of the A-7's existence, several aircraft were lost due to departures from controlled flight. Developmental test pilots led by D.D. Smith at the Naval Air Test Center studied these accidents, and concluded that A-7 pilots needed to be trained to recognize these departures and practice proper recovery procedures, which mostly involved letting go of the flight controls. Up to this time, pilots were basically taught to stay away from the edge of the envelope or risk getting into a potentially non-recoverable stall/spin. This attitude changed when a pilot who ejected from what he thought was a spin, sustained injuries associated with a high-airspeed situation. In a true spin, airspeed is near zero, which meant that he was actually in a high-speed spiral, which he should have been easily able to safely fly out of.

Our mission, at first, was to demonstrate departures to the students. This was an absolute blast, as we got to purposely fly the aircraft past the flight envelope into a full departure from controlled flight. We would climb the A-7 to around 25,000 feet with the student flying on our wing a safe distance from us. We would slow down, pull the nose up, initiate a left or right turn and then stomp the opposite rudder. This would create a huge adverse-yaw movement of the nose and a corresponding tumble of the now out-of-control aircraft.

To recover, we would simply drop the control stick (grab the glare shield) and take our feet off the rudders. Within a short time, the aircraft would end up 90-degrees nose-down. We would monitor the airspeed, which would soon begin to build up, at which time we would get back on the controls and recover the aircraft. Two things became apparent from this demonstration: the aircraft will recover, but it will take a significant amount of altitude to complete the recovery. We also demonstrated high-speed/high-G departures, which were more violent, but took much less altitude to recover. After a couple of months of demonstrations only, we got to the point where the students performed the departure and recovery themselves.

Those of us who flew these tactics flights everyday soon became very comfortable at extremely high angles of attack, learning to maneuver the A-7 beyond flight envelopes that adversary aircraft and fighter aircraft in fleet air wings were used to seeing. During this time frame, there was a classified flight program that was taking place involving captured MIG-17 and MIG-21 fighters.

Two of our tactics phase instructors, Jim Larkin and Don Simmons, had recently joined the RAG after serving combat tours flying in the very challenging North Vietnam air defense environment toward the end of the war. Both these guys could fly the heck out of an A-7 so we sent them to participate in the program facing the MIGs. They performed much better than expected against the MIGs, receiving accolades from the fighter pilots leading this training. *"These two guys are not typical A-7 fleet pilots. If every A-7 pilot was this good, they might not need a fighter escort."*

***

REAL WARRIORS ARRIVE AS NEW INSTRUCTORS: Immediately following the end of the war, we got an influx of new instructors who had participated in the resumption of operations in NVN. *Operation Rolling Thunder* was the initial campaign from March of 1965 through November of 1968, and many of those folks were my instructors in the training command and the A-4 RAG. Although those of us who operated in South Vietnam and Laos during the bombing pause of NVN still saw significant combat, it was nothing in comparison to the sophisticated and deadly air defense system encountered over NVN when bombing resumed with *Operation Linebacker 1 and 2* in 1972–73.

Three new LSOs joined our team. Don Simmons from *Coral Sea* (CV 43), Tom Demarino from *Kitty Hawk* (CV 63) and Mule Mulholland from *Midway* (CV 41). These three were great LSOs and had combat stories that made us sweat just listening to them. We also got Mike Penn who was shot down, captured and repatriated in the last six months of the war, and Jim Larkin who, as far as I know, is the only A-7E pilot to recover from a night stall/spin while avoiding a SAM missile. Unlike the long-term POWs, who endured up to eight years of captivity and torture, the shorter-term guys, although often captured in extremely harrowing circumstances, were ready and able to get right back into flying. There were several more-senior arrivals, but these five were my close friends and contemporaries.

***

MOVE TO 1560 MULBERRY LANE: Shortly after my return from Vietnam, Barb and I began to look for a house in town. We were also in the process of expanding our family so we decided that a four-bedroom would be ideal. We found a typical Beck house (the main builder in Lemoore in the 1960/70s). I will never forget signing that 30-year loan with a "giant" mortgage of $25,000. Even more daunting was the requirement for a monthly payment of $219. It was equipped with a typical swamp cooler system, so we purchased a real air conditioner to replace that inefficient system.

I was waiting at the end of the street to help guide the installers for the air conditioner when I saw this giant crane coming on the back of an 18-wheeler. They stopped and asked if I was Gil Rud. I said, "Yes I am, but I just have a little three-ton unit coming for my house." The dude in charge said, "No, you ordered a much larger and more powerful unit." I showed him my receipt and he said, "Well, you are correct. This one was supposed to go to a restaurant in Hanford, but this is your lucky day, because it costs so damn much to rent this crane that we are going to install this big mother for the same price as the smaller one."

They did, but when they turned it on to check it out, it blew out a transformer in the power system (in spectacular fashion I might add) and our entire neighborhood went dark. *Let's welcome the Rud family to our neighborhood! They upgraded the local power grid and we had the best A/C in Lemoore.*

***

Goodbye Muscle Car: Barb loved to drive the W-31, but after she got nailed for going 95 in a 65 zone, we decided to grow up and power down. We traded it in for a 1972 Datsun 510. The kid who sold us the new car was enamored with the W-31. I cautioned him on being careful as he drove it off to test it out. But he stomped on it and promptly spun out in the gravel parking lot. Since we needed two cars, I also bought a used 1969 Ford pickup that we converted into a camper of sorts. I was now a fully domesticated husband/father with maturity issues now confined mostly to flying.

***

January 30, 1973: Joining the Rud family at a whopping 9 pounds 9 ounces, we welcomed Ryan Theodore Rud. It was a challenging birth and my first and only opportunity to witness this blessed event. Eventually, he became a very handsome boy, however due to the difficulty of the delivery, at first glance, he looked a little rough and as the doctor initially held him by his legs, he appeared to not be breathing. Not to worry, he was just holding his breath so that he could let us know what a great voice he had. In those days, no one knew what the sex of your unborn child was so as soon as he was delivered (believe me there was absolutely no doubt that he was a boy), Barbara shouted, "Oh my God, it is a boy!" For some reason she had been certain it would be a girl.

This is also about the time that we were lucky enough to acquire the best baby sitter in Lemoore. Denise Rosenthal, who at the time was just a high school kid, came into our lives as a trusted sitter for our growing family. She later married a local farmer, Hill Fuller, and started a family of her own. To this day, my children still consider her a very special person who positively influenced their lives.

***

HOW TO GET THE FIRST OVERHEAD FOR CQ ON USS *LEXINGTON* (CV 16): Bill Bodenweber was my contemporary as the VA-174 RAG LSO at Cecil Field, Florida. Both VA-122 and VA-174 normally had very large groups of students to get qualified so it was a huge advantage to be assigned to be the first group to begin flight operations for that at-sea period. Bill was curious as to why we always got that honor. I fibbed a bit and told him that it was because we had to travel all the way from the west coast.

Really, it was because I had made a special deal with the operations officer on *Lexington*. He loved Coors beer, and in those days it was not available in Florida, so I would fill a cruise box full of Coors, label it "spare parts" and send it on the C-9 with the maintenance troops. The operations officer would meet me at Sherman Field, check out the contents of the cruise box, pay me for the beer and schedule us for the first overhead the next day. As I previously mentioned, this was a huge advantage and often resulted in us completing an entire qualification process in a 48-hour time frame. *The Leadership Lesson here is that there is always a path to success; you just must find it and execute.*

***

TEACHING NIGHT CARRIER LANDINGS: Although the A-7E was a challenging aircraft to fly aboard the carrier, I am very proud to say that every single pilot that we worked up and took to the boat successfully carrier qualified. We did have a few that needed to repeat the field carrier landing practice (FCLP) a second time prior to being considered ready for the boat, but eventually they all made it. What was the secret? It was all about them being predictable and responsive to the LSO. We would test their reactions during FCLP by asking them to add power when they did not actually need it, just to make sure that they did not second guess us. We also waved them off unexpectedly when they were flying a perfect pass.

In 1974, we added a Night Carrier Landing Trainer (NCLT) to our training program. Although quite rudimentary compared to today's sophisticated simulators, it was very effective in exposing the students to the procedures involved in the night environment. We LSO instructors acted as shipboard radar controllers simulating the carrier-controlled approach (CCA). The flight control system in the NCLT did not accurately reflect the real aircraft, but everything else in the approach was realistic, and we found that the students were much better prepared than pre-NCLT groups. Old timers with lots of night landings absolutely detested the NCLT so we only gave them one session each. *I firmly believe that my early struggles with carrier landings gave me the necessary patience to teach to all skill levels of students. My sense is that a pilot who never was challenged during his own training, or even worse won't admit to having struggled in some phase of training, often makes a lousy instructor.*

***

BAGGING TRAPS ON USS *LEXINGTON* (CV 16): Don Simmons and I really worked well together as a team. We had great success on all our detachments. On our last one together on *Lexington*, Don and I had walked aboard and were on the platform where we had just finished working our first group of A-7 students. As we prepared to leave the LSO platform, we got a call from the Air Boss. He said, "Are you guys qualified to wave S-2 *Trackers*?" We replied, "No sir, but we are qualified to wave C-1s, which is a COD version of the S-2." He said, "Hell, that is good enough for me, because the aircraft their LSOs were in just had to return to the beach due to a maintenance issue and here come their students."

It looked like a scene from the movie *Tora! Tora! Tora!* as the sky around the carrier was soon filled with S-2s. We must have waved for at least an hour before their LSOs finally arrived to take over. No big deal we thought, until a couple of days later when we finished up our A-7 class with a final "day" landing by the last student. At this point, the plan was for Don and I to man-up two A-7Es and fly to Pensacola, then to NAS Lemoore. Just as we were about to leave the platform, the Air Boss called and said, "Boys, I owe you for helping out with those S-2 students. Start those two A-7s up and the deck is yours for the next 30 minutes!"

We looked at each other and thought, "Shit! We don't have an LSO to wave us." What we did have was a fleet squadron trainee that was just out there writing comments for us. I looked him in the eye and said, "OK kid, for the next 30 minutes you are a qualified LSO to be revoked upon completion of our last trap. Don't say a word other than 'Roger Ball' and 'Wave Off' if the deck is fouled." And we were off to man-up and catapult into the pattern.

Since it was just the two of us, we launched, made a climbing left turn, which got us to the abeam position just in time to continue the turn to pick up the ball. We each got nine carrier landings in just over 30 minutes. When we finished, they refueled us and we were off to NAS Pensacola. *I know, I know, totally unprofessional and grossly unsafe to anybody but Simmons and Rud. Remember the T-28 carrier qualification caper when the AIRPAC Chief of Staff found out who was behind it and promptly threw the request in the shitcan? Some risks are worth taking. By the way Don Simmons soon departed the RAG for a tour as* Blue Angel *3 and 4.*

***

THE FOG BOWL: Most Navy pilots participated in team sports at the high school level, and many also played in college. Flag football was a very popular sport at NAS Lemoore with every squadron and other support organizations on the base participating. The most-prestigious event was a sort of super bowl held every January during the height of the Tule Fog season. During the regular season, the teams were a mix of enlisted and officers, but the "Fog Bowl" was limited to the pilots assigned to VA-122 and VA-125.

Initially it was the A-4 community against the A-7 community, which already was a great rivalry. Later it progressed to the A-7A/B community against the A-7E folks. When I say it only involved these two squadrons; that was not the case. It was so competitive that many college-level players would get temporary orders back to the RAG from the fleet for a month or so just to participate in this event.

Although certainly not a top-notch flag football player, I did play four years of high school and a year of college, so I participated in several of these Fog

Bowls. I played initially for VA-125 and then three times for VA-122. The first one I participated in was in 1971. I was playing defensive back and matched up with this athletic- looking wide receiver. He sort of lopes out of his stance and turns up field. I am playing off him a bit in perfect position for an interception, when I see the pass coming our way. I relax a bit when I notice that it is way over our heads and uncatchable. That is about the time this guy leaps into the air and grabs the football with one hand, hauls it in and proceeds to run for a long touchdown. "Geez! Who the hell is that dude?" *It was Rob Taylor an all-American wide receiver from the U.S. Naval Academy.*

The last Fog Bowl I played in was for VA-122 in 1974. We were blessed to have John Ross going through as a student. Talk about a stud. He had been the quarterback and a track star for Slippery Rock University in Pennsylvania (he was inducted into the Slippery Rock Athletic Hall of Fame in 1991). Everyone took this game seriously, so we practiced for a couple of weeks prior to the event. Unlike most of the participants, I had played the whole season for the VA-122 squadron team, so I was in great shape. I had been kicked out of one of our squadron games for exchanging punches with an opponent. Barb happened to be watching that game, so my behavior cost me a night on the couch. I seem to remember her saying something like, "I am not sleeping with a misbehaving high school kid!"

I was the center linebacker and defensive captain. We each wore a special belt with two flags attached in the hip area on either side of our body. A tackle was recorded when one of those flags was successfully ripped off by the opponent. During a practice a few days before the event, I was in the process of ripping one of those flags off when I caught my right ring finger in the runner's football jersey. It appeared to me at first to be jammed so I asked one of my teammates to pull on it and get it back in joint. He took one look at it, almost threw up and said, "Christ Rud, your finger is broken and twisted around so I am looking at the bottom side of it!"

So, over to the hospital I went (it was located right across from the football field), checked in and got a flight surgeon to reset my finger. He placed a long metal splint around it, chewed me out for playing football and walked away. The significance of this was that he never issued me a "down" chit, so I continued to fly, albeit in a bit of discomfort. Also, the T-28 maintenance folks noticed that there was a ring of paint missing around the Magneto switches after I flew the airplane. That was caused by the metal part of my brace scratching that area every time I flipped the Mag switch.

You would think that this injury would end my participation in the Fog Bowl—however, I was able to wrap the iron brace sufficiently so that the coach for VA-125 decided it would be safe for his players. He probably also figured that a one-armed center linebacker would not be much of a threat. I ended up playing every defensive snap in that game. Adrenalin took over to the point that I felt no discomfort in the finger and was soon using both hands to pull flags. Due mostly to the incredible play of John Ross, both passing and running, we managed to pull off a win.

After the game, as the adrenalin started to wear off, I finally unwrapped the brace and took a look at my finger. Not good. It was totally out of joint and twisted again. I headed over to the hospital and unfortunately got the same surgeon that had performed the initial procedure. I was covered in mud and probably some blood, still in my football uniform. He was so pissed at me for playing with the injury that he reset the finger without even numbing the area. Once again, his anger and frustration caused him to forget about a down chit, and I continued to fly.

16. CAG-14 LSO and the First Deployment of the F-14 Tomcat

THE GOAL OF EVERY Junior Officer pilot on his first shore tour is to avoid non-flying shipboard duty, officially called a "disassociated sea tour" for his second sea duty assignment. My goal was to be assigned as a Carrier Air Wing (CAG) LSO. As one might expect, these assignments were highly competitive. So much so that one needed to be a bit flexible on timing to be ready to fill one of these billets.

I got a call from the type commander (COMNAVAIRPAC) LSO, who was in charge of these assignments. He was also very familiar with our performances as RAG LSOs since he was present at most of the RAG CQ events. He gave me an offer that I could not refuse. "Rud, I would like you to roll out of shore duty six months early to relieve Tom McClelland as the Assistant CAG LSO in CAG-14 on USS *Enterprise* (CVN 65). They just deployed for the initial operational cruise of the F-14A *Tomcat* and the first S-3 *Viking* west coast cruise. The orders will include additional duty with the VA-97 *War Hawks*, which means you will get to keep your family in Lemoore." I did not hesitate. "Yes sir, I would love to take those orders. Thank you for picking me." *And I was off on another adventure that in many ways was my favorite sea duty in a 28-year career.*

Since I was "current" in the aircraft the only requirement that I had was to achieve night carrier landing currency by getting eight night traps, which was the rule for any pilot "must pump" directly to an operationally deployed carrier. I got those on the next CQ detachment to *Lexington* and flew my last flight in VA-122 on November 1, 1974. A week later, I checked into CAG-14 aboard *Enterprise*, which was operating in the South China Sea just west of Subic Bay in the Philippines; and I began flying with the VA-97 *War Hawks*.

My initial boss, CAG-14, for just a couple of weeks, was CDR Skip

Furlong. CDR Furlong was followed as the new CAG by CDR John R. "Smoke" Wilson. Smoke was the greatest pilot and Naval Officer I ever worked for in my 28-year Navy career. He was a fighter pilot and a developmental test pilot with hands of gold. He flew every type of aircraft in the air wing, day and night; and he flew them all flawlessly. As an example, during workups before the deployment, the senior CAG LSO and my roommate aboard ship, Monroe Smith, related the following about Smoke's unmatched flying skill: "I was waving the VS-29 pilots, all of whom were new to the S-3 *Viking*. Although it was a great carrier landing aircraft that was relatively easy to fly, it did have some aerodynamic quirks that the squadron was struggling with to get consistent "OK" passes.

I had not given out an OK pass to anyone until a newcomer came into the break and proceeded to fly an absolutely "rails" pass on his first attempt. Astounded at this accomplishment, I asked for the pilots name." The answer in a rather terse tone: "CAG!" *Talk about setting the example, this is exactly how great leaders gain the respect and admiration of their subordinates.*

Since this was the first deployment of the F-14A *Tomcat* and the first west coast deployment of the S-3A *Viking, Carrier Air Wing 14* was loaded with talent. The two F-14 squadrons were VF-1 and VF-2. Their pilots and Radar Intercept Officers (RIOs) were pretty much all handpicked to be members of the first two squadrons to deploy this fantastic aircraft. It did create a bit of a problem though in the sense that they had more senior officers than a normal squadron. Normally the CAG LSOs are the most senior LSOs in the air wing. That was true in the case of Monroe, but not me. In fact, since I had rolled early from my shore tour, in addition to Monroe, there were five squadron LSOs senior to me. The good news was that despite my relatively junior status as a mere mid-grade LT, during my shifts on the LSO platform, I was the boss.

Enterprise was an awesome ship, and at the time, the only nuclear-powered carrier in the fleet. With Admiral Hyman Rickover heading up the nuclear power training pipeline, all the folks selected for this program had to be super smart and sometimes to the point, in my opinion, that they might not necessarily be the most charismatic leaders. We were extremely fortunate to have as our commanding officer, Captain C.C. Smith who was both intellectually brilliant and a charismatic leader. *We would follow him anywhere including on liberty, which was somewhat of a challenge, I might add.*

***

NIGHT FLYING ADVENTURE: On the evening of November 13, 1974, which was less than a week after my arrival aboard *Enterprise*, I launched on what turned out to be a very eventful night sortie. As most naval aviation "this is no shit (TINS)" stories begin, it really was a dark and stormy night. Although the sea state was not terribly rough, there was a swell that caused *Enterprise* to pitch and roll so badly that Monroe decided to go to the use of the manual, LSO-controlled meatball called the MOVLAS. He made this decision because the Fresnel lens meatball, although computer stabilized, is only accurate to a total of 20 feet of deck movement. Anything more than that you either quit flying or you go to the use of the MOVLAS.

By using the MOVLAS, the LSO now controls ball movement. It is something that, as an air wing, we practiced pretty much every day. "Day," being the key word here. The reason we practice it so often is two-fold. One reason is for LSO training and the second is for pilot training. In other words, the LSO is not showing you where you are on the glide path as much as he is showing you what corrections you need to make to land safely. For example, if you have added too much power in close, the LSO may show you a rapidly rising meatball even though you are still below the glide slope at that time. The proper pilot reaction will be to take off some power before flattening out and missing the wires. He will do the opposite if you need power by lowering the meatball until you respond and add power. The most important function of the MOVLAS is to stop the pilot from chasing the Fresnel lens meatball, which in these conditions is going from off the top of the lens, to off the bottom of the lens in a very short time period.

Remember, that I mentioned we practiced using the MOVLAS in the daytime? Well, when you must use it at night, in my humble opinion, you should not be flying. But we were, and I was one of those folks working my ass off trying to get aboard safely. The good news was that Monroe was doing a great job of keeping all of us safe by waving us off if the deck cycle was too severe and powering us into a safe Bolter if a Wave-Off was not appropriate. He was also able to catch about every third aircraft when the deck cycle allowed.

With these conditions, the radio was jammed with requests for fuel states and subsequent directions to rendezvous with the tanker to get enough fuel to try another approach. The weather was also a factor with low ceilings and layers up to about 10,000 feet. This required the KA-6 tanker to be higher than the normal 5,000 feet overhead the ship.

I finally got an opportunity to shoot an approach, which resulted in a Bolter. At this point, I was getting low on fuel, so without waiting for direction from the ship's controllers, I headed for the tanker. When I joined on the tanker's wing, I was number five in the queue to get gas, and one of the A-7 nuggets (I remember vividly who it was but will forgo his name to protect the guilty) was stabbing away at the refueling basket with no success.

I checked my fuel state, and without hesitation, departed the cluster around the tanker and headed for the "bingo" field, which was NAS Cubi Point. On my way to Cubi, I stayed on the ship's approach control frequency hoping for a chance to update my intentions prior to switching to Cubi Point tower. Finally, they called me and asked, "*War Ace* 310 say your fuel state." I replied, "*War Ace* 310's state is .8 (800 pounds)." This is critically low if you are still attempting to get aboard the ship. They replied, "Say again your fuel state!" I answered, ".8—but fortunately I am about to land at Cubi Point!"

I made an uneventful landing at Cubi along with several other air wing birds. We refueled and then received a call from the tower telling us that the rest of the night operations on *Enterprise* were canceled. We were given an overhead time for around noon the next day.

Upon returning to the carrier, I was summoned to go see the air operations officer, CDR Bob Coffee. Bob was a former commanding officer of VA-97 reluctantly filling a senior disassociated sea tour. I assumed that he would be interested in my perspective of the previous night's debacle. Well, he was in a sense. "Rud, you asshole, you scared the crap out of me last night when you reported that you only had 800 pounds of fuel. Next time, tell us where you are before you report that level of fuel."

Bob was an accomplished pilot and Air Operations Officer, and although I certainly did not make a good first impression, when I was on the LSO platform we worked as a great team through many dark and stormy nights. *Leadership lesson—before making a smart ass/untimely comment, think about how it might affect the folks doing their best to keep you alive.*

***

MY PERSONAL HERO: As I mentioned earlier, *Carrier Air Wing-14* was really loaded with handpicked talent except for yours truly, of course. For example, the air wing operations officer, Rick Hauck, was like CAG Wilson, a developmental test pilot. He came from the A-6 *Intruder* community; however, he had been flying a little bit of everything during his carrier suitability tour at the Naval Air Test Center Patuxent River. While on board *Enterprise*, Rick flew A-6s, A-7s and F-14s day and night. There may be other CAG Ops that have done this, but Rick is the only one I know to have accomplished this feat.

How good was he? Shortly after this tour, he was selected to be an astronaut. As an astronaut he flew several shuttle missions both as the pilot and as the Mission Commander. Following the Challenger disaster in 1986, Rick was selected to command the return-to-flight mission aboard Space Shuttle Discovery. *That flight of Discovery holds a special place in my heart because I got the opportunity to talk to Rick from the Houston Space Center during that flight. I will cover that wonderful event in an upcoming chapter on the* Blue Angels.

Dale Gardner and Hoot Gibson, from VF-1, and John Creighton from VF-2 were also selected to be astronauts so, as I previously stated, we were loaded. Now, just because an aviator is accomplished enough to be an astronaut, does not mean that they cannot hoot with the owls while on liberty. In fact, Rick and I were often times a toxic combination ashore.

During one in-port period, we were competing in some sort of booze-fueled activity at the Cubi Officers' Club when Rick got a call from the CAG 14 Duty Officer. "USS *Coral Sea* (CV 43) will be arriving in the operating area tomorrow, and they are going to be flying a bunch of F-4 *Phantom IIs* into Cubi at first light. The weather is not good, the runway will be wet, and CAG says that they will all need to make field arrested landings. He says for you to find Monroe and Gil Rud and get them down to the runway to help talk these F-4's into the short field arresting gear." Well, we were easy to find since we were sitting next to Rick. However, we decided that a more elaborate "welcome" was required for our *Coral Sea* compatriots, so Rick bought several bottles of champagne.

Since we only had a couple of hours until they would arrive, we decided that it was best if we did not actually go back to the ship, but rather just hung out on the chairs behind the officers' club. Just before daybreak, we jumped in the 1959 Plymouth CAG LSO car with the giant fins on the back, which was a traditional hand-me-down that Monroe and I had purchased from the previous air wing. We had paid the traditional price, which was $55. The car's transmission only had one functioning position, which was second gear. Yep, no reverse or first gear so you had to be careful where you parked. Also, it had virtually no interior. Rather, it had six metal folding chairs, and of course, no seat belts. It was painted with a Marine Squadron logo, so when it was impounded for various violations they could not trace it back to us. The good news was that we knew where the impound lot was, so we would just go, hot-wire it and drive it back to the ship (obviously security knew all about the tradition and just let it slide because they considered us to be a bunch of nut-job pilots letting off steam from flying off carriers at night).

I dropped Rick off near the refueling pits where the F-4's would be parking and headed for the LSO shack to join up with Monroe (he was F-4 qualified) to wave the incoming *Phantoms*. The weather was nasty, so they happily agreed to use the short-field arresting gear. The Fresnel Lens was set up to put them just short of the arresting gear, so we did not have to do anything but watch them land. In the meantime, as each of them parked, Rick presented each crew with a bottle of champagne. *Talk about a class act, that performance was and still is typical of my good friend and personal hero, Rick Hauck.*

***

GULF OF TONKIN PROVOLATERS: Just as the cruise was winding down and we were passing through Subic on the way home, the shit hit the fan in South Vietnam. Things had been going badly ever since we left in 1973, but now the North Vietnamese were advancing on Saigon and the fall of the regime was imminent. We were ordered into the Tonkin Gulf with the expectation that we would soon be supporting an evacuation of embassy personnel and other folks from Saigon. Once we entered the gulf, we initially operated a normal flight schedule. As part of that flight schedule, I was paired with Tom Smith, a very aggressive and super-competitive young pilot in VA-97 for a Dissimilar Air Combat Maneuvering (DACM) flight against two F-14's.

All our flights were, in effect, a show of force to the bad guys that we were ready to kick some ass if necessary. We did the usual split with a turn-in to start an engagement against the *Tomcats* from about 50 miles. Of course, their radar was far superior to ours, as was the performance of their aircraft, unless we somehow saw them first. I was astounded that neither of them were able to pick us up on radar and we saw them first and got a couple of simulated *Sidewinder* shots off before being dispensed with by their vastly superior maneuvering capability. We set up for a second engagement with similar results, which really pissed off the fighter crews.

Standard procedure upon returning to the ship was for each flight to stop by the ship's intelligence center (CVIC) to debrief with the Intelligence Officers. Tom and I entered CVIC only to be met by the Chief of Staff for Admiral Oberg who was our Carrier Group Commander. I said, "Good evening, Captain, what can I do for you sir?" He had this big smile on his face and he saids, "You guys did pretty well against those F-14s didn't you?" Tom and I puffed out our chests a bit and I said, "Yes sir, we had some *Tomcat* ass out there today." He burst out laughing and said, "You also had some carrier, destroyer, cruiser and North Vietnamese ass!" "What do you mean Captain?" We just checked your chaff dispensers and LT Smith's is empty."

Apparently in his exuberance to keep the F-14s from picking us up on their radar, Tom had deployed all his chaff (foil strips used to confuse radar), which had effectively befuddled radar all over the Tonkin Gulf and along the coast of NVN. The COS went on to tell us that North Vietnam had officially protested our aggressive behavior by labeling us Brazen Provolaters." They never were very good with their English translations, but apparently, they were referring to our behavior as a brazen provocation. Of course, this episode eventually resulted in a patch for our flight jackets. *As I recall it was a vulture sitting on a tree branch with the inscription: "Patience my ass, I am going to go kill something. Tonkin Gulf Brazen Provolaters, 1975."*

***

OPERATION FREQUENT WIND: There were four aircraft carriers involved in *Operation Frequent Wind*, the evacuation of Saigon. USS *Midway* (CV 41) and USS *Hancock* (CV 19) off loaded their normal air wings and replaced them with a whole deck load of heavy lift Air Force CH-53 helicopters. USS *Enterprise* (CV 65) and USS *Coral Sea* (CV 43) maintained their normal air wings with the role of potentially engaging the North Vietnamese to protect the rescue helicopters and evacuees. During the previous month, thousands of Americans and the South Vietnamese who had supported our efforts were evacuated by airlift from Tan Son Nhut AB in Saigon. Unfortunately, as the North Vietnamese got closer to Saigon, South Vietnamese defectors began disrupting operations at Tan Son

Nhut rendering it unusable to fixed-wing aircraft.

In addition to the four carriers, several amphibious ships with Marine helicopters on board also participated in the event. With their relatively shallow draft they could operate much closer to the coast, so they also rescued many evacuees who escaped by small boats as well as by helicopter. The carriers *Hancock* and *Midway* got as close as they could safely operate, and on the morning of April 29, 1975, they commenced the operation.

There were two main pick-up points for the helicopters: the roof of the American embassy and Ton Son Nhut AB. Attack aircraft from *Enterprise* and *Coral Sea* were assigned to initially orbit the city of Vung Tau, which was located on a peninsula protruding into the South China Sea. As the helicopters passed by Vung Tau on their way to the designated pick-up areas, sections of A-7s or A-6s would pick up escort duties to make sure that they were protected from enemy fire. There were also F-14s from *Enterprise* and F-4s from *Coral Sea* patrolling for any potential enemy air activity (which could be ARVN defectors flying against their former ally).

The rules of engagement allowed for those of us flying escort to retaliate against anyone firing at the helicopters. To minimize collateral damage that would most certainly be considerable, if we used bombs, we were instead loaded with *Zuni* rockets and our 6,000 rounds-per-minute Gatling guns. These weapons are much more precise than dropping bombs in a crowded city like Saigon. In preparation for this operation, *Enterprise* and *Coral Sea* had loaded the heretofore banned-from-aircraft-carrier *Zuni* rocket pods, while in port in the P.I. The *Zunis* were banned because of their instability, and their involvement in previous fires on *Enterprise* and *Forrestal*. The big caveat was that once we launched with the *Zunis,* we were not allowed to bring them back aboard. In other words, if not used over the beach against the enemy, we would need to fire them into the South China Sea prior to landing back aboard the carrier.

There must have been some high level of communication between Washington, D.C. and Hanoi because the main road coming down the coast from Da Nang was bumper-to-bumper with bad guys all stopped just north of Saigon. We are talking tanks, trucks and soldiers. This was certainly a juicy target for those of us who had been mostly blowing over trees in Laos during our combat time, but unless they shot at us, we were to leave them alone. I do not remember who my wingman was, but we picked up a group of CH-53s that were on their way to Tan Son Nhut AB. I would love to say we got shot at and then struck back, teaching those North Vietnamese a lesson, but that did not happen. Nobody shot at our helicopters either on the way in or back out, so it was a milk run as far as combat is concerned. The most excitement we had was firing our *Zunis* into the South China Sea prior to landing.

Now, not only is a *Zuni* Rocket dangerous aboard ship, but it also has a fragmentation pattern like a 500-pound bomb. This required us to use a 30-degree dive with a 3,000-foot minimum pullout altitude to evade the fragmentation pattern when we shot them into the water. Well, not everyone followed this procedure, and sure enough, a VA-27 pilot shot himself down. Once again, I will leave off the pilot's name to protect the guilty. He had forgotten to get rid of his *Zunis* as he prepared to recover aboard *Enterprise*, so he just armed them up pushed over his nose and fired them into the water. The fragmentation pattern literally destroyed his aircraft and he was lucky to eject successfully. *For a very short period, he was considered a combat loss, soon to be accurately revised to what it was: a dumb-shit accident.*

During the two days of *Operation Frequent Wind, Enterprise* and *Coral Sea* stayed some 50 miles farther out to sea than *Hancock* and *Midway*. Since by this time, there were no friendly airfields in South Vietnam, our emergency bingo field was Cubi Point, some 800 miles to the east. With no alternative for landing, we could not tolerate any of the numerous South Vietnamese helicopters and even small fixed-wing aircraft, escaping from Saigon fouling our landing areas while we had aircraft in the air.

The good news was that many of them successfully landed on *Hancock* and *Midway*. When their decks became too crowded, they simply pushed helicopters over the side to make room for more. *Midway* even landed a small Cessna O-1 *Bird Dog*. The pilot flew over the flight deck and dropped a note asking the flight deck crew to move some helicopters so that he could land. He further related that he had his wife and five children, including a baby on board. The Captain of *Midway*, Larry Chambers, authorized the action and the Air Boss, Vern Jumper, and his flight deck crew pushed the South Vietnamese helicopters (that had recently flown aboard filled with refugees, but were now clogging the landing area), over the side into the sea. Shortly thereafter, Buang-Ly successfully landed to the cheers of the crew of *Midway*. *It was certainly disheartening to witness the end of a free Vietnam, but I was proud to be a small part of the successful evacuation of at least some of these folks who had remained faithful allies of the United States of America.*

Despite the sadness associated with this operation, Navy pilots will always find some humor in even the darkest hour. Since the shooting war had been over for two years, most of the nugget pilots had no combat experience. Any flight over South Vietnam during *Operation Frequent Wind* was a combat flight so these youngsters finally got themselves some green ink (combat flight time in their flight logbooks). To commemorate these first combat flights, they came up with a patch for their flight jackets with the slogan: "When I die, I know I am going to heaven, because I have already been to hell at Vung Tau, One Combat Mission." All in fun of course since many of them never left the holding pattern area over Vung Tau.

Following the completion of Frequent Wind, *Enterprise* made a brief stop in the P.I., then we headed for home. Finishing up my third cruise, I was fortunate to be included in the VA-97 fly-in to Lemoore. As per tradition, we all taxied into the flight line, shut down our aircraft on signal, got out of our cockpits and headed for the roped-off area that held our precious families.

Suddenly, out from under the rope sprinted this little six-year-old girl, my Valerie. I dropped my flight gear and grabbed her into my arms for a giant hug before proceeding to a joyful reunion with Barb and my little boy, Ryan. Unbeknownst to us, a photographer from the *Hanford Sentinel* newspaper had snapped a photo catching Valerie, moments before she leaped into my arms. In fact, the kneeboard that I dropped to pick her up had just hit the ground.

Opposite page, top: Photo taken by a Hanford Sentinel *newspaper photographer of my six-year-old daughter, who had jumped over the crowd line rope to come greet her dad as we returned from evacuating Saigon. This photo, together with the one below that Barb snapped of me running out to greet Val 23 years later, when she was returning from* Operation Desert Fox, *eventually graced the pages of a father-daughter article in the* Tail Hook *magazine.*

Opposite page, bottom: Me greeting Val on her return from a combat cruise on Enterprise *in 1998. We are the only father-daughter combination to achieve centurion status (100 landings) on the same aircraft carrier, USS* Enterprise.

***

A Shout-Out to My Many Friends Who Served in Vietnam: Although I have made light of my experiences during the Vietnam War, it is appropriate for me to acknowledge several of my close high school and college friends who endured a much darker side of the war. Some of these folks were drafted and some volunteered, but all served honorably, earning the coveted Combat Infantry Badge. The Combat Infantry Badge is awarded for performing duties while personally under fire while assigned to a unit engaged in active ground combat. My very close friend, college fraternity brother and teammate from Portland High School, Roger Erickson decided to attend Army OCS and eventually ended up spending six months in the bush (front lines in Vietnam) as a leader and advisor to a South Vietnamese Army unit. A fraternity brother, Al "Pinto" Johnson stayed enlisted and was a radioman with a unit that participated in the infamous battle known as "Hamburger Hill." He was the first radioman to reach the summit and one of the very few who walked away from that awful experience unscathed. Rick Sys, another fraternity brother, was also involved in close combat with the enemy.

My cousin, Donnie Leland, was an A-6 *Intruder* pilot flying out of Da Nang for six months followed by six months as a forward air controller serving with a very tough group of South Korean troops. Jim Grandalen was a Marine, a couple of years older than me, who would not talk about his experiences because they were so harrowing. He did mention emptying his .45-caliber pistol during a firefight, which basically meant he was face-to-face with the enemy.

Jerry Swift was an Army helicopter pilot who flew combat missions into hot landing zones to resupply troops and evacuate wounded. Peter Paulson who was a couple of years behind me in high school was badly wounded as an Army infantryman, as was a fraternity brother, Daryl Lutovsky. My second cousin, Eric Lunde, flew Air Force F-105 *Thuds* into North Vietnam, and Doctor Glen Thoresen was an Air Force Flight Surgeon, but also flew as a Weapons Systems Officer in the F-4 *Phantom II*. Many others from our community also served. My best friend Fugie, who probably would have been the most capable soldier of us all, suffered from football knee injuries and was also running a critical farming operation that reluctantly kept him on the home front. *Even though the Canadian border was only a couple of hours drive, despite the unpopularity of the war, defecting to Canada was never considered an option from this ultra-patriotic community.*

***

TURNAROUND HIGHLIGHTS: The CAG staff was headquartered at NAS Miramar; however, I remained attached as additional duty to VA-97 in Lemoore. VA-97 was at that time the best Light Attack squadron in the Navy, recipients of the coveted "Triple Crown", which consisted of the "Battle "E," the Safety "E" and the "McClusky" Trophy. We had great leadership in the quiet warrior, CO, Bert Terry and the XO, Pat Nicolls. Our department heads, Tank Bledsoe, Paul Otto, Major Charlie Brame (Air Force Exchange), Mike Sullivan and Bud Orr related well to the junior officers and set great examples with their superior airmanship skills. During the turnaround we picked up LCDRs Goryanec, Miller and Kasting, all top-notch officers and aviators. CDR John "Digger" Murray also joined as CDR Nicolls' XO.

But as is the case with most outstanding squadrons, the JOs were the heart of the organization. They were a wild and woolly group that loved to fly and raise hell. There were several characters amongst this group, including a former enlisted Marine Grunt, who survived the siege of Khe Shan prior to becoming a naval aviator. Ken "Ike" Eisenhart was also some sort of gymnast/acrobat. He would enlist the help of the biggest/strongest JO, Mike Malone (yes, the same Mike Malone who eventually became a three-star Admiral and the Air Boss of all naval aviation) to help launch him into various events via multiple flips and twists *almost* always landing on his feet. Ike was also infamous for coming up with call signs for newly arrived nuggets, most of them terribly politically incorrect.

Despite all these awards and accolades, VA-97 was most famous for the Wives Group. They were by far the most attractive group of ladies (and believe me there were many beautiful women married to naval aviators) in any squadron in all of naval aviation. Not only were they attractive, but they also loved to party, and Barb and I had an absolute blast socializing with this very fun group of people led by the incomparable Norma Nicolls.

***

WEEKLY TRIPS TO MIRAMAR: CAG would hold staff meetings every Thursday that he expected me to attend. No problem, I would grab a VA-97 jet and fly to Miramar every Wednesday evening, just in time to attend their famous Wednesday night happy hours. Since running was very popular in 1975, NAS Miramar would host a three-mile run around the golf course first-thing Thursday morning. I was a serious runner by this time, and it was a great way to get rid of a hangover, so I would run the race, shower, and spend the rest of the day with the CAG staff carrying out my collateral duty, which was the Air Wing Human Resources Officer.

One of several VA-97 donkey carts heading for some fun in Tijuana. Barb and I had a blast while I was assigned to the best Light Attack Squadron in the Navy. Barb and I are the second couple from the left. Other names omitted to protect the guilty.

Following the racial riot on *Kitty Hawk*, the Navy got very serious about addressing discrimination issues. I went through some very intense training that really opened my eyes as to why minority sailors were so dissatisfied with how they were being treated. *In this job, I had an opportunity to influence/change the behavior of not only the Wing staff, but also all the squadrons in the air wing.*

***

FALLON DEPLOYMENT: In preparation for the next cruise on *Enterprise*, all the squadrons in the air wing deployed to NAS Fallon for a two-week training evolution. The training involved basic bombing and strafing flights followed by competitive exercises called COMPEXs and culminating in large practice Alpha Strikes. There were certain minimum proficiency goals that had to be met, and of course, head-to-head competition to determine the top squadrons and individuals. The CAG staff was tasked with observing these events so

I convinced CAG that the ideal platform to accomplish this would be a T-28. Unlike some of my previous attempts to bag more flight time, CAG Wilson agreed with my proposal.

I borrowed a T-28 from VA-122 along with a First-Class Radial Engine Mechanic, and he and I flew to Fallon. We parked the T-28 right below the window of the CAG office. I found an old WWI style aviator's scarf that looked really cool flying this beast. I never took off without a passenger. Sometimes it was a fellow staff aviator helping to observe a competitive event, but more often it was one of the numerous Squadron Sailor of the Month awardees.

Most of the events I observed were flown over a target area known as B-17. The great thing about this target area was that the border fence defining the Restricted Area ran right down the ditch of a well-traveled highway. I often flew the first event of the day. Just imagine a bunch of sleepy/hung-over jet jocks attempting to pay attention to a morning brief when it is suddenly interrupted by the beautiful sound of a 1425-horsepower radial engine coughing a cloud of smoke as it came to life.

It took me a bit longer to get to the Restricted Area, so I would take off a few minutes ahead of the jets. With the sun just beginning to pop over the mountain on the east side of B-17, I would take the T-28 down to 50 feet or so above the ditch that ran alongside the highway (my wing was probably only 100 feet from the shoulder of the highway), come out of the sun on an east to west run and scare the shit out of whatever vehicle was on the road. It was great fun and actually legal versus many of the fun, but illegal things that we did. *The amazing part of this two-week event was that I also flew the same number of A-7 flights as the rest of the VA-97 pilots. I ended up with over 60 hours of flight time in only two weeks. It doesn't get any better than that.*

***

VF-126 Adversary Training: As part of the many lessons learned from the Vietnam War, it was determined that dissimilar air-to-air training would be very beneficial for not only the Fighter community, but also for those of us in Light Attack. VF-126 was established at Miramar and outfitted with the A-4 *Super Fox*, which could out-turn and out-accelerate the A-7E. It also proved to be a good match for the F-14 *Tomcat*. LT Tom Smith and I (the Provolaters) were selected to be the first VA-97 pilots to receive the training syllabus. VF-

126 deployed to NAS Lemoore to conduct the training, which was awesome. Due to our relatively limited performance capability versus enemy fighters, especially when fully loaded with ordnance, the training concentrated on defensive maneuvering to escape from the enemy fighter.

The final graduation flight consisted of Tom and me attacking a simulated target in the mountains just east of Bakersfield. The target was defended by two VF-126 Adversary A-4s. Our mission was to avoid the adversary fighters, simulate a bombing run on the target and then egress without being shot down. The lead adversary instructor made it realistic by allowing us to use any attack profile that we wanted. This was a big mistake on his part because there were not two pilots in Lemoore that were more competitive than Rud/Smith.

We filed an IFR flight plan at an ingress altitude of 30,000 feet. We canceled the flight plan passing 24,000 feet and checked in on the shared tactical frequency. Of course, they were expecting us to be coming in at low level for a pop up-type of attack. Instead, coming in extremely high, we now accelerated to almost 600 knots for the simulated bomb run. The other advantage this gave us was that we easily spotted the two adversaries. After simulating dropping our bombs, we called "tally ho" on the A-4s. We were carrying training heat-seeking *Sidewinder* missiles so with the incredible airspeed advantage we both got off "Fox 2" shots prior to pushing our noses over and escaping the area. Yahoo! They trained us really well.

Not so fast. After landing back at Lemoore, we headed for the ready room expecting accolades for our successful graduation flight. Instead, we received a tongue lashing for gaming the event. The fighter pilot egos initially came into play, but it was not long before we were all laughing up a storm about our extreme competitiveness and the suggestion that they need to put in some rules of engagement for future training. *It doesn't matter how you win an air-to-air engagement because losing is not an option.*

***

F-14 TOMCAT CROSS-TRAINING: As I recall, Smoke Wilson said, "Gil, as we work up for this second cruise of the *Tomcat*, I think it would be great for your credibility as an LSO to be qualified in the F-14." I replied, "I could not agree with you more, but the RAG is nine months long." "RAG, screw the RAG, I am talking in-house training with VF-1." He then handed me an F-14 NATOPS manual and said, "Study the manual and fill out the open-book exam. Come back down to Miramar in two weeks and take the closed book NATOPS Exam. We will then give you a simulator check of the emergency procedures and schedule you for two flights. The first one will be with a pilot in the back (there are no flight controls in the back) and the second will be with a RIO to work all of the systems."

Keeping in mind that this was 1975, and the aircraft was still pretty much considered to be departure and spin resistant, it was extraordinary for a non-test pilot A-7 driver to get this opportunity. *No problem though because Smoke Wilson was an extraordinary leader.*

I did a good deal of studying during those two weeks, so I was ready to go on December 18, 1975. My first flight was with a former F-8 *Crusader* pilot, "Rocky" Rockwell. We started the aircraft and he attempted to get the radar and other stuff working. He even had troubleshooters plugged into the aircraft intercommunications system (ICS) so he could get help with the avionics. They did not seem to be making any progress so I said, "Rocky, I really don't care that much about you demonstrating the systems, I can get that on my second flight with a RIO. Let's just go out and fly the shit out of this fabulous machine." And that is exactly what we did.

As we arrived at the hold-short area, just before taking off, I noticed an F-8 waiting for his wingman. At closer scrutiny, I realized it was my old AOCS classmate and fellow CAG LSO, "Bug" Roach (he had this wildly decorated helmet that was unmistakable). I told Rocky that this was an opportunity that I could not resist and to bear with me for just a minute. At that point, I took off my helmet and waved at Bug. Here we had Bug, the ultimate fighter pilot, looking at his old friend, a Light Attack puke, flying the F-14 before he had a chance to. *Bug would end up with thousands of F-14 hours to my two flights, but, by golly, I flew it first.*

As we taxied onto the runway, Rocky suggested that since it was my first flight, maybe I should limit takeoff power to Zone 3 afterburner. I replied, "Rocky, I only get two flights, so I am going to make the most of it and use Zone 5." "OK, but I bet that you will not be able to stay within the confines of the "Blacks Beach" departure parameters." He was correct, of course. Since altitude violations are a lot easier to detect than speed violations, I stayed within the altitude restrictions, but I was probably about 100 knots above the speed limit.

With Rocky Rockwell, following my first flight with VF-1 in the F-14 Tomcat.

The flight was an absolute blast. We got the supersonic stuff out of the way (no big deal other than to say you did it) and then proceeded to accomplish some pre-briefed performance maneuvers. A "slick" *Tomcat* is one hell of a machine, and even though I ended up with 1,100 hours of F/A-18 *Hornet* time, this was a unique experience. As I recall, we did a high-G, level turn in basic engine, lit the afterburners to Zone 5 and pulled the nose up 60 degrees to show the incredible climb rate.

The next step was to deselect afterburner, roll inverted and bring the nose back down to the horizon. A rather straightforward and simple procedure when accomplished properly, which I did.

I was not the only person cross training, however, and one of the other guys scared the shit out of his pilot instructor. During their attempt to accomplish this maneuver, the ICS failed, so the instructor lost communication with the pilot. Instead of bringing the power back to "military" and rolling inverted to stop the climb, this guy (again his name is left out to protect the guilty) stayed in Zone 5 and continued his climb until the following happened via a direct quote from the instructor, "We climbed so high that it started to get dark, and I could actually see the curvature of the earth before he finally stopped the climb. I was in fear of engine failure, and without a pressure suit, that we would die a horrible blood-boiling death!" *Of course, this situation was exaggerated some, but it does make for a great story.*

Since the main purpose of my flight was to become familiar with the landing qualities, we proceeded to the bounce field at San Clemente Island. My pilot logbook shows that I performed eight FCLPs. After the completion of the last one, we had briefed that I would light the burners, accelerate down the length of the runway and then pull up into an "Immelmann" maneuver, ending at an altitude appropriate for our transit back to Miramar.

We checked in with approach and got vectored to the initial point for the VFR carrier break on runway 24. Rocky advised that we needed to watch our airspeed until we got to the initial, then I could add it up to achieve an appropriate airspeed for the carrier break. He made the mistake of not defining "appropriate." Just inside the initial he said, "OK Gil, you can add the power for the break." I immediately pushed the throttles into afterburner all the way to Zone 5.

During the debrief, he said, "I almost pissed in my flight suit when you went to zone 5 because we ended up entering the break at over 500 knots. (The admiral at Miramar at the time was the notorious "Field Day" Fellows who had his own personal radar gun, which he would use to check airspeeds in the break.) I rolled that baby into 90-degree angle of bank and pulled like hell as I reduced the throttles. From the back I hear a muffled transmission, "Wings." "What?" "Never mind." By this time the wings were programming forward into the landing configuration, which was what he was worried about. We landed, and as we were taxiing into park, the low fuel light illuminated. We had been airborne for 1.3 hours. *Now that is what I call a great use of flight time.*

My second flight, also flown on December 18, was with a VF-1 RIO by the name of John Uelses. He got all the systems working, but we were restricted to limited maneuvering by a gear door issue. John was very famous as he was the first man to clear 16 feet in the pole vault, and he held the world record with the old fiberglass pole. *Smoke was right about the* Tomcat *flight experience increasing both LSO skills and my credibility with the Fighter community.*

***

CHANGE OF COMMAND DINING OUT: "Dining Ins" and "Dining Outs" are formal events meant to honor achievements, honor guests and honor fallen comrades within the organization. They are meant to be fun and in good taste. A Dining In is a closed event without guests or a speaker. It is often an unruly affair, with charged grog and a chance for everyone to let their hair down, so to speak. A Dining Out on the other hand, includes wives and girl-friends, a formal speaker, the mess president and a "Mr. Vice" to help keep decorum at the event.

As I have previously mentioned, Smoke Wilson was an extraordinary leader, so we, as his loyal staff, decided that we would put on an extraordinary Dining Out. We did not surprise him with our plans, but rather included him to make sure that this night would be one that he would cherish and never forget.

I remember the initial meeting that we had at the staff headquarters at NAS Miramar. Monroe, who could have easily made a living as a script writer/producer/director, oversaw the preparations. Once we decided to go rogue with the event (not follow prescribed Navy protocol), everyone agreed that "Mr. Vice" would be the key to a successful event and would also be the individual whose career following the event would more than likely be in jeopardy. Suddenly, they all looked at me, "Rud, you are the guy! You are the junior aviator on the staff and don't really have a career to lose anyway." And so, just like that, I was unanimously chosen to be the sacrificial LT.

I was thrilled and honored to get the job. Monroe, Rick Hauck and I had put together several end-of-line-period "Foc's'le Follies." These events were held to celebrate completing a specific at-sea period, and to honor the top individual pilots and squadron teams with the best landing grades. The only space large enough to hold these events was the ships forecastle, which is pronounced "foc's'le." Each squadron was encouraged to perform a skit as part of the entertainment and, at least at-sea, political correctness was not a consideration. Of course, we made fun of everyone including ourselves, so a thick skin was helpful, especially if you were a more-senior target.

With this shared experience driving our imaginations, we went to work scripting the best, or worst, (depending on your point of view) Dining Out, ever. Smoke was the President of the Mess, and I was Mr. Vice. All requests for toasts and other frivolity are directed through Mr. Vice. Tradition is quite clear that at the completion of the meal, the smoking lamp is lit. Well, Smoke thoroughly enjoyed a good cigar, so we started off the evening by Mr. Vice (me) lighting the smoking lamp before we even started the first course of the meal, which of course delighted the boys and thoroughly disgusted the ladies—and unfortunately RADM Coogan—who was the guest speaker.

I might mention, about now, that there was a great deal of pre-event imbibing among the primary participants, so things went downhill quickly. It is traditional for Mr. Vice to sample the beef prior to it being served. The chef, in full chef's cooking uniform, brought the beef out on a rolling table. He cut off a piece of the meat and I took a bite, chewed, swallowed and would normally pronounce, "Mr. President, the beef is fit for human consumption." Instead, I took a bite, chewed, swallowed and pronounced, "Mr. President, this beef tastes awful!" Mr. President replied, "That is totally unsatisfactory Mr. Vice, what action do you recommend that we take?" I replied, "Sir, I recommend that we shoot the chef!" Mr. President replied, "I agree, carry out the deed Mr. Vice." At that point, I pulled out my .38 survival pistol and aimed at the chef, who screamed, "NO! Please don't! I proceeded to shoot him in the chest with the full noise, flame and smoke of a .38 inside a room (with blanks instead of bullets, of course). The chef had this movie-quality bag of fake blood that he slapped onto his chest in a manner so realistic, that for a minute I wondered if I had fired a real bullet. He fell to the floor, and the waiters picked him up and carted him off to the kitchen.

By this point, the women, at least the ones who had never been to one of these events, were a bit hysterical, and of course the guys were laughing heartily. In other words, it was a chaotic scene. But it was not over yet, as a new chef quickly appeared with a drop-dead gorgeous blonde (wife of one of the VS-29 junior pilots) laid out on the rolling tray table. "Mr. Vice, would you like to taste this beef?" I got up to accomplish that potentially pleasant task. At that point, my wife, who was part of the plot, stood up and said, "Mr. President this Mr. Vice is done tasting beef. Let's eat!"

Eventually we completed the meal, and it was time for the introduction of RADM Coogan, the guest speaker. Of course, the last thing that any drunken bunch of naval aviators wants to do at this point is to pay attention to some admiral drone on about whatever. I am sure he sensed this, so he kept his remarks quite succinct, but he did say this, "LT Rud, you are the worst Mr. Vice I have ever seen in my 35-years of Naval Service!" And he was not smiling when he said it.

Following the conclusion of the event itself, we retired to the bar area of the North Island Officers' Club for a nightcap. Smoke and I had arm wrestled a few times over the last year or so. He always won when we used our right arms, but old lefty here would take him left-handed. He asked for one more opportunity, so we got on the floor and locked arms for the big contest. At that moment, a woman came into the bar, leaned over and said, "LT Rud, what do you think of the guest speaker, RADM Coogan?" I replied, "I think he is a bit of an asshole!" She replied, "I do too!" I then asked Smoke who that woman was. "Oh, that was Mrs. Coogan, the admiral's wife." *It was a good thing that Smoke and not the admiral was writing my fitness report, because this was just another one of many reasons why I should have never made it past LT.*

***

NEW LEADERSHIP AND A SECOND DEPLOYMENT ON *ENTERPRISE*: Another awesome fighter pilot replaced Smoke as the new air wing commander. CDR Roger Box took over as the CAG and Bill Snyder replaced Monroe as the senior (fighter) CAG LSO. Bill also became my roommate for my second *Enterprise* cruise.

We already knew each other well since Bill was a plow back (assigned to be an instructor right after receiving his wings) and was my primary instructor in the F-9 *Cougar* back in the advanced training command. Obviously, he too, was way senior to me, so like the first cruise, several other LSOs were senior to me, but not on the LSO Platform.

***

THE GREAT VELCRO SLIDE: As a CAG LSO, I was not eligible to officially compete as an individual, or to be counted for the VA-97 team in the carrier landing grades competition. I was, however, included on the VA-97 "Greenie Board" in the ready room. In addition to all the pilots being listed on the Greenie Board, to step up an already heated competition, they included what was referred to as a "Velcro" Ladder. Each pilot's call sign was then placed on the ladder in the order of best landing grades, to worst. My call sign at the time was "Rudy" (I would earn a more permanent one a bit later in my career) and during this particular line period, I was doing great. In fact, I was at the very top of the ladder with the best landing grades in the squadron. I launched on the very last night event, completed my assigned mission and headed back to land. For whatever reason, I screwed up the pass, over-corrected for being a little low, missed all the wires and got a rare Bolter. Then everything went to shit. I had two more Bolters before finally getting a "no grade one wire" on my final pass. This is what naval aviators refer to as a "Night in the Barrel." And it wasn't over yet.

Traditionally, once the last event launches, the squadron skipper will authorize the SDO to start the movie. The PLAT television which shows the aircraft landing is still on over to one side of the movie screen so if the last recovery gets colorful, all the pilots are in the ready room to watch. I was the one making it colorful that night. Did I get any sympathy? Hell no, they were all laughing at the plight of the cocky CAG LSO suffering through his Night in the Barrel. Feeling totally deflated, ashamed, and of course pissed off, I headed for the ready room.

As I approached the door to the ready room, I noticed my call sign on the deck outside of the ready room. They had stapled together what was probably at least 50 feet of Velcro all the way from the Greenie Board in the front of the ready room, under the door into the passageway, and placed Rudy at the very end. I opened the door to a rousing round of guffaws of all sorts, including mock prayers of thanks for my safe return. I don't care how good you think you are at flying aboard an aircraft carrier, you will have your night in the barrel, and you'd better have a thick skin when it happens. *By the way, they would never do this to a youngster who was struggling with his landings. To him they would offer encouragement; me, they made fun of. It was a good feeling to know that they respected me enough to pull the Great Velcro Slide caper on my sorry butt.*

***

SAFARI IN MOMBASA, KENYA: Mombasa was certainly the most unique liberty port on this deployment. I had been in touch with a childhood idol, Dave Simonson, who was a Lutheran Missionary in what was then called Tanganyika. Dave had been a standout football player at Concordia College in Moorhead, Minnesota. While he was a student, he was dating and eventually married Eunice Norby, who was the daughter of our parish minister. The parish parsonage was right across the road from our farm; so, Dave would come over and play catch with little Gil. My most-vivid memory is when he was showing me how to properly punt a football, and he proceeded to kick it over the top of our barn, scaring the crap out of our milk cows.

Since Tanganyika bordered on Kenya, I had submitted all the necessary paperwork/visa, etc. to visit him. He had been there for many years, so I was really looking forward to getting a feel for the real Africa. It was not to be however, since due to political turmoil that eventually resulted in the renaming of the country to Tanzania, my visit was denied. To replace that disappointment, I joined some of the other CAG-14 staff members on a camera safari into Tsavo National Park in Kenya. We rode in VW minibuses, which were equipped with sliding roofs that we could stand up in to view the wildlife and take photos. The driver of each minibus was also a guide of sorts, pointing out where to look as we literally drove through hundreds of miles of an open wild animal park.

We saw all sorts of native African animals as they exist in the wild. This included getting quite close to a large pride of lions. We were able to do this safely, according to our guide anyway, because they had just killed and eaten a zebra, so they were not interested in skinny, tasteless humans. Ed Nieusma, the staff Air Intelligence (AI) officer and a fellow North Dakota native, and I, did have a bag of potato chips heisted out of the mini-van by one ugly, nasty, baboon.

We spent the night at a resort that had a tunnel running from the lodge to a large watering hole. At the end of the tunnel was what I would best describe as a WWII pillbox, with slits that we could look through to view the various elements of the food chain. The lower portion of the food chain drank quickly, while nervously checking for predators. Later the big cats and elephants sauntered up without a care in the world. *It was a wonderful experience that I will cherish for a lifetime.*

***

RESPECTING THE CREW CONCEPT: As the assistant CAG LSO, I considered that an important part of my job was to learn to know every aviator in the air wing. This included the Naval Flight Officers since they were part of a team working together with the pilots to complete the mission. In some cases, an experienced Naval Flight Officer would be paired with a junior pilot who was struggling with shipboard landings. In one case, a newly arrived F-14 *Tomcat* Nugget was having a night in the barrel (not able to get aboard the ship, with several Bolters and/or Wave-Offs). Not having much success talking to the pilot, I addressed the RIO, Willy Driscoll, who just happened to be an *Ace*, who was awarded a Navy Cross flying with Randy Cunningham in F-4s a couple of years earlier. Willy was able to calm the pilot down and help get him aboard. That being said, what I did not appreciate was an NFO questioning a grade given to his pilot by one of my LSOs. Fortunately, that rarely happened. I also knew all the flight deck and air operations guys on the ship. *These friendships were important when it came to port visits, since I could hang out with whatever group was having the most fun.*

***

INTERACTIONS WITH CAPTAIN C.C. SMITH: We were so lucky to have this man as the Captain of *Enterprise*. He was both a great leader and a wonderfully entertaining character. Being from Alabama, C.C. combined a deep/authoritative voice with a distinct southern drawl that was unique and effective. Whenever he addressed the crew with the 1MC (communication system that can be heard in every space on the ship), he always ended the address by saying, "Thaas Awl." Upon our arrival in the western Pacific, a new admiral and his staff came aboard *Enterprise*. RADM "Little" Bill Harris (there was also a "Big" Bill Harris, and they had very different leadership styles) considered Captain Smith's "Thaas Awl" to be inappropriate, so C.C. had to stop ending his addresses to the crew

223

with that phrase.

During the later stages of the cruise, as we were about to go into port in Hobart, Tasmania, an A-3 flew out from the Philippines to take the Admiral and his staff off *Enterprise*. I was on the LSO platform to wave the A-3 aboard, when suddenly the direct line from the captain's bridge began to buzz. I picked it up and said, "Yes Sir, what can I do for you?" C.C. said, "Ruuud, that A-3 is coming to pick up the Admiral, so don't let that sum-bitch Bolter!" Following the usual pomp and circumstance of side boys and honors for the Admiral, the A-3 finally got catapulted off *Enterprise*. As soon as the noise from that departure quieted down, C.C. came on the 1MC with the following message to the crew, "Thaas Awl!"

Another memorable interaction that I had with C.C. occurred during some bad weather operations in the Indian Ocean. During that day, the deck had been pitching to the point where we were using MOVLAS to safely recover aircraft. Since we were in a "Blue Water" situation (no bingo fields), I felt rather strongly that we should cancel night operations. I asked CAG, and he recommended that I go directly to Captain Smith with my concerns.

C.C. was an avid runner, but since he needed to stay near the bridge, he would often jog in place on the bridge wing. Rather than stop his workout, he would ask folks who needed to talk to him to join him on the bridge wing and jog in place with him while we discussed the issue.

C.C. said, "What can I do for you, Ruuud?" "Well, Captain, as you know the deck has been moving to the point where we have had to use the MOVLAS to safely recover aircraft. With that in mind, I recommend that we cancel night operations." He continued to jog in-place, seeming to be in deep thought, then he turned to me and said, "Ruuud, I will compromise with you, I won't launch the *Vigilante*." It was a long night, but I certainly appreciated the small concession he gave because the RA-5 *Vigilante* was tough enough to land in good weather (by the way, C.C. was a *Vigilante* pilot).

Although C.C. was a typically brilliant nuclear power guy, he possessed a great sense of humor, and he loved to hang out with and address the air wing, whenever he had an opportunity. He would begin his address to the air wing: *"There are three things that you do not do. You do not piss into the wind. You do not step on Superman's cape. And you do not mess with Carrier Air Wing Fourteen!"*

***

HOBART, TASMANIA: During a long at-sea period, we were given the uplifting news that our next port of call was going to be Sydney, Australia. This news was greeted with great exuberance from the entire crew. As we worked our way south, we got some bad news. Only the ships other than the carrier were going to be allowed into Sydney; *Enterprise* had been re-routed to Hobart, Tasmania. "Noooo!" we all thought, where in the hell is Tasmania?" This is when an Australian exchange pilot came into the CAG office with a big smile on his face. "Cheer up you blokes, Hobart is our favorite holiday destination." C.C then let him address the crew about Hobart and the unique opportunity it was going to afford the crew. We were certainly not disappointed.

What was not so great was operating in the Tasman Sea on our way to Hobart, which is located 370 miles south of the Australian mainland. Operating that far south was similar in many respects to operating way up north in the Bering Sea, with very quick and nasty changes in the weather, producing very rough seas. I recall having an event of 15 or so air wing aircraft airborne when one of them reported a particularly bad rain squall nearing the ship.

We canceled the launch and immediately recalled the aircraft to get them on deck before entering the squall. We got them all aboard except for the KA-6 tanker, which was being flown by an experienced pilot (I think it was Bob Whittenberg, former star offensive lineman at the U.S. Naval Academy). As the CAG LSO, by this point in the cruise, I would normally be watching my LSO team perform or maybe backing up the controlling LSO. However, when a situation such as this arises, it becomes my task to get the aircraft aboard.

The rain and wind were such that the visibility was near zero. It was also cold as hell, just above freezing. I took the pickle (device the LSO uses to control the Fresnel lens) and told everyone else on the LSO platform other than the team leader, who was backing me up, to get into the safety net. I then walked out onto the flight deck as far as I could with the communication equipment. I told Bob to turn on his landing light, so that I could see him through the rain. He did this and I picked him up at about a quarter mile. He had flown a great approach even though he did not see the carrier even at that distance. I gave him a couple of power calls (because of the high wind conditions), a little right for lineup and he landed, darn near blowing me over the side, since I was still on the flight deck away from the blast shield that protects the LSO platform. He was awarded an 'OK, no comment', 5-pointer for that landing.

Enterprise anchored in the Derwent River, which was basically in downtown Hobart. The great news was that instead of enduring long lines waiting to go ashore via liberty boats, two giant 600-person ferries from the mainland were available for our use. They were in Hobart because the main bridge across the river had been damaged when hit by a ship, so it was not useable.

The people of Hobart had almost no exposure to the United States Navy and to this point in 1975, as far as I know, absolutely none to an aircraft carrier. I had the staff duty the first day in port, so of course, I quickly volunteered to be one of the sailors hosted for dinner by a local family. The patriarch of my host family met me at the pier. He was a retired rugby player, so we immediately headed for his rugby club bar. After a couple of beers and introductions to some very imposing rugby players, we headed for my host's home.

He and his lovely wife were the parents of ten wonderful children ranging in age from 20 to a toddler of about two. Of course, they peppered me with questions about the United States and what it was like to fly from an aircraft carrier. I, in turn, had a chance to learn a great deal about Tasmania. By the time he brought me back to the pier, we had set up a day for me to host all of them aboard *Enterprise*. I did that to the absolute delight of the whole family. *We often underestimate the effect of civilians having a chance to come aboard an aircraft carrier to see and touch the world that we take for granted.*

The liberty for the sailors in this port was unmatched anywhere in the world. The older folks remembered WWII and thanked us for saving them from a potential invasion by the Japanese. The younger folks were thankful that we were putting a halt to the spread of communism, so our crew rarely had to pay for a drink. *Enterprise* had a crew the size of a small city, so we had several musicians on board, and many of them had formed bands to entertain other sailors. Those bands now were in huge demand in the city of Hobart. Whether Rock and Roll, Country Western, or Soul, they all found work for the seven days we spent in Hobart.

On my second duty day, which was toward the end of the port visit, I received a call from the ship's public affairs officer. He said that he was in desperate need of a uniformed officer to assist him in helping to coordinate a photo shoot for an Australian Government Tourist Bureau production that was highlighting the *Enterprise* port visit. Once again, I was saved from a day on the ship as I put on my dress blues and headed for the pier.

I was given a ride by the camera crew to a nice restaurant in downtown Hobart to join several other naval officers from both the air wing and ship's company. The producer of the film was a classic character from the 1950s in that he was smoking a cigarette using a cigarette holder and pointing it around at people just like you see in Hollywood scenes from the 40s and 50s. He began pairing up the dashing (his words) naval officers with these beautiful models/actresses he had brought with him from the mainland, with the intention of all of us having a lunch that he was going to film. Apparently, these ladies were big news in Hobart, because there was a huge crowd of local folks gathered in the street outside the restaurant, and the restaurant itself was packed.

Since I was the last naval officer to arrive, and certainly the least dashing, the producer/director ran out of ladies. No problem. He brought me to the center of the restaurant where there was a small stage and then he said in a very loud voice, "All of you local ladies out there, this young man is now going to pick one of you to have lunch with, and you will be in our movie." It was like a bunch of single girls at a wedding trying to get in position to catch the bride's bouquet. Of course, I was enjoying the moment, too. Finally, I noticed this gorgeous red head, who did not seem the least bit interested in participating. I pointed at her and said, "I would like to have lunch with you!" She just rolled her eyes and reluctantly joined me for lunch. I soon found out why she was reluctant to join me, she was the wife of the United States Navy Attaché to the Australian Embassy. After a glass or two of wine, she settled into her movie actress role, and we had a wonderful lunch as I got to learn a lot more about the country of Australia.

As I left the restaurant to head back to the ship, a couple of teenagers bumped into me and one of them grabbed my cover and took off running. Since a replacement cover would be tough to find in Tasmania, I took off after them to the cheers of several locals. I really was in terrific shape and I was much faster than my adversaries. They started up this steep hill where I caught up to the one without my hat. I tapped him on the shoulder, he turned and screamed, "Bloody hell, the old boy is a fast bugger Tommie, you better give him back his hat before he gets upset with us!" Tommie gave me my hat, I put my arm around them both and we walked back down the hill to the clapping and cheering of the locals. *What a great experience—and it could only happen in Hobart.*

Since *Enterprise* was anchored in the middle of the city, and the winds were coming right down the River Derwent, on the morning of our departure, C.C. put the "pedal to the metal" and we launched 15 aircraft while basically still in the city. They made several flyovers, to the delight of the locals, and we recovered them when we reached the open sea. When C.C. next addressed the crew, he divulged that he had taken out a full page add in the Hobart newspaper with a photo of *Enterprise* and the words: "To the City of Hobart from the crew of USS *Enterprise*, thank you for your wonderful hospitality. We will never forget our time here with you."

***

ORDERS TO SHORE DUTY: I had been in touch with my detailer about my next set of orders. Initially, I was thinking that a tour of duty as an instructor in the "Advanced Training Command" would be my first choice. Then I began thinking about what I had accomplished in my first seven years as a naval aviator. At this point, I was still an O-3. I already had 450 carrier landings and I had been basically a flight instructor for several years. However, I had accomplished that instruction, almost exclusively in single-seat aircraft.

The last thing I wanted to do was to become a nervous screamer sitting in the back seat of a TA-4 while some student attempted to kill us both. I inquired as to what else was available that would still include the opportunity to remain in a "flying billet." Following some discussion with my detailer and my wife, I decided to become a Navy Recruiter. *Although certainly not my favorite tour of duty, I strongly believe that this choice was a key factor in my eventual selection as the Flight Leader of the* Blue Angels.

17. Selling the Navy for a Living

ON NOVEMBER 20, 1976, I flew my last flight with VA-97 and was relieved as the assistant CAG-14 LSO. My replacement was a great pilot and LSO, LT John "Skid" Roe. I absolutely loved, and knew that I would miss this job, but I was also excited to get back to my family and I was looking forward to a new adventure as the Head of Officer Programs at Naval Recruiting District (NRD), San Antonio, Texas.

Since we were pretty sure that we would end up back in Lemoore for our next sea tour, we kept our house and eventually rented it out for enough to cover our house payment. Prior to reporting to NRD San Antonio, I was assigned to attend a recruiting preparation school in Pensacola, Florida. I took some leave to spend time with my family prior to the start of the recruiting school in January of 1977. By this time, Barb was teaching full-time so we decided that she and the kids would stay in Lemoore until the semester break.

I took my old '69 Ford truck and headed for Pensacola. Since I was going to pass through San Antonio on the way to Pensacola, I decided to see if I could find a house for us to rent for the next 2.5 years. What I found instead, was a brand-new home for about $40,000. Connie Weigert, the Realtor whose husband Sid was a pilot stationed in Beeville, was quick to point out that we were now in Texas, which had a whole lot more reasonable housing market than California. I purchased our new abode and set up the move before continuing to Pensacola.

The school was really effective. It consisted of a full-blown sales course that would have cost a fortune to go through as a civilian. They also provided opportunities for realistic role-playing and a solid ethics foundation to make sure we did not succumb to the temptation of making a recruiting quota at the expense of an honest appraisal of the recruits.

For the aviators, the last week of the school consisted of getting qualified to fly the T-34B. There were three instructors running the qualification operation. Two of them were young LTs with P-3 backgrounds. The third was a "salty" old commander, who at one time had been the commanding officer of VT-1, the T-34 training squadron. I was the only tailhooker going through the training and the other guys were scared to death of flying with the old salt. My goal was to become the best T-34 pilot that I could, so I chose (much to the delight of the other trainees) to fly all my flights with the old timer.

It was a great move on my part because he really taught me how to fly that machine. He even showed me prohibited maneuvers like snap rolls and inverted spins. The inverted spin really got my attention. The most important part of that demonstration was how not to get into one in the future. He also stressed precision aerobatics, teaching me how to stop a spin with my wing intersecting a particular spot on the ground, and how to accomplish literally perfect aerobatics across the board. *With all the tactical jet time that I had it may seem a bit weird to say this, but the flight time that I was to get in the T-34 over the next 2+ years really helped me become better at precision aerobatics, because that is pretty much all we did on our flights with potential recruits.*

***

HIGHLIGHTS FROM THE RECRUITING TOUR: Following seven years of tactical tailhook flying and the necessity for critical split-second decision-making on a regular basis, the transition to what was important in this job was hard for me. Don't get me wrong, I had some wonderful people working for me, it just took a while for them to get used to my priorities. My predecessor was a great guy that everyone in the office hated to see leave. I never really got to know him, but I found that he had been very active in the local community and had also traveled to virtually every two-and four-year college in our south Texas area of responsibility. I looked at the results of these visits and decided to concentrate on the schools that had produced viable officer recruits.

My official title was Head of Officer Programs. LT Fred Crecelius was responsible for the Aviation Programs. Fred was outstanding in every respect. He was a gifted pilot who had, up to this point, very bad career timing. He entered the pilot training pipeline right at the end of the Vietnam War. Since there were limited, active duty flying slots available, he ended up in a Fleet Composite (VC) squadron. The good news was that while in that squadron he flew everything they had, including helicopters, C-130s and even had some A-4 time. It did not take me long (one flight) to realize that I was blessed with a very solid T-34 partner. Fred and his family also lived just a short distance from us, so we became close friends and running partners.

We also had a Restricted Line (1100) woman officer, Kay, who was good at recruiting other than aviation. Since the University of Texas, Austin, was by far our most-lucrative market for recruiting, my predecessor had stationed Kay permanently in Austin. She was great at her job, but there was just too much of a market, especially aviation-related, for one person to handle, so I brought her back to San Antonio. She then joined Fred and me on our traveling team when we visited the University of Texas, which we eventually did, as often as once a month.

***

THE SAN ANTONIO ROAD RUNNERS: During my tour at the NRD, running had become very popular. I was now running close to 35 miles per week with plenty of free time to do it. I joined the San Antonio Road Runners and began to participate in competitive events. Most of the large events consisted of 10,000 meters or 6.21 miles. Al Becken, an accomplished distance runner, taught me how to be more efficient with my running style. Initially, I struggled with the longer distance, but still had a lot of fun because I knew that wherever I was in a pack of runners with 400 meters to the finish, nobody was going to pass me.

I eventually shaved my 10,000-meter time down to a best of 36 minutes and 50 seconds, which equates to just under six minutes-per-mile. There were over two thousand runners in that race. The first mile was run on a paved road. We then had to cross a footbridge basically single-file before continuing a cross-county course through a park for four plus miles. We then re-crossed that same bridge prior to finishing up with a mile of paved road. The first 100 folks to finish were awarded this awesome beer mug and all the beer you could drink from a keg (of course this was going to elicit some bad behavior on my part).

My strategy was to run the first mile as fast as possible to get to that bridge where we knew there would be a backup to get across. I did that, running what was probably about a 5 minute 15 second mile, way too fast to sustain, but perfect to arrive in time to make a quick crossing of the footbridge. I struggled through the cross-country portion and re-crossed the bridge. Just on the far side of the bridge, a fellow was counting off what place we were in. As I passed by, he said, "You are number 82." Yes! Now I just must hang on for one mile and a beer mug is mine. That was not easy, since many better runners than me had been held back at the bridge and were now picking up the pace to try get into the top 100. Somehow, I managed to hold-off enough of them to finish number 93! The bad news is that I came home before noon on a Saturday morning, legally intoxicated. *That race and beer mug cost me a couple of nights on the couch.*

***

RECRUITING COMPETITION STRATEGY: Recruiting has an ugly word associated with it. That word is "quota." Every NRD had a quota of new recruits to sign up each month. Also, every NRD was then ranked initially within their region, and eventually against every NRD in the entire nation. To reach the overall Navy goals, the competition was weighted toward the more important and difficult-to-recruit programs. We took a very close look at those programs to decide which ones we should emphasize, and of course, which ones we could find recruits for.

What we discovered was that NROTC two-year and four-year scholarship programs were the most heavily weighted in the competition. I then convinced CDR Stacy Clardy, the NRD commanding officer, to offer all newly graduated and recently commissioned Ensigns from the San Antonio area who were in a "pool" status awaiting aviation, surface warfare, or nuclear submarine training, six weeks of recruiting duty in their hometown.

We eventually ended up with four of them. Knowing that they were newly commissioned and very fired up about the Navy, I then sent them to the high schools in the area to obtain application packages. The results were spectacular, as we went from last of the eight NRD offices in our region, to first in officer recruiting and ended up number one in the nation in NROTC recruiting. *These recruiting quotas took most of the fun out of the job, and frankly, tempted us to push forward some marginally qualified recruits near the end of each month. We never did that, but it was stressful in any case.*

***

Use of the T-34: Our bird, when we were in San Antonio, was in a Confederate Air Force-owned and operated hangar less than five minutes from our office (in 2001 they changed their name to Commemorative Air Force which is a more accurate description of the organization). There was also a gap in the San Antonio International air space that we could use for aerobatics, which was only 5 miles northwest of the airport, so we flew as often as we desired, which resulted in me getting almost 400-hours of T-34 flight time in my two and one-half years as a recruiter. And most of these were very short flights.

Officially, we were only supposed to fly aviation-related applicants. Fred and I decided that this aircraft was our most effective recruiting tool for any and all programs. We even went well beyond that by flying people who we felt would help us with our recruiting efforts. This included secretaries that could get us the best interview rooms and spots on campus to set up our recruiting booths. The other service recruiters could never figure out why we were so well treated, especially on the University of Texas campus. We also used the familiarization flights to discourage aviation applicants that we felt had no chance of making it through AOCS.

We had one case where the applicant scored the maximum on the Academic Qualification Test/Flight Aptitude Rating (AQT/FAR) exam. The minimum requirement to even process an application was a 5/5. This young man scored a perfect 9/9. However, he was totally out of shape, had a nasty personality, and would never have made it through the first day of AOCS. Fred and I had a great system set up for our flights. While one of us briefed and prepared the applicant for the flight and strapped him into the back seat, the other one would fly him. In this case, Fred was the pilot and I was the briefer.

We had already decided that if this kid wanted to apply, we would have to at least forward his application. So, our opportunity to discourage him going to be the flight itself. I strapped him in rather loosely, gave him an air-sickness bag, checked to make sure he had aircraft intercommunication (ICS) with Fred, and they were off. Thirty minutes later, Fred pulls up to the Fixed Base Operation (FBO) that we were operating out of and there is nobody in the backseat! Holy crap! I thought. Did he fall out or jump out?

Taking a closer look and noticing that Fred was smiling broadly, I approached the back cockpit. There he was, slumped down into a corner of the cockpit, white as a sheet (not sick, thank God) and obviously scared to death.

I said, "Did you enjoy the flight?" His answer, "I don't ever want to get near an airplane ever again! You guys are all crazy!" Fred had spent the entire 30 minutes either upside down, pulling Gs, in a stall-spin, or doing aileron rolls. The kid had not said a word because his ICS became disconnected during the first stall/spin. Since he was not complaining, Fred assumed that he might be enjoying the experience. *I know, we were not very nice to this guy, but we saved the Navy some time and money by getting rid of him early.*

***

FAMILY EVENTS IN SAN ANTONIO: San Antonio had several major Air Force and Army facilities, but absolutely nothing Navy other than our recruiting office. Commander Clardy was not that into social events, so he would often delegate various commitments to the other officers on his staff. Barbara and I would jump at these chances since the other service representatives were all General Officers and I had just made LCDR. During one of these gatherings a local theater group performed a singing skit, and they were awesome!

Later, we got to know a couple of them while attending an outdoor theater presentation, and again thoroughly enjoyed it. Although there was no active-duty Navy presence outside of our office, what did exist was a very vibrant retired Navy community. These people loved to have a good time, and they sort-of adopted Barb and me into their group. They set up a Navy Birthday celebration and asked Barb if she knew of any entertainment that they might hire. Bingo! She got hold of the theater group and they agreed to perform *Pensacola*. These kids were terrific. *Barb was about eight months pregnant with our third child at the time, and every one of the retired wives treated her like it was going to be their grandchild. Since both of our families were a great distance away, it was a wonderful experience for us to have a Navy family, albeit a bit older.*

***

SALLY AMANDA RUD JOINS THE FAMILY: On March 20, 1978, our third child, Sally Amanda was born. Since there were some complications, the obstetrician at Fort Sam Houston Army hospital decided to perform a cesarean section delivery. Unfortunately, that meant that unlike the birth of my son, Ryan, I was not allowed into the delivery room. When I was allowed to see Barbara, I walked by a nurse who was tenderly carrying this absolutely beautiful, dark-haired baby, who appeared to be a month or so old. The nurse smiled at me and inquired, "Are you Gil?" "Yes, I am." She held this beautiful baby so I could see better and said, "Meet your newborn and very healthy daughter!"

Both of my other children were redheads at birth so I just assumed Sally would be too. She was not, and the fact that it was a cesarean delivery, she had not gone through the usual trauma, so she looked older than the few minutes she really was. What a blessing. We were now a family with three healthy children, who fortunately would all inherit their mother's good looks and superior intellect.

***

SOUTHWEST AIRLINES: Our regional headquarters were in Dallas, just a few blocks from Love Field. Southwest was just getting started, but they flew to Love Field versus Dallas/Fort Worth (DFW) International Airport, which was much more convenient for us. Their rival was Braniff Airlines, a much more traditional airline that was also more expensive than Southwest and only flew into DFW. I will never forget showing up in the morning at San Antonio International to a big table full of free donuts and other delicious pastries, laid out for those of us riding the daily commuter flight to Dallas Love Field. The flight attendants were decked out in "hot pants" and boots and appeared to all be selectees from a beauty contest. It was only a 50-minute flight to Dallas. We would conduct business and return on the afternoon commuter flight back to San Antonio. The difference on this flight was that it included the "drink chug-a-lug" option. As I recall, if you purchased a beer, they would continue to refill it for the entire 50 minutes. *No wonder so many naval aviators opted to fly for Southwest. In those days, it really was an awesome job.*

***

VIP PROGRAM: To help reduce the attrition rate in Aviation Officer Candidate School (AOCS), a program was established to make certain that a potential selectee would be fully aware of what to expect, and most importantly, what kind of physical condition he or she should be in, before beginning the training. Once a candidate had officially been selected, they were grouped together with all the other selectees from across the nation. The Navy then paid for them to travel to Pensacola where they would be met by two Navy pilot recruiters. Those two pilots would lead them through a cross-country run, and the infamous obstacle course, both of which they would eventually be required to complete within a minimum time requirement. They were also exposed to the swimming program, and finally, they met a real-live drill instructor (DI).

I was assigned as one of the two pilots for a VIP group. I got to Pensacola a day early and headed for one of the gyms to get in a workout. When I arrived, there was a "five-on-five" basketball game in progress, with four guys waiting to play the winner. They asked if I would join them, so I did. Although not much of a basketball player, I was in great shape, and I had played enough high school-ball to hold my own. It did not take long to realize that we had a superstar on our team. He was a deadly shooter, so I just started setting screens for him. It was weird because he looked so familiar, and his style reminded me of a kid I played against in high school.

We quickly dispensed with the first team we played. In between games, I asked the superstar what he was doing in Pensacola. He replied, "I am a P-3 pilot, and a Navy recruiter from NRD Kansas City, and I am here to escort a VIP group." "What a coincidence, I am the other recruiter assigned to the VIP group." I then asked him if he had played college basketball. "I sure did. I played for the U.S. Naval Academy." "You look so familiar, but I never watched a Naval Academy game so that can't be it. Where are you from?" He replied, "I am from Cooperstown, North Dakota."

Aha! His name was Gary Bakken, and he set the record for most points scored in a Class B State Tournament game. I think it was 52. We had played against each other. He was only a freshman when I was a senior, but I remember how impressive he was even at the age of 14. Gary was a great guy, but he was a bit on the cocky side, especially when it came to basketball. When I reminded him of the 52-point effort he replied, "I would have scored a heck of a lot more than 52 if the three-point shot had been in effect back then."

The next day, we set up the cross country-run and the obstacle course for the 20 or so VIP folks. I led them on the cross country, and Gary led them through the obstacle course. None of them were able to keep up with us so they were very impressed by a couple of old guy pilots (I was 33) kicking their college butts. We then set up a meeting with a DI who was one scary dude. Even though I was now a LCDR, I found myself snapping to attention when he entered the room. He smiled broadly and said, "Mr. Rud, I can tell that you are a graduate of AOCS." He then explained the program, what role the DI would play, and ended with this comment: "You better want to be a pilot really bad, and you better get your sorry asses in shape, or you will never make it through this program!" *We found that the VIP experience was very effective because it got rid of the folks who were shocked by how difficult it was going to be before the Navy spent any more money on them. It also motivated the others to get into better physical condition prior to beginning the training.*

18. Academic Meathead Goes to School

Seminar group from the Armed Forces Staff College Class. I am fourth from the right in the back row.

ONE OF THE SMARTER THINGS that I had done when I accepted orders to the recruiting command was to negotiate an assignment to a mid-grade service school prior to going back to the fleet. There were several choices including the prestigious Naval War College. Being a mediocre academic, I chose what my friends all referred to as the easiest option. The Armed Forces Staff College, which was (and still is) located in Norfolk, Virginia. The course is six months long, versus a full year for most of the others.

I checked in as a member of Class 66. There were a total of 257 of us broken down by service to 95 Army, 72 Air Force, 35 Navy, 18 Marines, two Coast Guard, 17 Civilians and 16 Allied Officers. The school was designed by General Eisenhower and Admiral Nimitz to turn out the best staff officers in the world. The first class started in 1947, so with two classes completed each year, in August of 1979 we began our studies as the 66th class. We were divided up into 14 Seminars with about 18 students in each. I was assigned to Seminar 12, which consisted of six Air Force, six Army, two Navy, one Marine, one Coast Guard, one German Air Force and one civilian.

What a great experience this school was. I learned so much about the other services, and most importantly, I learned how to work with them in a joint environment. It was also a family affair. The compound that the school was in was very secure and patrolled by Marine guards. We could literally allow our children to roam freely about the grounds without worrying about their safety.

The Rud family enjoying life on fertility row at the Armed Forces Staff College in 1979.

Barb and I were assigned to quarters on what was known as "fertility row." It contained the larger quarters that were only available to families with three or more children. The wives were invited to hear all the unclassified guest speakers, as well as other socially stimulating events. *It really was a little paradise, especially for those of us who regularly deployed away from our loved ones.*

To enhance the experience, each service was granted a week in which we would have a chance to show off, so to speak. With Naval Station Norfolk located within a few blocks, Jim Sullivan (who was a Surface Warfare Officer) and I arranged tours of ships including an aircraft carrier. For the Army and Air Force students, it was a very competitive experience as they were rated against one another for class standing and a fitness report. For the Navy and Marine Corps students it was a six-month break from the grind as we were in a "not observed" fitness report status. Because we had no pressure, we performed better in a lot of areas.

In addition to the normal course load, we had a chance to pick an elective subject taught by professors from Old Dominion University. I chose creative writing (also known as "bonehead English"). From the first day of the class, the goal of the professor was to get us away from the normal military writing style. She succeeded, since this one class changed me from what I would call a struggling, paperwork-avoiding meathead into a pretty decent writer. *I would end up working for a commanding officer with a degree in English from Brown University, so this added skill set was a key to my success in my next sea tour and eventual selection for operational command.*

Sports competition was a huge factor in the rivalries between the Seminars. Probably the most intense was in "slow pitch" softball. We had a very good team, with several of us having high school and a bit of college background in baseball. One afternoon, we were matched up against Seminar 14, playing on a field with no fences. I was playing rover, a position normally set up between left and center field to pick up the balls hit over the shortstop's head.

Fortunately, I recognized the "cleanup hitter" from their team. Not from my military contacts, but rather from my high school/college days. He was Marine Major Emil "Buck" Bedard, who was a standout football and baseball player at Mayville State University near my hometown. As he came up to the plate, I headed out to deep—and I mean really deep—left-center field. Sure enough, he hit this monstrous fly ball that would have been a home run on any field with a fence, but now became a routine fly ball.

I did a similar thing on his next at-bat before he got smart and stopped trying to go for the "fence." After the game I walked over to him and said, "Go *Comets!*" (*Comets* were what Mayville State athletic teams were called). Although he did not remember me, he did remember my best friend, Richard "Fugie" Fugleberg, a teammate of his in college football, so we had a lot of memories to share. *Buck went on to become a three-star General in the Marine Corps. He has continued to be a very active supporter of Mayville State and we have remained friends all these years.*

Although I have lost contact with many of my classmates, they were a great bunch of guys. Most of them were Vietnam veterans, with Air Force fighter pilot, Major Bill Scherwertferger (call sign "Shortfinger") having been shot down and serving time as a POW. Major Spurgeon Ambrose was an Army combat vet, who related an experience he had while home on leave from the war. He was attending a movie, and in those days, they would play the *Star-Spangled Banner* prior to the movie. He stood up and placed his hand over his heart and no one else in the audience did. That is until he stepped up on the stage in front of the movie screen and threatened to kick the shit out of anyone who did not stand. They stood! *Spurgeon had an interesting perspective on Air Force and Navy pilots. "It must be a lot easier to get the respect of your enlisted men, when instead of telling them to go take a hill while getting shot at, you are the warrior that goes in harm's way."*

***

GRADUATION: We had many wonderful speakers that addressed us throughout the curriculum. I must say though that the one we were most excited about was our graduation speaker. Senator John Warner from Virginia was scheduled for that special occasion, but in all honesty, the excitement was all about his beautiful wife, Elizabeth Taylor. Guys our age had all grown up worshiping this lady so, albeit a bit of a shallow attitude, we continued to be fired up until the Senator showed up sans Elizabeth. We managed to swallow our disappointment, received our diplomas and headed for our next duty assignments.

19. Back in the Cockpit

WHILE WE WERE IN SAN ANTONIO, the housing market in Lemoore convinced us to sell our place on Mulberry Lane. We did that at a reasonable profit, which we immediately turned into two new cars that we paid cash for. They included a 1978 Volkswagen Jetta and a 1978 Dodge stretch-cab pickup truck. So now we needed to get back into the housing market. We did so by purchasing a four-bedroom house at 501 East Hazelwood on the north side of Lemoore, very close to the Meadow Lane Grade School. As I recall, we paid $86,000 for it, setting up payments under a 14% loan, which basically meant that most of the monthly payment went to interest. Oh well, it was a great house, and perfect for our family.

Shortly after moving in, we decided to put a hot tub in the rather-confined back yard. During that installation process, two workers got into a fight. One hit the other over the head with a block of wood, almost killing him. The badly injured victim approached Barb about getting some help, scaring the crap out of her with his badly bleeding head, while his attacker took off. Of course, like everything else that happens to Navy wives, I was not home. The victim ended up with a long hospital stay and the attacker with a long jail sentence. The good news was that the company quickly finished our hot tub, which we thoroughly enjoyed over the course of our stay in this house. Barb returned to a full-time teaching position and the kids settled into what is probably the best Navy town to raise a family.

Since it had now been three years since I had flown the A-7E, I checked into the RAG, VA-122, for a category-2 syllabus, which is just a little shorter than the normal training evolution. It was an absolute blast. I was so happy to get back into the tactical jet community that I really studied hard, prepared for all the flights and did exceptionally well. I also took the opportunity to mentor the category-1 nuggets, especially when it came to weapons deployment and carrier landings.

***

EARNING A PERMANENT CALL SIGN: Unlike some naval aviators, up to this point in my career, I had not really acquired a call sign that stuck. Initially in the training command it started as "Farmer," then in the fleet it progressed to "Rufus," then "Norwegian Two," and finally, "Rudy." With a rather unique first name like Gil, most of my squadron mates simply used that moniker instead of a call sign. That is until a fateful Saturday as a RAG student, practice bombing on B-16 at Fallon.

Since it was a weekend, the spotting tower that called our bomb hits was not manned. I am not positive of who my wingmen were, but I believe it was Jack "Jocko" Chenevey and Bob Sprigg. All three of us were experienced veterans heading to department head tours, so the instructor just sent us out to the range and told us to call each other's hits. For whatever reason, the shotgun-type shells that set off the smoke charges in our MK-76 practice bombs were not working properly. Normally we could easily spot the hits of the aircraft in front of us in the pattern and relay that information over the radio. Without the smoke charges, however, each pilot would drop his bomb, pull his nose up above the horizon, roll inverted and spot his own hit. Since there was no smoke involved, the hit was called a "duster."

I was number three, so the chatter in the pattern sounded something like: *Three is in hot, three is off safe, my hit 50 feet, 6 o'clock, Duster.* We continued until we had completed twelve bomb runs and then headed back to Fallon to debrief. Of course, there was money on the outcome with the bet being that the pilot with the best circular error probable (CEP) (overall closest to the bull's-eye) would get his first two beers paid-for by the losers. My called "Dusters" were quite obviously superior to the other guys so after a quick debrief, we headed to the club for my richly deserved free beers.

On Monday morning, the three of us were again briefing for a flight, but this time, together with an instructor. Just prior to the instructor starting the briefing, the squadron Ordnance Chief entered the ready room, walked up to us and—obviously a bit distressed—announced, "Mr. Rud, my sincere apologies for not being able to get any bombs on your aircraft on Saturday!" And so… from then on, I was "Duster." Some folks think it was because I was a crop duster, but I was only a "wanna-be" crop duster and most likely would have killed myself in pursuit of that title. Luckily, I was drafted into a little less dangerous (at least better training) career as a naval aviator now with the call sign "Duster."

VA-122 had some great flight instructors that accompanied us on the weapons detachment. My fading memory recalls "Shifty" Peairs and "Smooth Dog" Vaughn as being two of the more colorful ones. They both went on to very successful careers, but on this detachment, they were still hell raising LTs. Now that I have shared the level of competition and how it resulted in my call sign, here is another one.

There really is nothing more embarrassing than forgetting to arm your system for a practice bomb run. Smooth Dog was leading a practice "nuclear-lay-down" bombing pattern at B-16. This flight was great fun as we were armed with Mk-106, "beer can" bombs, which simulated a retarded (parachute-equipped) nuclear device. The run-in was 450 knots at 100 feet. As soon as your bombsight passed over the target you pressed the pickle to release the Mk-106. The hits were generally very close to the target. I was following Smooth Dog and noticed that on his first run a bomb did not come off his aircraft. As we were on the downwind leg setting up for a second run, I spotted a puff of smoke in the desert. Aha! Smooth Dog was getting rid of his Mk-106 because he forgot the master arm (switch that allows ordnance to be dropped) on his first run.

I did not say anything in the debriefing, and we finished up the weapons detachment. As if we needed any excuses to party, the tradition of all weapons detachments was to have a class party to celebrate the successful completion of the event. Barb and I hosted this party. I had a MK-106 ashtray in my garage, which included the top-half of the bomb with the readily distinguishable fins. Barb constructed a dummy cowboy. We laid him on his belly in the yard, with the Mk-106 firmly implanted in his back. As the guests arrived, they were mostly incredulous, until Smooth Dog showed up. He immediately turned red as a beet and said, "Darn it! You saw me drop that 106 on the downwind." *Naval aviators don't really cheat, we just bend the rules a bit to make sure we do not lose a bet.*

***

JULY 21, 1980, CARRIER QUALIFYING WITHOUT A HEADS–UP DISPLAY: Our carrier qualification class was a bit extraordinary in that there were several of us that were experienced and were on our way to department head assignments. The LSOs were "Gib" Godwin and "Gyro" Knight, and the ship was once again USS *Lexington* (CV 16).

Now that the A-7E and its ground-stabilized Heads-Up Display (HUD) had been around for several years, pilots had pretty much become addicted to using it. We were now so dependent on it that not having it working was almost considered an emergency, especially at night. To ensure a full navigation system and working HUD, the A-7E needed to get a full alignment. This was no problem ashore, and it was also accomplished aboard the aircraft carrier, using their Ship's Inertial Navigation System (SINS). Unfortunately, the venerable old *Lexington* was not equipped with SINS. No problem though, because we just made sure to get alignments ashore and then hot refuel aboard the ship. If you did not shut down the aircraft, you maintained your full system, including the HUD.

To ensure full systems for night operations, the plan was that after day operations, the nuggets would fly the aircraft back to Pensacola. They would then start them up, get a full alignment with a functioning HUD, fly out to the ship and get three night landings, then we more-experienced folks would hot-seat and get all six of our night traps.

It was a good plan and we old guys soon headed up to flight deck control to prepare for the hot seats. Suddenly, a wicked thunderstorm hit. The Air Boss decided that he couldn't hot refuel in a thunderstorm, so instead of keeping the aircraft turning until the weather passed by, he ordered them all shut down!

Shit! Now the best we could do was to get a basic alignment with no chance for a working HUD. This was like going back to flying A-4s or A-7A/Bs. I was fortunate enough to have flown those, but some of the other guys had been brought up on the A-7E. Following launch, instead of turning downwind directly into the pattern, we were all sent to marshal (designated area to hold prior to commencing an approach to the ship) until the weather improved.

During this period, one of our would-be department heads literally disappeared for 20 minutes in a vain effort to achieve an airborne alignment.

It was déjà vu for me: back to Vietnam and *Oriskany*. I absolutely worked my ass off to get six night traps, even putting the gun sight on dim at zero mils as sort of a "HUD pacifier." *Pilots of other non-HUD equipped aircraft may scoff at this situation, but there is not a single A-7E guy who will read this and not sympathize.*

20. The VA-147 Argonauts and Some Great Flying in the Philippines

ORIGINALLY, I HAD ORDERS to be a Department Head in VA-27. However, all of us are acutely aware that the needs of the Navy will ultimately determine where we end up. Although we were not engaged in a shooting war, due to the aggressive and realistic training flights, Light Attack Aviation was still a risky business. Just prior to my completion of the RAG, LCDR Ben Tappen, a department head in VA-147 was killed while participating in an exercise in the Philippines.

VA-147, the *Argonauts* had been off-loaded as part of the change in *Carrier Air Wing Nine's* complement for an extended Indian Ocean deployment of USS *Constellation* (CV 64). In the past, air wings had off-loaded a mixture of all types of aircraft, which resulted in a very inefficient/ineffective beach detachment. By off-loading an entire squadron versus the mixed bag, the beach detachment was now being run by the Commanding Officer of VA-147, CDR David "Huey" Le Herault. CDR Frank "Tank" Bledsoe was his XO. Tank and I had been squadron mates in two previous squadrons, so I think he convinced Huey that I would be a good fit to replace Ben.

This was a very different situation than being deployed aboard the ship. The good news was that the squadron was filled with great guys, who all loved to fly. And the flying was, by far, the most tactically challenging and enjoyable that I ever experienced outside of actual combat.

The squadron received the same amount of operational flying budget as our sister squadron, VA-146 did on the ship. This meant that we had enough money to fly up to almost 50 hours per pilot, per month. And fly we did! Surrounded by open air space, realistic targets and adversary aircraft from both VC-5 and Air Force squadrons stationed at nearby Clark Field. We were, without doubt, the most tactically proficient A-7 squadron in the fleet. Even during the monsoon season that produced heavy rains and even the occasional typhoon, we continued to fly. We found that despite the cloud layers and rain, it was normally VFR with good visibility below the clouds.

***

MY INITIAL DEPARTMENT HEAD ASSIGNMENT: I joined Russ York, Rich Nibe and Dave Wallace as the fourth department head. As the new guy, the skipper assigned me to be the Safety Officer. He also gave me another job, which was the Human Resources Officer. The squadron had failed a recent Human Resources inspection and frankly, we were a bit under the gun to get prepared for another one. Since I had served in this position on the CAG-14 staff, it was the perfect fit. We cleaned up our act and passed the inspection with flying colors. *Although I did not realize it at the time, the skipper deeply appreciated this seemingly innocuous accomplishment and generously rewarded me for my effort.*

The deployment was about half-over when I arrived, so it was not long before *Constellation* came back to the South China Sea and we flew back aboard. Our JOs struggled a bit with the carrier-landing phase, but soon they were back in the groove. What was really fun, though, was to fly against our sister squadron and even the F-14 squadrons in air-to-air tactics. We were incredibly proficient, to the point of embarrassing a couple of arrogant F-14 nuggets.

Doug Ashley, who although just a nugget, was by far the best air-to-air tactics A-7 driver I ever flew with, or against. The kid was an energy management genius and kicked my ass every time we went head-to-head. Obviously, we could not compete with experienced *Tomcat* drivers with their tremendous performance advantage, but an over-confident JO was a different story.

I recall coming back from one of these sorties where Doug and I had gotten in position for gun runs on a *Tomcat* JO. As they were the fighter experts, they always led off the post-flight debrief. I remember who this was, but as I have throughout this publication, I will, leave out the name to protect the guilty. He initially claimed victory via a "Fox One" with a radar-guided missile, and totally left out the rest of the engagement. This is when his Radar Intercept Officer (RIO) chimed-in and said, "Duster, I am not sure how you did it, but you guys kicked our ass!" *I might add here that by this time, I was an avid runner. In fact, there was only one person in the squadron, either officer or enlisted, that could beat me in the 1.5 mile-fitness test. You guessed it, Doug Ashley.*

***

TIGER CRUISE: This wonderful opportunity to bring a relative aboard in Pearl Harbor to ride the ship from there to San Diego is continued to this day. My son Ryan was just a smidge under the minimum age of 8, but he was a tough little guy who was totally up and ready for the adventure. Since I had only made half of the deployment, I volunteered to ride the ship into port, versus flying off. I then took over responsibility for the other squadron boys whose fathers had flown off the day prior to our arrival in San Diego. It was a wonderful bonding experience that I will never forget.

***

AIR SHOW FOR PRESIDENT REAGAN: During our turnaround prior to my second deployment with VA-147, *Constellation* and *Carrier Air Wing Nine* were selected to host a visit from President Reagan. The visit was to take place while we were at sea near San Diego. It was scheduled for August 20, 1981. Our Air Wing Commander was CDR Bill Newman, who had just come from a tour as the Flight Leader of the *Blue Angels,* so he knew how to put on an aggressive, but safe event.

My role in the show, besides being a member of the A-7 diamond formation, was to perform a John Wayne-style gun firing demonstration. Normally, when we fire our guns in combat, we do so with several short bursts. For airshow purposes only, I was to continue the strafing run, pulling the nose of the aircraft along the run-in line for maximum distance and effect. We practiced the show several times with CAG Newman in the tower directing the event until we had perfected it.

On the other side of the world, on August 19, the day before our scheduled airshow, two F-14 *Tomcats* engaged in a dogfight with two Libyan SU-22 fighters. The final score, not surprisingly, was *Tomcats* 2, Libya 0. "Oh no," we all thought. Now the President will cancel his visit to *Constellation.* He did not. In fact, he used his visit as a platform to show Libya and any other potential adversaries that they had better not mess with his boys.

We had manned up all the participating aircraft, which were set up on the flight deck ready to start up, when the Presidential helicopter, and a second one with press on board, landed. This is something I will never forget. We were all so darn proud to be hosting our Commander in Chief. Yet, as often happens in these events, a beautiful reporter getting out of the second helicopter stole the show from the President. How? Because when she disembarked from the helicopter, the combination of rotor wash and wind on the flight deck blew her red dress up over her head, much to the delight of all of us sitting in the cockpits of our jets. And yes, there were wolf whistles and other terribly politically incorrect comments prior to CAG Newman calling for some communication discipline.

The air show went perfectly with perhaps one small hiccup. During a *Tomcat* supersonic pass close aboard the viewing area on the flight deck, the President's knees buckled a bit when the sound wave hit the ship. The Admiral quickly grabbed his arm and related that the President had just looked at him, smiled, and said something like, "*Shit hot!*"

***

SECOND DEPLOYMENT IN VA-147: During the workups, I had moved from Safety to Administrative Officer. We had a change of command so that Tank Bledsoe was now the commanding officer and Eric Vanderpool became his XO. Normally, we would have swapped aircraft with VA-146 (whichever squadron was on the ship had the newer Forward Looking Infrared Pod (FLIR) capable aircraft), and they would have become the Beach Detachment in the Philippines (P.I.). For whatever reason, this swap did not occur and we became the first back-to-back Beach Detachment squadron.

VA-147 Argonauts *on second beach detachment in 1982. I am third from the left.*

Once again, it afforded a wonderful opportunity to do meaningful training daily versus boring holes in the sky in the Indian Ocean. The bad thing was that the junior pilots did not accumulate enough carrier landings to be competitive with their counterparts for future assignments.

Shortly after off-loading, I became the squadron Operations Officer. This to me was the ideal job, playing to most of my leadership strengths. On the first cruise, we had to contend with the monsoonal wet season. This time we were blessed with the opposite, and made use of great weather to participate in several joint exercises with the Air Force and Marine Corps. We were also able to build flight time with cross-country flights to Okinawa, mainland Japan and Korea.

I planned and executed a flight to Osan AFB just outside of Seoul, Korea. I do not recall whom my wingman was, but our mission was to buy athletic shoes for anybody in the squadron who wanted them. We got all the sizes (many wanted them for their children), wrote them down, hung a blivet (a large wing gas tank reworked to carry cargo) on my aircraft and took off. We had to stop for fuel in Okinawa, so on the way back we were prepared to pay duty on our purchases.

We went to a shoe factory and literally sifted through what they referred to as the Blemish Department. This included shoes that had some minor defect that would not be accepted by a brand name retailer. The good news was that if we bought 1,000 pair of shoes, they would charge us only $2 per pair. We bought 1,000 pair of shoes, which filled the blivet.

As I mentioned earlier, we did not sweat the Philippines charging us duty, but we did expect it from Japan when we stopped for gas in Okinawa. Sure enough, the customs official met us at the aircraft as we began to refuel. He asked me if I had anything to declare. I said, "Yes sir, I have 1,000 pairs of shoes in that gas tank hanging on my wing." He looked at me, grinned broadly and said, "Very funny. You have 1,000 pairs of shoes in your gas tank?" He then broke into a raucous laugh, slapping his thigh as he headed back to his office. "Crazy Americans! 1,000 pairs of shoes in the gas tank, ha, ha, ha!" So, we successfully outfitted the entire squadron and most of their children for $2,000.

***

COPE THUNDER EXERCISE: This exercise involved the Navy, Air Force and Marine Corps. Navy participants from Cubi Point included VC-5 flying adversary

A-4s and VA-147 with A-7Es. It also included A-7E, A-6E, EA-6B, KA-6 tankers, F-4s and E-2s from CAG-5, which was the *Midway* air wing. *Midway* was undergoing some maintenance in their home-port of Yokosuka, Japan, so part of the air wing deployed to Cubi Point to stay current. A Marine F-4 squadron from MCAS Iwakuni, Japan also deployed to NAS Cubi Point for the exercise. In other words, we had a shit load of aircraft flying single runway operations from the same location.

The Air Force seemed to have participants from all over the Pacific including F-4s, brand new F-15 *Eagles*, F-5 adversaries and even an E-3 AWACS to pretty much run the war for them. All the Air Force aircraft were flying out of their own single runway environment at Clark AFB.

How do you accomplish this event safely? We did it by operating the Naval Air Station exactly like an aircraft carrier. We developed a daily air plan that included takeoff and landing windows. If you did not make your take off time frame, you were scrubbed to ensure that the folks scheduled to land could do so on time. We also had a KA-6 tanker airborne filling the role of a recovery tanker. We really did not expect anyone to run low on fuel with the missions we were flying, however, we wanted to make sure that we had emergency fuel in the air in case the runway became closed due to a blown tire or some other landing incident. We really needed to keep our operations Blue Water (no bingo field), because the only bingo field was Clark AFB and, as mentioned earlier, they were as saturated as we were with all sorts of participating Air Force units.

The basic idea of the exercise was to divide into a Red and a Blue Force. This was not necessarily adhered to for the entire exercise, but the point was to engage with the enemy. Very basically, I would describe it as the Air Force protecting Clark AFB from an attack by the Navy and Marine Corps. Rather than actually attacking the base, we flew sorties against remote target areas in the Crow Valley Range that were defended by the Air Force.

Of course, we had rules of engagement, but they were mostly safety related and, quite frankly, often ignored. The result was some of the most aggressive and realistic training I ever experienced. And of course, there was a great deal of pride on the line.

The Air Force E-3 AWACS was effective in giving the Blue Force defending Clark AFB accurate information as to where we were, how many of us there were and other vital information. It was a huge advantage for them. We tried all sorts of tactics to try mask our intentions, to no avail.

That is until the EA-6B arrived from CAG-5. Their senior Electronic Countermeasures Officer (ECMO) asked if we had a good mission for them. I replied that anything they could do to help us confuse the AWACS would be helpful. They took off as part of a large strike force, basically on their own to see if they could help us out. The strike was much more successful than our previous attempts with several attackers getting through the Blue Force fighter screen.

Shortly after we landed, I got a call from a full bird Air Force Colonel. He was furious to the point where I had a hard time understanding what he was saying. "Colonel, with all due respect, would you please calm down so I can understand what the problem is?" "Rud, you asshole, your EA-6B followed my E-3 AWACS around for the entire evolution and they not only jammed, but literally fried our multi-million-dollar radar!" *Of course, the Rules of Engagement were immediately revised for them to regain the use of the E-3, but we sure as hell proved that a real enemy could shut their arrogant butts down.*

Some of the events were purposely designed to instigate a full-blown air-to-air engagement between the forces. These "fur balls" produced some incredible duels, especially between the VC-5 A-4 adversary-trained pilots and the Air Force F-4 fighter pilots. I was the strike leader for a large-scale attack force protected by VC-5 A-4s. During our ingress to the Crow Valley target area, a giant fur ball erupted over a little coastal village called Iba. Radio discipline was quickly lost with all sorts of shots called, etc. Suddenly, this wonderfully sweet female voice said, "Fox Two on the F-4 in a right-hand turn at 10,000 feet over Iba." Followed shortly from the obviously female aviator, "Guns! Guns on the F-4 now at 8,000 feet over Iba." With this very specific description of the encounter, there was no doubt as to the identity of the lady's victim.

As the strike leader, immediately following the event, I landed at Clark versus Cubi so that I could represent the Red Force at the mission lessons-learned session. Arriving in the auditorium that was large enough to accommodate all the Blue Force participants in the recent event, I was immediately accosted by questions as to the identity of the lady fighter pilot that had just shot down the F-4.

Although I had no idea who the lady was (I believe it turned out to be the VC-5 Aviation Intelligence Officer riding in the back seat of a TA-4 with the squadron CO), I addressed the group with the following statement: "Boys, I recommend that you put a suicide watch on that F-4 crew because the lady who shot them down is a drop dead gorgeous LTJG on her first flight in an A-4 with VC-5. If you put that same crew up on four more events, she will become an ace in her first week in the squadron."

The whole auditorium erupted in laughter and applause. *As Paul Harvey used to say, "… and now for the rest of the story." LTJG Crystal Lewis, who fits my made up description perfectly, actually did begin flying with VC-5 a few months after this event. She spent four years battling Air Force fighters prior to getting orders to VT-4 in Pensacola as a flight instructor. VT-4 shared a hangar with the* Blue Angels *where she met and eventually married LT Mike Campbell who was number 8, the Events Coordinator for the team when I was the Flight Leader. She went on to a very successful career as an airline Captain for Delta, and she served as a great role model for my daughter to become a naval aviator. It really is a small world.*

***

Eagle Shot Down by a SLUF (Short Little Ugly Fellow): Toward the end of the deployment, Dave "Magic" Wallace and I had an opportunity to participate in an exercise involving the brand-new F-15 *Eagle*. The two of us flew to Clark AFB, where we briefed with four Air Force F-5 adversary pilots for the mission. The mission was to provide opposition for a 4-ship of *Eagles* that would be flying out of Okinawa. The total distance to engage was just short of 900 miles. This was no problem for the A-7 (we would land at Okinawa to refuel after the mission), but the F-5s were very short range, so the Air Force provided them with a KC-135 tanker.

During the briefing, we had written down all the F-5 pilots' call signs. Just an hour or so into the flight, they lined up for their first refueling, only to discover that the tanker was sour (unable to transfer fuel). The F-5s had no choice but to cancel their participation in the mission and return to Clark AFB. Their flight leader recommended that we continue with the mission, and to check in using all four of their call signs, plus our own. He said that this would cause great confusion, since the F-15s will be looking for six targets.

We did exactly that. Also, I climbed to 43,000 feet, which was pretty much as high as the A-7 could climb. It was high enough to create a vapor trail, while Magic descended to wave-top level. As soon as they reported a radar lock on me, I rolled inverted and pulled to 90-degrees nose down, and headed for the water. By doing this move, I showed no Doppler closure on them (more confusion). Also, throughout the evolution, we were making all sorts of bogus calls changing our voices and using the F-5 call signs. The greatest part of all these moves is that Magic popped up behind one of the *Eagles* and got a valid Fox Two shot. Of course, over time, these magnificent machines were able to dispatch the two lowly *Corsair IIs*, but not without significant embarrassment.

Magic and I landed and headed to the after-mission briefing. We were met by what turned out to be an instructor pilot. He was one of the four *Eagle* pilots; however, he was in more of an observer role as the other three were on a graduation flight from an air-to-air weapons syllabus. He was super excited saying, "You were the only two aircraft in this entire event, weren't you?" I said, "Yes, the F-5s had to return to Clark so we just picked up their call signs and pressed on." He replied, "What a great lesson for these guys. They are still looking for the F-5s to land. But the most important lesson was that one of them was shot down by a SLUF! Talk about humiliating!"

I don't think I have ever had so much fun in an after-mission briefing. They thought that the F-5s were married up with us in some sort of masking formation, and of course they did not see Magic approach at 50 feet above the water. Lots of beers were consumed and these three *Eagle* drivers were humbled, but also as the instructor stated, *"You three guys are now a lot better fighter pilots having gone through this experience."*

***

THE GREATEST PILOT IN *CARRIER AIR WING NINE:* Having spent the entire deployment once again in the Philippines, when *Constellation* came back from the Indian Ocean, we, VA-147 *Argonauts*, hosted some of our fellow *Carrier Air Wing Nine* buddies for a party in the town of Olongapo. To get from the base to the town, we crossed a bridge on the Olongapo River. The river was pretty nasty, but that did not prevent local kids from swimming in it. They often did so from small skiffs, showing off their diving skills, singing and whatever else would grab the attention of folks crossing the bridge. Their goal was to get us to toss money to them from the bridge.

LT Craig Thomas, a very colorful pilot with a penchant for showmanship, did some preparatory work with the divers. As we got to the middle of the bridge, he suddenly shouted to the kids, "Who is the greatest pilot in *Air Wing Nine?*" All of them together chant, "Box of Rocks! Box of Rocks is the greatest pilot in *Air Wing Nine!*" That was Craig's call sign. He really was a good pilot and I have no idea where he earned the call sign, but he hit a home run with this little show.

Craig went on to a very successful career as a FedEx pilot and made Captain in the Naval Reserve. Craig and one of our sister squadron pilots, George "Elwood" Dom, often performed a perfect impression of the "Blues Brothers." Elwood eventually served as the Flight Leader of the *Blue Angels* and as a Carrier Air Wing Commander. *You may begin to see a pattern of colorful Junior Officers that lived a bit on the edge ending up with very successful and impressive careers. I worry that the same opportunity for these types of leaders may not be possible in today's Navy.*

PROMOTED AND SELECTED FOR OPERATIONAL COMMAND: Thanks to VA-147 Skippers Huey Le Herault and Tank Bledsoe providing the Military Personnel Command with glowing reports on my performance as a department head (someday I will have to try and live up to all the great things they said about me), I was promoted to Commander (05) and selected on my first look for operational command. Once you are promoted to 05, you are given three windows of opportunity to be selected for command. The first and second looks are considered primary looks, with the third—and last—look more likely to result in selection to head up a training command squadron, which was also a great opportunity.

21. Gold Socks and Command of the VA-192 Golden Dragons

SHORTLY AFTER OUR RETURN to NAS Lemoore from the second VA-147 *Constellation* deployment, as the Operations Officer, I was briefing the Commander, Light Attack Wing Pacific (COMLATWINGPAC), RADM Corky Lennox, on lessons learned from our deployment. This included the status of some of the more-senior department heads. I mentioned that I had been selected for command and expected to be leaving VA-147 for an Executive Officer position in about six months. At this point, RADM Lennox stops the brief and said, "Commander Rud, there has been a slight adjustment to the six-month time frame; you will be taking over as Executive Officer of VA-192 on Monday!"

Two tragic events had recently occurred that prompted this rather immediate action. Two current Commanding Officers and friends of mine, George Garner and Mike Gary were killed in separate aircraft accidents. In at least one case, VA-94, the XO had just gotten to the billet and was not ready to take over as the commanding officer. To fill this leadership gap, Jack Zerr, the current CO of VA-192 was ordered to VA-94 after only a year or so as the CO of VA-192. The good news was that he got two CO tours. The bad news was that instead of a laid-back six months of family time, I was now going to be starting a second consecutive sea tour on Monday, 28 June 1982. Some more great news though—I was going to be working for my old VA-215 squadron, four-man bunkroom roommate, Harry "H.T." Rittenour.

***

THE WORLD-FAMOUS *GOLDEN DRAGONS:* I was incredibly fortunate to get this assignment. The squadron, commonly known as the Super Shit Hot World-Famous *Golden Dragons* (SSHWFGD), really was one of the very best. The maintenance department was the best by a long shot, with a cadre of hard-charging Chief Petty Officers leading some exceptionally talented sailors. H.T. was a natural leader. He was a great pilot, and probably the most efficient manager of resources that I ever worked for in my entire Navy career. Of course, the squadron was also blessed with the perfect blend of experienced department heads and gung-ho junior pilots, all of whom loved to fly, and deeply appreciated the opportunity to be a member of the *Golden Dragons*. The only problem working for H.T. was that he was so darn smart and efficient that I literally had nothing to do as the XO but fly. *Go ahead and throw me in that briar patch.*

***

Maple Flag Exercise at **CFB Cold Lake, Alberta, Canada:** Like the *Cope Thunder Exercises* in the Philippines, *Maple Flag* was a very realistic and effective training exercise. We were fortunate enough to be invited to participate with a few other U.S. units, and virtually all the Canadian Forces fighter aircraft, for a total of 20 squadrons. The Cold Lake base and operating area is located 180 miles northeast of Edmonton, with half of the range in Alberta and the other half in Saskatchewan. It is massive, covering 1.6 million acres of trees, lakes and grass. Despite its remote location, it is not only the largest, but also the busiest air base in Canada. In October of 1982, there were literally no flight restrictions to worry about. You could fly as low or as high as you wanted and you could also fly supersonic.

The missions and the roles we played varied from one event to the next. Obviously, our primary role was to be Blue Force strike aircraft, however, we also played a role as Red Force Adversary fighters. I recall one event where we were assigned to oppose a flight of A-10 *Warthogs*. The Canadian forces had set up some incredibly realistic plywood targets closely resembling tanks, trucks, AAA guns and even SAM sites. If we stuck to practice bombs, these targets would easily last the whole two weeks of the event.

Not so fast, the *Warthogs* from a reserve unit in Syracuse, New York, rolled in with their 30 mm cannons and absolutely shredded all these targets on one mission. These guys had a pattern set up, whereby, they always had at least one bird pointing that nasty gun at us, as we attempted to defend the hapless targets.

Since that did not work, we next turned our attention to attempting to intercept the Canadian F-104 *Starfighters* that were making 700 knot, (supersonic) low level runs, on the target area. I saw one of them a good distance out and rolled in on him. With him having a 200-knot speed advantage, my attack was totally ineffective. "OK boys! We have one more attacker still coming, so let's see if we can get him." It was a B-52 flying a mission all the way from Ellsworth AFB, in South Dakota. He was level at 500 feet and 360 knots; we could see the engine smoke trail from 50 miles, so we took turns making *Sidewinder* and gun runs on this lumbering bomber.

During the post mission briefing, the B-52 representative, who was nothing short of a stand-up comic, related that the crew passed to him that they estimated themselves to have been shot down over 20 times. They still considered their mission a success, because they got some great pictures of their numerous attackers and the moose that they scared out of the marshes.

***

COLD LAKE LIBERTY: We were lucky enough to enjoy two Thanksgivings in one year. Canada celebrates their event the second Monday in October, and our Canadian hosts put on a tremendous feast. On the weekend, we got the squadron into a Canadian Forces van and traveled to Edmonton to watch a Canadian Football League game between the Edmonton *Eskimos* and the British Columbia *Lions*. The quarterback for Edmonton was Warren Moon, who later became a standout NFL quarterback.

At halftime, as part of the crowd, we participated in the filming of "Running Brave" the Billy Mills story. They basically set up the final lap of his thrilling 10,000-meter win in the 1964 Olympics. It was amazing to see how they prepared the actors playing the runners, so that they all appeared to be on the last lap of a brutal 10,000-meter race. The set-up showed Billy Mills coming from behind on that last lap to win, and since the stadium was packed with fans, we made lots of noise to make that part of the movie pretty darn realistic. *This was another amazing opportunity that could be featured in a Navy Recruiting commercial.*

As a unit, we were introduced (then, of course, challenged) to a game of CRUD. The game is played on a pool table using the cue ball as the shooter ball and one normally striped object ball. Pool cues are not used; the shooter ball is launched across the table surface with the hand. The game involves moving around the table and other players, trying to grab the shooter ball and either strike the object ball or sink it before it stops moving.

Sounds innocuous enough, however, the version of the game played at Cold Lake, was Combat CRUD. Now, the rules allowed for full body checking and blocking like what one finds in ice hockey. Rank is left out of the game, so it is perfectly within the rules to nail a more-senior officer without any repercussions. What got me, as the XO, into trouble, were the boys putting together what they considered to be an appropriate competition uniform. The uniform consisted of a bandana tied around the head with a *Golden Dragon* sticker affixed to the front of it. Bandanas were not to be found at Cold Lake, so they decided to simply rip up a few bed sheets to make the bandanas.

This destruction of Canadian Government Property was frowned-on by the local authorities and earned me a face-to-face with the XO of the base. I offered him a bandana with a *Golden Dragon* sticker affixed to the front and all was forgiven. Well, not all since I did have to pay for a couple of bed sheets. *Combat CRUD did tend to leave players with some significant bruises, most of which were just about waist level. There were a few inquiries from the wives and girlfriends when we got home, about how in the hell we got those.*

The Golden Dragons *following award as the best unit in the* Maple Flag Exercise *1982. That's me, third from the right in the front row.*

***

BASE COMMANDER'S TROPHY: We may not have won the CRUD tournament (or we may have but since it ended at 0300 the next morning, nobody remembers), but we were presented the Base Commander's Trophy as the best-performing squadron of all 20 participating in the exercise. Hats off to our maintenance department for keeping our aircraft in a flying status to the point where we did not miss a single sortie. *This super-competitive attitude permeated all throughout the squadron making it very easy to be successful in a leadership role.*

***

NUCLEAR TECHNICAL PROFICIENCY INSPECTION (NTPI): In 1982, the Light Attack community was still tasked with being part of the Nuclear Triad Mission. The Triad consisted of land-based intercontinental ballistic missile (ICBM), submarine-based ICBM and aircraft strike capability. Readiness was a huge part of the mission. We practiced loading and delivering the various nuclear practice devices, as well as constant monitoring of the personnel involved in any phase of the program. LT Pat "Sponge"' Walsh was assigned as the officer in charge of preparing the squadron for the dreaded NTPI. Sponge was relentless, tireless and meticulous in carrying out this task. He was a perfectionist and his efforts resulted in the squadron scoring an unprecedented "zero-deficiency" score, which, to my knowledge had never been achieved by anyone. What makes it so amazing is that the inspectors absolutely nitpicked the crap out of everything. I am not positive of the exact quote from the chief inspector in the official debrief, but I seem to remember him saying to Skipper Rittenour, "Skipper, I am almost ashamed to say that my inspection team were unable to find a single discrepancy in any phase of your program. Congratulations on an Outstanding, Zero-Discrepancy NTPI!" *Sponge went on to be a* Blue Angel, *a White House Fellow, a Carrier Air Wing Commander, a Battle Group Commander and finally a four-star Admiral. I was incredibly lucky to have him as a squadron mate in both VA-192 and the* Blue Angels.

***

THE MIRAMAR O' CLUB CAPER: Our JO contingent was a force to be reckoned with. Of course, having a now 38-year-old XO with the maturity level of a 24-year-old LTJG, the combination could be rather risky. The *Golden Dragons* were a bit cocky because, by golly, we were good and we knew it.

We were members of *Carrier Air Wing Nine* deployed aboard USS *Ranger*. The other squadrons included VF-24 and VF-211 flying F-14 *Tomcats*; VA-165 flying A-6E *Intruders*, VS-33 flying S-3 *Vikings*, VAQ-138 flying EA-6B *Prowlers*, HS-8 flying SH-3 Helicopters and VA-195, our sister squadron, flying A-7E *Corsair IIs*.

During the workups, *Ranger* pulled into port over a long weekend. The JOs all headed up to Miramar to attend the infamous Friday night happy hour. During this period of socializing with our fighter brethren, there was a good deal of competitive banter tossed about. To the point where on their way out of the officer's club, the *Golden Dragon* boys decided to remove a very famous painting of the F-14 *Tomcat* from the wall in the entry to the club.

I might add that this was a very public area, so of course there were numerous witnesses to this brash caper. Still, they might have gotten away with it, if all three of them had not attempted their getaway in a little Fiat Spider convertible, with the top down. The painting was so large that they had to hold it over their heads as they headed for the gate to make their escape with this worthy prize. They were stopped at the gate, promptly arrested for theft and possibly a few other charges. In any case, they had violated enough laws to put them into the Miramar Brig for the rest of the evening.

It wasn't long before I got a call from the XO of Miramar, CAPT Boyd Repsher. He briefly described the situation (in between giggles regarding the utter incompetence of the *Golden Dragon* boys) and stated, "I am going to scare the shit out of these nitwits, and then I am going to release them with orders to report to you in the morning." He then added, "I have to admit your boys do have some balls. Brains, not so much, but balls, they have."

Bright and early the next morning the three bandits reported to me in the ready room on USS *Ranger* (CV 61). LT Rich "Padre" Bruce stepped up as the ringleader and the driver of the getaway vehicle. The other two, Barry "Bear" Brocato and Brian "Waldo" Forsyth, were standing there looking like they were about to be put to death.

This is when I addressed them with the following, "I am so damn proud of you idiots that I just want to give each of you a big hug. However, due to the seriousness of the charges, I am immediately placing all of you on "Double Secret Probation." Now go enjoy your liberty and stay the hell away from Miramar." *I think I mentioned earlier that a 38-year-old XO with the attitude of a 24-year-old JO might have some difficulty disciplining his boys. Not really, they received plenty of discipline by getting tossed in the brig for a couple of hours; now it was time to let them know how much I appreciated their incredible team spirit, and, no matter how poorly planned, almost awesome caper.*

***

EARLY CHANGE OF COMMAND: My boss, H.T. Rittenour was one of the few naval aviators smart enough to be chosen for Nuclear Power. After only a year as the commanding officer, he was selected to attend Nuclear Power School. I felt bad for him that he did not get a chance to fill out his entire 18-month tour, however, the timing was perfect for me, as I now took over just in time to lead the squadron through a full deployment.

This was the opportunity that I had really been looking forward to. Changes of Command are very traditional and emotional. It is an opportunity for the outgoing commanding officer to highlight the squadron's achievements and for him to thank the members of the squadron for everything that they have done to support him.

It is also very much a family affair, with both immediate and extended family members being recognized throughout the ceremony. Of course, the most important people recognized are the respective spouses. Being a Navy wife is often referred to as the "hardest job in the Navy." Since we are gone for long periods of time, they are stuck with raising the kids, balancing the checkbook and making all the hard decisions. We had an ombudsman assigned to officially coordinate family affairs, but as the skipper's wife, Barbara now took on the responsibility of watching over all the officers' wives. This included several newlyweds who had no experience with the hardships of being separated for seven plus months.

It is traditional that the speech given by the outgoing commanding officer be of whatever length he desires. However, the incoming guy's speech (mine in this case) should be of much-less duration, since to this point anyway, he has not accomplished anything. I followed that tradition much to the delight of the entire audience, especially the squadron standing in formation in the 100+ degree heat of a June day in Lemoore.

Before dismissing the troops, LCDR Jerry Arbiter, the adjutant, brought all the officers to attention, then gave the following command: "Officers, present socks!" They all reached down and pulled up their left pant leg to reveal the traditional, incredibly ugly gold socks. Much to the ire of the Commander, Light Attack Wing Pacific, we also defied the Navy's all dull-gray combat paint scheme by adding gold colored landing gear struts. My argument to his chief of staff was that if we were going to be fighting the enemy with our landing gear down and locked, yes, they might see us. *If I had asked to do this, there is no way it would have been approved, so this move was a perfect example of the fact that it is sometimes better to ask forgiveness than permission.*

***

NEW EXECUTIVE OFFICER: Commander Dewey Englehardt now took over as my second-in-command. Once again, I was lucky to get the perfect partner to share in the leadership of the *Golden Dragons*. Dewey was like H.T.—a great pilot and a super smart, effective and efficient leader. He was exceptionally strong in the areas of responsibility that I was not, so, once again, I was free to fly my ass off, which of course, I did.

***

GOLDEN DRAGON TRACK CHAMPIONS: I was blessed with not only having a great group of pilots and the best maintenance department, but we also had a group of talented athletes. Steve Moreau and Chuck Wright had both been members of the U.S. Naval Academy track team. We also had a young, enlisted sailor who had been a state champion high jumper and pole-vaulter. I was still fast enough to participate in the mile relay and of course the much-anticipated Commanding Officer's, 60-yard dash. My good friend and sister squadron CO, Shedd Webster, had been a sprinter in high school, and he was faster than I was. So, I bought a pair of track shoes that were especially made for the state-of-the-art, rubber-type track at Lemoore High School (they cost me a small fortune). Since you were not allowed any sort of regular spikes, the rest of the contestants wore regular running shoes.

We had enough old-fart skippers to fill all eight lanes. We used starting blocks, and with the huge advantage of the track shoes, I got a big jump on Shedd and managed to hang-on by a nose to win that event. Chuck won the mile and the half-mile, Steve won the 440, the kid won both the high jump and the pole vault, we won the mile relay and I won the CO's 60-yard dash. The Lemoore Marine Detachment was the defending champion. They had four times as many competitors as we did, so they had several point-scoring finishers.

I seem to remember the top five in each event getting points; however, with a breakout of something like 10 points for first and significantly less for the others, we became the first fleet squadron that I know of to win the NAS Lemoore Track Meet. *Events like these are great morale boosters for the entire squadron. We also set a good example regarding the importance of physical fitness, especially with me being a relatively "old man" of 38.*

***

Deployment Aboard USS *Ranger* **(CV-61):** Just prior to our departure on *Ranger* from San Diego on July 17, 1983, CAG Chapman called all the squadron COs into the classified area of CVIC. Our original plan was to head to Hawaii, the Philippines and then leisurely to the Indian Ocean operating area, via several exotic port calls.

CAG opened the meeting with the following statement: "There has been a change of plans. Due to the threat imposed by the potential alliance of Nicaragua and Cuba, we have been ordered to steam southeast versus southwest, to set up operations in the Pacific, just off the coast of Nicaragua. This is "Top Secret" so you will not share this with anyone, even within your squadron, until we are out to sea."

And so, we were off on another adventure that we referred to as "Banana Station." It was not a good start. Just a couple of days after getting underway, *Ranger* collided with USS *Wichita* (AOR 1) during an underway replenishment. The collision was severe enough to put one of the ship's four flight-deck elevators out of commission for the entire cruise. Unfortunately, that would not be the last problem that we faced, but our deployment was underway. The 45-day delay at Banana Station cut out all our planned port calls, except for short provisioning stops in Hawaii and Subic Bay.

Since I had been selected for command on my first look, sent immediately to be an XO and now moved to CO six months earlier than normal, I was by-far the junior CO in the air wing. What that meant was that our squadron was considered the junior squadron and would therefore be last in line for any good deals, and first in line for all the undesirable assignments from the CAG Staff. *We decided that the best way to deal with this situation was to volunteer before being assigned. Having been on a CAG Staff, I knew that in the long run, the squadrons that always stepped up to the plate and reliably supported the staff would be the ones receiving the good deals, no matter what their seniority was.*

***

BANANA STATION: This unusual course of action was very interesting in that the threat that we were sent to deal with was a potential deployment of Cuban *MIG* fighters to Nicaragua. It is amazing what four acres of U.S. territory, 12 miles off a country's coast, with 75 aircraft on board, can do to discourage aggressive behavior.

Our almost-new A-7Es were equipped with Forward Looking Infrared (FLIR) pods that were very effective locating even the smallest fishing boats, in the darkest of nights. On one of these sorties, I spotted this good-sized fishing boat and flew over to get a better look. I passed directly over them at about 200 feet, turned around and made a second run, just in time to see the crew throwing all sorts of contraband over the side. Although we had no arrest authority to go after drug runners (*posse comitatus*), we did scare the shit out of them and reported their location to Coast Guard units.

***

CONTINUING WEST: It was not long before Nicaragua agreed not to base any Cuban *MIGs*, and we were off for a short stop in Hawaii, then on to the Philippines. From the Philippines, we proceeded through the Straits of Malacca and on into the Indian Ocean. We spent most of the deployment operating in the North Arabian Sea. In those days, we did not enter the Persian Gulf, but we did fly continuously over and near the Straits of Hormuz.

The threat was Iran and their rhetoric about closing the Straits, which is the only way in and out of the Persian Gulf. We planned several contingency strikes, so that we were ready to retaliate if they did close the Straits or attack allied shipping. Our battle group commander was RADM Stan Arthur, a true war hero with over 400 Vietnam combat missions. He instilled total confidence with his leadership style, to the point where we were hoping that Iran would give us an excuse to engage.

We practiced those contingency strikes on Masirah Island and other practice areas in Oman. My wingman for those strikes was Randy "Claw" Causey. He was just on his first cruise, but he was one hell of an aviator. He was our Weapons Training Officer (WTO), and the kid finished in the Top Nine of the 100 or so air wing pilots in landing grades. *We always carried 1,000 rounds of 20 mm and a live* Sidewinder, *so the two of us felt as though we were good enough to go head-to-head with the Iranian F-4s if the opportunity arose.*

***

FIRE! FIRE! FIRE! I would like to preface this incident by reminding everyone who is not familiar with the Navy, that every single sailor and officer, including the embarked carrier air wing, is a trained fire fighter. On the morning of November 2, 1983, I was working in my stateroom, which was located on the 03-level about mid-ship, just below the flight deck. I am not positive of the exact sequence, but I seem to remember hearing the following over the 1MC, "Major fuel oil leak, major fuel oil leak, number 4 Main Machinery Room." This was followed shortly by a rather frantic announcement, "Fire! Fire! Fire! Class Bravo Fire in number 4 Main Machinery Room. General Quarters! General Quarters! All hands man your battle stations."

By the time I opened the door to my stateroom, smoke was already making its way into the passageway leading to the flight deck. I hauled ass banging on doors as I went by (just in case someone was asleep) until I got to the catwalk and then up onto the flight deck. Our General Quarters station was in our ready room, which was Ready Room Nine. It was located on the 03-level aft and was well away from the fire and smoke affected area of the ship. Because of its relatively safe and convenient location, and the loss of the normal medical section of the ship, our ready room became a triage area for treatment of the injured.

Unfortunately, the initial explosion and resulting fire took the lives of six engineers working in Main Machinery Room (MMR) number 4. From that point on, it was a battle to save the ship. This meant that fire teams were sent into that space to battle the blaze using Oxygen Breathing Apparatuses (OBAs) and fire-retardant suits. These brave heroes consisted of both ship's company and air wing personnel. It was amazing to watch, how, without hesitation they entered the space to battle the fire.

Others helped replace used OBAs and transported the injured (mostly smoke inhalation/physical exhaustion cases) away from the affected area to triage spaces like our ready room. It took several hours to get the fire totally under control, but we continued to operate as a full up operational aircraft carrier. *Fire is the greatest enemy of an aircraft carrier. The fires are classified as Class Alpha, which is the most common and least serious. It includes flammable materials such as paper, cloth, wood, etc. Class Charlie refers to an electrical fire. Class Bravo, by far the deadliest, refers to flammable fuel and lube oil. In any case, there is no place to go so you better be able to put the fire out. Hence, we were all trained fire fighters.*

***

BARRICADE ARRESTMENT: Most of our operations were "Blue Water", which means we had no bingo fields. It was land aboard the carrier or eject. For certain emergencies, in addition to the normal four arresting gear wires, a 5th arresting gear engine was located between the #3 and #4 wire. However, instead of a wire, two 20-foot-high stanchions were raised up from either edge of the landing area to support a specially designed webbed barricade. This could be used in conjunction with the four wires, but more commonly was used with the four wires removed. LT Wade "Stanky" Tallman was the senior *Golden Dragon* LSO. On returning from a routine mission, he discovered that he could not lower his right main landing gear. He tried all sorts of positive and negative-G maneuvers, to no avail.

Wade Tallman's emergency barricade with his wing still being held off the deck. His ability to flawlessly perform this dangerous evolution resulted in minimal damage to the aircraft.

Eventually, it was determined that the best option was to remove the wires and rig the barricade. Stanky flew a perfect pass, to the point where he engaged the barricade with amazingly little damage to the aircraft. Of course, due to not having his right landing gear down his right-wing tip eventually contacted the deck, but we were able to repair the aircraft and it flew again in just three weeks. *It was fortunate that one of our best "ball" fliers was in the cockpit for this event. He did a masterful job of controlling the aircraft, keeping that right wing off the deck, whereas a more inexperienced pilot may have been at risk of damaging the aircraft beyond repair, or even worse, injuring himself or somebody on the flight deck.*

***

DALLAS COWBOY CHEERLEADERS VISIT: This was really a tough cruise for the crew of *Ranger*. We ended up setting a record of 121 consecutive days-at-sea without a port call. As a pilot, you don't notice it so much, because, frankly, you are having fun flying every day. However, for the average sailor or airman, it gets really old. To break up the monotony for them, the captain of the ship held Steel Beach Picnics every 45 days. This included an allowance of two beers per person and a full day off from flight operations and maintenance.

The old days of smuggling booze aboard were long gone. Instead of a slap on the hand or possibly a verbal lashing for getting caught consuming alcohol aboard ship, the penalty in 1983 was severe, including ending of one's career. To get the attention of the JOs before we left on this cruise, rather than being a pompous ass, I shared my Vietnam era stories of rampant boozing aboard ship, much to their delight and amusement. I then added, "Now, we cannot do it anymore. If you get caught, I cannot come to your rescue, because it will be taken out of my hands, and you will go directly to the captain of the ship for adjudication. Do not bring booze aboard or imbibe while we are at sea, period." *If they did do any shipboard boozing, they did so sneakily enough not to get caught.*

After 100 or so days at sea, I am not sure who decided it would be a good idea to bring the Dallas Cowboy Cheerleaders, some of the most beautiful women in America, aboard the ship, but God bless the United Services Organization (USO). As part of their visit, they were invited to participate in a FOD walk down (this is an all-hands evolution walking elbow to elbow from the bow to the stern to check for and pick up any foreign objects that could potentially be ingested by a jet engine) prior to a launch. As the junior squadron commander, I was put in charge of the walk down.

Imagine a flight deck crew averaging about 20-years old, none of whom have seen a woman for over three months, and now they are expected to walk literally arm-in-arm with a Dallas Cowboy Cheerleader and totally behave themselves. The good news was that these ladies were dressed very modestly, in warm up suits, versus their usual game day attire. The other good news was that they were all wonderfully polite and charming to these awestruck kids who risked their young lives on that flight deck daily. The culmination of that Steel Beach Picnic was a wonderful dancing show that same evening, with the ladies now dressed in their fabulous game day attire. It was wonderful. *I overheard one sailor say to another, "Hell, I will stay out here for another three months if they guarantee these ladies will come back to visit."*

***

VA-192 Aircraft Accident: Toward the end of the cruise, LCDR Dave "Grunt" Kendall and LT Carl "Chach" Abelien were flying an air-to-air training mission. As part of the training, Grunt set up as the defender with Chach climbing to an altitude above and a little behind Grunt. We referred to this position as the offensive perch. On signal from Grunt, the fight was "on," with Chach rolling in on Grunt, looking to achieve a practice firing solution for his *Sidewinder* missile (we carried one live *Sidewinder* on the shoulder station of the A-7). Since it was a live missile, we always double-checked the master arm switch in the "off" position, then we could still use the tone of the missile safely to indicate a firing solution.

Just as Chach got a tone, Grunt's aircraft EXPLODED! And I mean exploded! The back half of the aircraft separated from the front in a massive fireball. Chach quickly checked to make sure his missile was still on his aircraft (it was), then watched in horror as Grunt, trapped in the cockpit of his wildly gyrating half of an A-7, careened toward the ocean.

In Grunt's cockpit he described the event from his perspective: "I hear Chach's Fox 2 call and start a hard left-hand pull to practice evading the missile. Suddenly, I am slammed into the top of the cockpit with such force that I almost lost consciousness. I was experiencing unbelievable negative-G, to the point where I was about to red out." *(The opposite of blacking out, which is the*

270

result of excessive positive-G. It is also far more uncomfortable and disorienting).

"I realized immediately that I was in deep shit, and that my only option was to eject. The problem was that I was pinned up against the canopy, in a position that kept me from reaching either the face curtain or the lower ejection handle. Finally, I managed to push myself off the canopy with my right arm, far enough to reach the lower ejection handle." (Grunt was a California high school wrestling champion, and easily the strongest person in our squadron). The ejection was, of course, a wild ride, but he survived, got in his raft, and was rescued in short order by an SH-3 Helicopter from HS-8.

It is amazing to see the toll on the human body that an aerodynamic event like this can cause. Thank goodness Grunt was in the terrific physical condition that he was prior to the event, because I don't think anyone else would have survived. The most obvious damage was to his physical appearance. His face would have made a great Halloween mask. There was no white left in his eyeballs. They were a reddish black color due to the negative-G rupture of all the little blood vessels. He could have easily picked up the call sign "Ghoul" or maybe "Zombie." Fortunately, there were no long-term physical effects and he was back on the flight schedule in a few days. *Since the aircraft crashed into very deep water, there was no opportunity for a hands-on investigation of the cause, other than to assume a massive engine explosion. Naval aviation is a dangerous business, but if we bring everybody home, it is a successful deployment.*

***

Golden Dragons **Dominate:** In a peacetime environment, the sortie completion rate and Tailhook Landing Competition tend to define the best squadron in the air wing. Usually, the competition is very close, with different squadrons winning the various line periods of operations. We did not let that happen. The *Golden Dragons* won every single line period and took the trophy for the top grades for the entire eight-month deployment.

How did we do that? We did not have any weak links. Even our most inexperienced nuggets worked extremely hard to become good ball fliers. Our LSOs, Wade "Stanky" Tallman, Carl "Chach" Abelien and Barry "Bear" Brocato, stayed in the teaching, versus criticizing, mode with the new guys, which brought them along quickly. Landing on a carrier is the most difficult and challenging evolution in aviation. A pilot makes an average of over 200 minor movements during the last 18 seconds of the landing approach to the carrier. He must balance out how fast he is going, whether he is on the glide slope, and whether he is lined up laterally on the moving airfield that is an aircraft carrier.

The Golden Dragons *showing off the award as the Top Hook (best overall squadron landing grades for the entire deployment). Back row, left to right: Dave Kendall, Randy Causey, Brian Binnie, Carl Abelein, Mal Branch, Dewey Englehardt and Steve Moreau. Front row: Tim Estes, Barry Brocato, Wade Tallman, me (holding the plaque) and CAG Austin Chapman. Eventually, nine of these awesome aviators ended up commanding their own squadrons.*

As a former CAG LSO, I absolutely forbid any of our pilots from arguing about grades given by the controlling LSO, and I set the example by taking whatever grade I was given and thanking the LSO for helping me get safely aboard (occasionally having to grit my teeth while doing so). Individually, we ended up filling three of the "Top Nine" with Dragons: Wade Tallman, Randy Causey and me. *From my experiences as an LSO, there is nothing more unpleasant than debriefing pilots from a squadron that tends to argue about the grades they are given. This attitude starts with the CO. If he is a weak ball flier who tries to cover up his relative incompetence by arguing/bullying the LSOs', the rest of the squadron will follow his lead. It doesn't work. In fact, if there is some debate on the LSO platform about a grade and the pilot is a known "whiner," it will probably go against him.*

***

500-KNOT BREAKS AT THE LSO PLATFORM: So how does one go about winning the individual Top Hook Award for an entire deployment? Consistency is the key. You cannot afford to have more than maybe one or two Bolters, no Wave-Offs and no "cut" or "no grade" passes. Basically all 100+ of your landings need to be at least a three-point "fair" with the vast majority being four-point "OKs."

In *Air Wing Nine*, we also had the opportunity of getting an "OK-no comment" pass, which resulted in five points. Normally, that would be a night landing where the ball stayed in the middle for the entire pass, also known as a "Rails" pass. Because we had a couple of shit-hot CAG LSOs in Dave Bernard and Pat Madison who enjoyed being occasionally entertained while on the platform, they would award five points for a high-speed early-carrier break, basically adjacent to the LSO platform, versus at the bow of the ship. There was great risk in this move however, since unless you performed it flawlessly, you could very well end up being waved off.

There was also a very limited opportunity to perform this during normal operations due to potential landing pattern disruption. I figured out that if I flew a yo-yo test flight (launch and land on the same event), I would be landing last and have the pattern to myself. Not wanting to surprise the controlling LSO, one of the *Golden Dragon* LSOs would put out the word that "Duster" was on a test flight and intended to perform a 500-knot break at the LSO platform. A normal break was accomplished parallel to the ships heading, from 800 feet of altitude, at 350 knots, to end up level at 1.25 miles abeam of the back of the ship, at 600 feet, in a landing configuration, which included your landing gear, tailhook and flaps down. This resulted in about an 18-second wings-level final approach. None of this was true in a shit-hot break.

Rather, I would enter the break with about a 30-degree angle outside (starboard) of being parallel to the ships heading. I would also be level at 600 feet instead of 800 feet so that I could perform the entire break at a level altitude. Most impressive and fun to watch (according to the LSOs) was that I would be at 500 knots. I would roll into a 90-degree angle of bank and pull about six Gs, pulling my power back to near idle with the big belly speed brake extended. To keep that speed brake extended, I would have to hold my gear and flaps up (if the flap handle was moved to lower them, the speed brake would automatically retract) staying in a significantly banked attitude, until I was probably nearing 45-degrees from rolling out on final.

I would then drop the gear and flaps, all the while being at idle power with an angle of attack a bit slower than "on-speed." I would usually reach "on-speed" angle of attack and add what often amounted to full power, when I was just a few seconds from touchdown. Instead of the usual 18-seconds of wings level on final, this approach resulted in only a few seconds of wings level time. If I did not get a one wire, I was awarded a five-point grade for the pass. *I accomplished this a couple of times which turned out to be just enough of an edge to squeak by the XO of VA-165, Gary Wasson, for Top Hook. Gary was also a former CAG LSO. As you read through this book you will find that many successful leaders in naval aviation did come with an LSO background.*

***

Last Night of Tailhook Competition: Just prior to pulling into Cubi Point where we would have the End of Cruise Foc's'le Follies and the award of the top squadron and individual tailhookers. I was scheduled for the last night event. While we were briefing, Wade Tallman informed me that he had been helping the CAG LSOs to tabulate the final grades. Wade said, "Skipper, you are a tenth of a point ahead of XO Wasson. If you don't fly tonight, you will win the top hook, no matter what he does. On the other hand, if you fly and get anything other than an OK, he will be the winner."

I thought about it for a minute or so before replying, "Thanks for letting me know the status Stanky; this award means a great deal to me, so I am going to go out and earn it!" I did not think too much about it again, until I was holding up in marshal. Then just for a minute, I thought, "Rud you dumb shit! Now you had better fly the best pass of your life." It was not the best pass of my life, but it was good enough for an OK, and I had now achieved a very coveted award, that I believe played a large factor in my eventual selection to be *Blue Angel #1. Amazingly, on this same evening back at home in Lemoore, my 14-year-old daughter was working with her math teacher mom to make sure that she would be able to solve the math bonus question, and thus beat a smartass-boy classmate who had told her that "girls can't do math." She did beat him, and 25 years later her squadron won the top hook, and she won the top individual hook award for an eight-month combat cruise on USS* Nimitz *(CVN 68). And yes, her overall grade was better than mine.*

***

CORROSION CONTROL INSPECTION: I had always thought that the Navy assigned too many people to inspection teams, and not enough to teams that could teach and train versus critique and terrorize. Now just imagine this scenario. We had just reached our first port in 121 days. Our kids had the opportunity to drink exactly four beers in 121 days. In other words, they were ready to party.

How do you utterly spoil this opportunity for them? A shore-based staff schedules a corrosion control inspection for the morning after our first day in port. I don't remember what staff they were assigned to (I think it was COMFAIRWESTPAC), but this team came aboard *Ranger* with the usual high-minded arrogance and bullshit to inspect the air wing. Some of my fellow COs were so concerned about this inspection that they canceled the first night of liberty for the maintenance folks involved. There was no way I was going to do that, plus I knew that we had, by far, the best-maintained aircraft and least corrosion problems of anyone.

As the sun rose the next morning, it was obvious that the whole squadron had immensely enjoyed their first night of liberty in four months. So much so that one of my AMH2s (Hydraulic Technician Second Class) assigned to accomplish the Hydraulic Contamination Test, passed out during that event. I heard about this from one of my maintenance chiefs and laughed heartily, because this was just one small part of the inspection, and I knew how well prepared our maintenance department and aircraft were.

At the end of the day-long inspection, the maintenance officer, his division officers, the XO and myself gathered in the ready room for the debrief. The Team Leader, I think he was a LCDR ground-pounder, stood up in front of the ready room, looked directly at me and said, "Skipper, I am sorry to tell you that your squadron failed the Corrosion Control Inspection. We just cannot tolerate the unprofessionalism exhibited during the hydraulic contamination test. But the good news is that although we are giving you a failing grade, we will be happy to conduct a re-inspection first thing tomorrow morning."

I replied, "Thank you for your comments. By the way, you can stick your re-inspection where the sun doesn't shine." I then added, "I will be calling the admiral in Lemoore in a few minutes. I will be asking him to send a team to meet us in Pearl Harbor, so that they can conduct the re-inspection. You know full well that they will discover the best maintained aircraft in the Pacific

Fleet. When they do, I will recommend that you and your team be disbanded as a useless waste of manpower. Now get the hell out of my ready room!" *OK, I know, control your Norwegian temper, Gil. Actually, I was not as angry as I appeared to be to this now very nervous inspector. I sensed an opportunity to make a difference for future squadrons, so that they would never be put into these ridiculous, morale-destroying circumstances. It worked, too.*

***

THE FLY OFF: As often happens when we transit from the tropics back to the States during the winter months, the changes in temperature end up setting off a flu bug that gives the whole ship the sniffles or maybe something a little worse. In my case, it was a little worse, bordering on walking pneumonia. Despite my health issues, there was no way in hell that I was not going to lead the squadron fly off to our long-awaited homecoming in Lemoore.

CAG was well aware of my medical situation, as was the Air Boss. On the morning of the fly off, my always-scheming JOs asked if I would go along with a bit of a ruse. Just when I was beginning to think that my mind was approaching the age of my body, I replied, "Sure what do you have in mind?" They got a stretcher from medical, put me on it, then had our flight deck maintenance guys carry me to my aircraft. I rolled off the stretcher, pretended to vomit under the aircraft and climbed in. You can imagine the reaction in Primary Flight Control and the Captain's chair on the bridge. Just to make sure nobody panicked, I stood up in the cockpit and did a little bow as did the stretcher crew. *The Super Shit Hot World-Famous* Golden Dragons *strike again. Of course, as usual, this is not the sort of behavior that promotion boards are looking for.*

***

BACK WITH THE FAMILY: Now well into my fourth straight year of sea duty, it was wonderful to be back with Barbara and the kids. Barbara was teaching math full-time. Valerie was a freshman at Lemoore High School, Ryan was a fifth grader and Sally was in kindergarten.

Val had started out as the student manager for the girls' basketball team, but once the coach got a look at her 5-foot 8-inch and growing frame, she said, "Get a uniform on girl, we are going to teach you how to play basketball."

Ryan was part of a very competitive soccer program under coach Mike Graham. Since they were sort of an unofficial select team, their uniforms consisted of some sort of (not necessarily matching) yellow shorts and plain white tee shirts. They were invited to a tournament in Fresno, California, where the first team they faced showed up in super-fancy uniforms including warm-up gear. One of the parents from the other team came over to me and asked about the lack of uniforms, and basically, "What are you doing here? You do know that this is a select soccer tournament, do you not?" This was stated in a very arrogant manner.

I replied, "Yes, we know that. I have been watching your boys in their fancy uniforms warming up." He said, "Very impressive, aren't they?" I said, "Our kids may not look so hot in those white tee shirts, but they are going to absolutely school your boys once the game starts." The final score was 10-0 as our small-town and country-boys ran rings around the fancy-pants, city slickers. They also won that tournament, which would be the first of many. *Ryan fell in love with soccer, and would continue to play in northern Virginia, and Florida, before returning to Lemoore to join this same team for his junior and senior years of high school. Thank you coach Mike Graham, for giving my son the start he needed to eventually end up playing Division 1 soccer for the University of California, Berkeley.*

***

Turn-Around Highlights: A couple of months after we got back from deployment, rumors started to swirl regarding what squadrons were going to be next to transition to the F/A-18 *Hornet*. We were still experiencing terrific availability and continued success with our almost-new A-7Es so we did not expect to be in the discussion. We were doing so well, that we had enough aircraft to loan a couple of birds to CAPT Joe Prueher and CDR Bill "Bear" Pickavance as they stood up the Naval Strike Warfare Center (STRIKE U) at NAS Fallon, Nevada. Bear was a long-time friend and fellow A-7 pilot from Lemoore and Joe was an iconic A-6 pilot. Both had experienced the worst part of the war, having flown hundreds of missions over North Vietnam. In other words, they were the perfect choice to establish a much-needed center for air combat training and tactics development.

They also established the command so that instead of being another dreaded inspection unit, they assisted in the training of the various carrier air wings and squadrons and provided much-needed standardization across all tactical naval aviation. During this same period of time, we hosted an east coast

A-7E squadron, VA-105 as they passed through on their way to a six-month deployment at Marine Corps Air Station (MCAS) Iwakuni, Japan. The CO of VA-105 was Don Weiss. Don and I would cross paths again as COs of aircraft carriers, but although we had not met before, we instantly became friends, due to shared stories of our mutual North Dakota roots. Don was an all-state football player from Jamestown, and even has an airport named after him. *Talk about successful; Joe became a four-star Admiral, while Bear and Don both made two stars.*

***

SURPRISE AND ANOTHER EARLY CHANGE OF COMMAND: As I mentioned, we were not expecting to be in the transition discussion, but suddenly we were right at the front. I got a call from COMLATWINGPAC telling me that the squadron would be transitioning starting in a couple of weeks. Wonderful, I am thinking as I called an all-officers meeting (AOM) to share the great news. The pilots were super excited to have this opportunity, but of course, it was not that simple. The Military Personnel Command and the budget boys soon entered the picture and quickly muddied the waters. Not all of us would transition. Those with less than six months remaining in the squadron would not be included in the transition. Unfortunately, that included me.

I was bitterly disappointed and expressed my frustration with my chain of command. "How am I supposed to lead a squadron if I am not even flying the aircraft?" The answer was, "Well, we can fix that issue with an early change of command." Not the answer I was hoping for, to say the least. And so, with only 15 months of a scheduled 18 as the commanding officer of the *Golden Dragons*, I was relieved by Dewey Englehardt. I was extremely honored to have one of my heroes, RADM Stan Arthur as the guest speaker at the change of command. He had lots of nice things to say regarding the performance of the *Dragons*, and I got the opportunity to personally thank the members of the squadron for making my tour as the commanding officer a wonderful experience. I also took the opportunity to thank Barbara and the kids for their support, and for putting up with our nomadic lifestyle. Then, as was the tradition, we were gone. *The Rud Family was off on another adventure.*

22. Avoiding the Pentagon

WITH THE FIRST-LOOK SELECTION for command and then two consecutive early changes of command, I was now a bit out of sync, seniority-wise, for a normal career progression. Fortunately, CAG Chapman recognized the achievements of the *Golden Dragons* and made me the beneficiary of some very competitive fitness reports. My detailer threatened to send me to a job in the Pentagon, which I had rather skillfully avoided for 18 years. Although unable to avoid Washington, D.C., my next assignment was to the Navy Military Personnel Command (NMPC) as NMPC-432, Head of Aviation Junior Officer detailing versus the five-sided wind tunnel.

Up to this point in time, the job had always been filled by an experienced detailer/someone who had been a detailer on a previous shore tour. I was to be the first, straight from-the-fleet operator, with no previous detailing experience. I arrived in time for a couple of weeks of turnover with my predecessor, CDR Jay Johnson. Jay and I had been shipmates on *Oriskany* during the war. He was an F-8 fighter pilot with VF-191 when I was flying A-7Bs with VA-215. This was Jay's second tour as a detailer, so he was an absolute wealth of knowledge. He was also very personable and seemed to know everyone in the building. He introduced me to all his contacts and did his best to prepare me for the responsibility I was about to assume. Despite his best efforts, I was in way over my head. *Yes, this is the same Jay Johnson who would eventually become a four-star Admiral and the Chief of Naval Operations (CNO).*

When you feel overwhelmed, you turn to the folks that you are supposed to be leading for help. I was blessed to have as my civilian assistant, Mary Nell Greer, known affectionately and respectfully by everyone as "Mom." What a wonderful lady she was. I simply looked her in the eye and said, "Mom, I have no idea what I am supposed to be doing, or, for that matter, what all these people working for me are supposed to be doing." She laughed and said, "Don't worry Gil, I will teach you the ropes, keep you out of trouble—and I bet that we will have fun while we do it." That is exactly what happened.

The personnel assignment system consists of a placement department, NMPC-433, which is responsible for determining what jobs are available, and a detailing department, NMPC-432, which is responsible for who fills the job. These two departments work for the same boss and must closely coordinate their respective actions. I was lucky to have CDR Mike Bowman as my counterpart as head of NMPC-433. *Yes, the same Mike Bowman who would eventually become a three-star Vice Admiral and be assigned as the head of all naval aviation, CNAP, the Air Boss.* Mike and I worked very closely together as did all the folks who worked for us.

MILPERSCOM was located up the hill from the Pentagon, just outside the entry gate to Fort Meyers, which encompasses the hallowed grounds of Arlington National Cemetery. The 432-office consisted of a very large space set up very much like an aircraft carrier ready room. Aviation communities such as Light Attack A-4/A-7/F/A-18, Medium Attack A-6E, Fighter F-14, Airborne Early Warning E-2/C-2, Electronic Warfare EA-6B/A-3, P-3 ASW, S-3 ASW, Helicopters and the Training Command were all represented with detailers from that specialty. The larger communities had both a shore and a sea detailer. Limited Duty Officers (LDOs) and Aviation Maintenance Duty Officers (AM-DOs) were also represented. The assignment process considered the preference of the individual, his/her career progression, fitness reports, and of course, the needs of the Navy. Each detailer was afforded a modicum of privacy via pukka office-dividers, because in 1985, computers were not sophisticated enough, so telephone conversations were the primary means of communication.

When an assignment met the preference of the individual and the needs of the Navy, I would not be involved. When there was pushback from the individual or possibly from a senior Navy officer lobbying on behalf of the assigned individual, that is when I earned my pay. Unlike the Pentagon, where most of the jobs entailed planning five years down the road and changing that plan to accommodate whatever political or budget related hiccup that emerged; our jobs were much more rewarding. We dealt with real, present-day issues, and we could get the satisfaction of helping people smoothly transition from one assignment to the next.

I worked for NMPC-43, which initially was Captain Henri B. Chase, and then for his replacement, Captain Dave "Snake" Morris. Both were great leaders and easy to work for. I did, however, have one major disagreement with Snake over the assignment of a Flag Lieutenant (Aide to an Admiral). The normal process for a career enhancing assignment like this was to put together a package of at least three candidates for the admiral to pick from. In this case, all three of them were exceptionally competitive. A couple of days after we had submitted the package, Snake called me in and said, "The admiral does not want any of the folks we submitted in the package, instead, he desires LT _____." "OK, I will get his record out and formally submit it as an addition to the package."

I looked at this guy's record and I was appalled. He was a below-average performer, not anywhere close to the typical quality of a Flag Lieutenant. I went in to show the record to my boss. He looked at it and said, "Yep, he is a marginal performer, but he is a good tennis player and that is what is important to the admiral." "Bullshit! I will not send this guy to be a Flag Lieutenant." Snake looked me in the eye and said, "Listen to me carefully, you stubborn Norwegian. You have two choices, you can go home right now and cool off, or I will fire you!" *I went home and cooled off, but I had a heck of a time putting up with this type of "good old boy" stuff. My boss didn't like it either, but he knew better than I did what battles to fight, and when to let the admiral have his way.*

As I mentioned earlier, I rarely got involved in the detailing process unless there was an issue. In one particular case, we assigned a rather average-performing aviator to a squadron forward deployed to Japan on *Midway*. He fought this assignment by using the chain of command to try and pressure me to change his assignment. Unlike the Flag Lieutenant issue, on this one, I had the full backing of my boss. Overall, though, this guy was a real pain in the ass, even having his wife call the detailer. The situation was common knowledge throughout my whole department, with lots of good-natured ribbing of me for being insensitive, etc. It all came to a head one afternoon when Mom said, "Gil, you have a visitor (which was not uncommon)." "OK, who is it?" "You are not going to like this—it is the wife of the Japan assignment whiner." "Crap! OK, send her in."

Oh my God, in walked this drop-dead-gorgeous woman. Not only was she extremely attractive, but she was dressed in a mini-skirt and a top that showed an extensive amount of cleavage. She sat down next to me and leaned in very closely to identify herself as LT Whiner's wife. I was having one hell of a time maintaining eye contact, or for that matter listening to what she was saying.

At some point during the conversation, she burst into tears and said, "CDR Rud, I will do anything—and I mean anything—for you, if you change my husband's orders." I was now quite flustered as she stood up and placed her arms around my neck as though she was going to kiss me on the lips.

That was when I finally noticed that the entire office had gathered around the two of us, as they all burst out in laughter and applause. I had been punked big time. The lady was an actress that the office had hired to play the part of LT Whiner's wife—and she played the part very well! *I like to think that when your subordinates know they can pull this sort of stunt without risking retaliation, morale in your command is where you want it to be.*

***

TAKING ADVANTAGE OF A GOLDEN OPPORTUNITY: Occasionally, the "manning" needs of the Navy and Marine Corps can get a bit out of sync with recruiting. This happened to the two aviation branches in mid-1985. The Navy did not have enough students to fill pilot slots in the training command, and the Marines had too many students in their pipeline. As a solution to both services problems, we were invited to address a group of student pilot-designated graduates of the Marine Officer Basic Course at Quantico, Virginia. John "Knobber" Sandknop from NMPC-433 and I were chosen for this mission. Our plan was to show them the opportunities available via the Navy pilot training pipeline. This included immediate transfer from the Marine Corps to the Navy, and assignment to a class in Pensacola within a few weeks. Their alternative was to be a Marine (non-flying) officer.

Our Marine counterpart was a very imposing Brigadier General from Headquarters Marine Corps. Brigadier General Jim Meade was an F-4 *Phantom II* pilot. He stood at least 6 foot 4 inches tall and had a voice that required no microphone. I estimate that there were about 120 First and Second LTs in the audience. General Meade addressed them with congratulations on graduation from Basic School. He then glowered over at Knobber and me, and said, "You probably wonder what these two 'squids' in their silly white sailor suits are doing here. Well, goddamn it, they are here because you signed up to fly airplanes, and the Marine Corps is not going to be able to provide you that opportunity. Now you and I are Marines first and pilots second (huge and rather scary "OORAH" from the crowd). However, if I was in your shoes, I would have to give serious consideration to what they have to offer." He then added, "Whatever that choice is, the rest of you will respect each other's decisions."

Our presentation covered every possible opportunity that was going to be available to them as a Navy pilot. At the end of the presentation, the obvious question was asked, "So this sounds good, but what sort of a delay will there be before we actually begin training?" I responded, "Today is Thursday. Those of you who choose to go Navy will be sworn in at an equivalent rank/pay longevity at the Navy Yard on Monday—and half of you will be in Pensacola starting a pre-flight class the following Monday. The rest of you will be just a month behind." This elicited a noticeable gasp from the audience followed by the reaction we had hoped for. I looked at Knobber and said, "I think we will probably get 15 or 20 of these guys." *We actually ended up transferring 89 out of 120, which was the perfect number to fill out the training command's student shortage.*

The Marine Corps manning issues also affected the operational end of their A-4 and F-4 RAGs. Our next trip was to Yuma where we managed to transfer another, already winged, 30 Marines to the Navy. We sent most of the F-4 guys to the F-14 RAG and the A-4 guys to A-7s and A-6s. One student went directly to Navy F/A-18s. *The commanding officer of the A-7E RAG in Lemoore was a former squadron mate of mine, CDR Bud Orr. He was not initially aware that he was getting these Marine transfers, so assuming that they were all coming from the training command, he remarked, "These are the most squared-away students I have ever seen."*

***

FAMILY ADJUSTMENTS TO LIVING IN NORTHERN VIRGINIA: We rented a nice home in the northern Virginia suburb of Springfield. Probably the biggest adjustment for the kids was the extremely competitive academic environment. Fortunately, they all took after Barbara when it came to intellectual capability, so although they started a bit behind their classmates, it was not long before they were settled in and doing well. Sally's first grade-elementary school, Orange Hunt, was right behind our home within a short walk. Ryan went to

Washington Irving middle school and Valerie enrolled at Lake Braddock High School. Barb started out substituting, and as usual, after a short stint as the most competent math teacher ever, she was soon hired full-time. The best part of my job was that although the workday was long, it did not include weekends and very little travel, so I was able to finally have lots of family time.

Valerie started her sophomore year at Lake Braddock High School, which had an enrollment of over 5,000 students. She made the junior varsity basketball team and the softball team. When we first arrived in Springfield, I went to the local recreation department to check on soccer opportunities. I mentioned that my son had played some select soccer in California. The reply was a rather arrogant, "He may have played select in California, but soccer here is much more advanced than out west. We will sign him up for regular recreation league soccer. If he is good enough, then he can try out for a select team after the recreation league finishes."

He absolutely dominated at that level and was selected to play in an all-star game. During that all-star game, the stands were full of coaches from all over the area. They all had clipboards taking notes on the players. Ryan had a great game, so the calls started coming in. Some of the teams that wanted him to try out were over a two-hour drive from Springfield. Fortunately, the local Springfield select team invited him to tryout. It was a two-day event on a Saturday and Sunday. I attended both days.

There were 30 kids trying out for just three slots. After the first day, Ryan was pretty fired up, as he seemed to have done very well. One of the drills I remember is that they had each kid shoot at a toilet seat hanging from the goal post (you must be kidding me). At the start of the second day, they put him in as a goalkeeper. He was obviously devastated. I inquired to one of the assistant coaches as to why they had done this. He smiles and said, "Ryan is the best player by far. We have already selected him to be a halfback, but there are other rival-team coaches watching this tryout, and we don't want them to see how good he really is."

I then said, "Coach, would you please tell him this, since he thinks you have put him in goal because he isn't a good-enough field player." I watched as the coach approached him and delivered the news. That was the biggest smile I had ever seen from my son. *Ryan's select team played almost 60 games a year. It was the perfect situation that gave him the experience he needed to eventually start all four years in high school, and end up playing Division 1 college soccer for the University of California, Berkeley.*

The whole family experience during this tour of duty was exciting to say the least. None of us had seen any of the wonderful Washington, D.C., sites, so every weekend was a new adventure. The transit system was safe and convenient, so we made the most of it. The two older kids were good to go to museums and other events on the Capitol Mall using the Metro. Valerie, in particular, fell in love with northern Virginia, and would come back to attend Virginia Tech along with several of her friends that she made at Lake Braddock High School.

23. "Who Me?" Selection to the Blue Angels

SOMETIME DURING THE SPRING of 1985, I got a call from a good friend and ex-*Blue Angel*, Jim Maslowski. He asked me if I had considered applying for the Flight Leader position. I honestly had not even given it a thought up to this point, as I had just been selected to be an air wing commander. The CAG job had always been my goal, and frankly I figured the odds of an ugly Norwegian becoming a *Blue Angel* were pretty darn slim anyhow. Jim's perception was that I should at least give it some thought, because he felt that I would be good at it. Wow! I was certainly flattered by his comments, but I still felt that it was a long shot and I did not want to give up the opportunity to be a CAG.

Then a twist of fate occurred. The Navy decided that they were going to upgrade the CAG position from the rank of CDR (05) to CAPT (06). This change was going to be gradual, still allowing me the opportunity to get the job as a 05. However, it now opened a window of opportunity to potentially do a two-year tour with the *Blues* and then get a 06 CAG job. I got a second call from another ex-*Blue Angel*, Randy "Pogo" Clark. Randy was strike-operations on *Ranger* during my cruise as the CO of VA-192. He was a great guy and did a tremendous job as Strike Operations. In that job, he had the opportunity to judge the leadership and airmanship of all COs in the air wing. He said, "Duster, you are the right guy to apply for the Flight Leader of the *Blues*."

OK, now I started to check into the application process. I approached my boss, Captain Morris, to get his opinion and permission to apply. I remember telling him that it would only entail some paperwork and if I was lucky enough to be a finalist, maybe a two-day trip to interview with the Chief of Naval Air Training (CNATRA) in Corpus Christi, Texas. I commented further that my chances of being selected were slim to none. He pretty much agreed with my assessment, which was not very encouraging, but at least, honest.

***

BLUE ANGEL FLIGHT LEADER SELECTION PROCESS: The Flight Leader selection process is very different from the selection process used to pick the rest of the pilots. Technically, if you are good enough to land a tactical jet on an aircraft carrier at night, you are good enough to eventually learn to fly a *Blue Angel* airshow. However, there is a lot more to being a *Blue Angel* than just flying airshows. So, to help ensure that the right pilots are picked (all but the Boss are picked by the current team), their process involves "rushing" the team by attending as many airshows as possible. By doing this, the *Blues* get a chance to get to know the applicant, and to judge how he might fit into the team.

Regarding the flight leader, the applicant must have completed a tour of duty as a commanding officer of a tactical jet squadron flying from the decks of an aircraft carrier to qualify for the Commanding Officer/Flight Leader job. Unlike the rest of the pilots, the flight leader applicants do not rush the team. They are encouraged to try and make at least one show during the process, but it is not always feasible since many of the Boss applicants may still be on sea tours. I was able to make one show at Naval Air Station Norfolk, Virginia, which was a real eye opener for me.

***

FINALIST: As I recall, there were a dozen Boss applicant packages submitted to the Chief of Naval Air Training (CNATRA). Initial review of career timing and individual service records narrowed this number down to six finalists. Due to a schedule conflict, one of the six interviewed early, which left five of us invited to an interview process at CNATRA Headquarters in NAS Corpus Christi, Texas. The interviewers consisted of the training command commodores and one ex-*Blue Angel*. This board, culminating in a one-on-one with CNATRA, RADM John Disher, interviewed each of us separately.

The interview was intense to say the least. Most of the questions were scenario-driven with no easy answers. Although most of us had mustaches, I was the only one who did not shave it off for the interview. My reasoning was that I had such a slim chance of selection, that it was not worth having to re-grow my precious red mustache. The favorite candidate, in my opinion, was John "Rat" Leslie. He met all the criteria. He was a "shit-hot" test pilot and the only F/A-18 *Hornet* experienced applicant. When asked by the board why I should be selected over the other candidates, I replied, "I believe that I am the only applicant with experience as a navy recruiter. If selected, I intend to work very closely with the Navy Recruiting Command to more effectively make use of the *Blue Angels* as a recruiting tool."

Other than this potential advantage, I was the least impressive when it came to the image factor, and the mustache did not help. When the interviews were completed, we all went to a nice dinner with the board members and the admiral. Unlike today, when the selectee is announced immediately following the interview process, we were simply thanked for applying. The admiral then said that within a month or so, he would be getting in touch with all six of us, whether selected or not.

***

The Call: I went back to a very busy schedule running the assignment process for some 14,000 naval aviators. In all honesty, I had pretty much forgotten about the *Blue Angel* interview. It was not unusual in my job to receive calls at home from flag officers. Usually, those calls were not all that pleasant since they were most likely intervening on behalf of a subordinate dissatisfied with an assignment.

On this evening, Barbara picked up the phone. I heard her say, "Yes sir admiral, he is in the other room, I will get him on the line for you." As I walked to pick up the phone, I asked her who it was. She said, "It is RADM Disher." I replied, "Oh of course, I know what this is about." I picked up the phone and said, "Good evening admiral. I suspect this is the call telling me that I did OK, but that you have picked someone else." He replied, "No, I have already made those calls. You are the selectee to be the new Flight Leader and Commanding Officer of the *Blue Angels*. One caveat. Shave off that mustache tonight!"

I almost fainted. I must have been white as a sheet after I hung up the phone because Barb seemed very concerned. "Another pissed off flag officer, I presume?" "Barb, please sit down." "Oh my God, have you been fired?" "No, but I will be leaving this job early, because I have just been selected to be the Flight Leader of the *Blue Angels*!" Now, she almost fainted.

***

Life-Changing Event: I was not in any way prepared for what happened once the announcement of my selection was made public. I am sure that most of my contemporaries and close friends were shocked because I had never indicated that I was even applying, so I fielded lots of calls that started with, "You have to be shitting me Duster, Flight Leader of the *Blue Angels*?" My usual reply was that I was selected to fill a Norwegian quota. Regarding my family and numerous friends from North Dakota, I would describe the typical reaction from them as one of shock. I now had just 60 days to get re-qualified in the A-4 and complete all the other prerequisites before officially joining the *Blues*.

***

A-4 *Super Fox* Training at the Naval Air Test Center: Since I had not flown the A-4 in 14 years and had never flown the *Super Fox*, the Navy sent me to the Strike Warfare Division at the Naval Air Test Center in Patuxent River, Maryland, for training. They had a two-seat TA-4J that I flew initially, including a cross country with LT Dave Kennedy to build up some hours, and to attend a *Blue Angel* airshow in Lake Charles, Louisiana. Now that I was a selectee, the team set me up for a ride in the spare #7 jet, which was a two-seat TA-4J. About halfway through the show, #4, Curt "Griz" Watson had a problem with his jet and indicated that he was going to land and jump in the spare. They quickly strapped me in the backseat. He paid no attention to me, other than to remind me that this jet only had a J52 P6 engine so he would be working his ass off to keep up with the other jets. This ride was another huge eye-opener for me, as it was my first exposure to how close these guys flew.

I eventually graduated to the single-seat Marine Corps A-4M, which was basically a *Super Fox* with a self-starter mechanism. Since Pax River was only a two-hour drive from D.C., I was able to accomplish this training while still holding down my job as NMPC-432. The clean A-4M was a literal rocket ship, and an absolute joy to fly. I also had an opportunity to get a ride in an F/A-18. The performance was impressive, but the flight controls were not. It became obvious to me that the upcoming transition would require modifications to the *Hornet* flight controls.

***

Joining the Team: About three weeks prior to taking command, I checked into the squadron as the newbie Boss. My fellow newbies (all first-year officers, no matter what their assigned position, are referred to and treated as newbies),

Donnie "Big Time" Cochran #3, Wayne "Leroy" Molnar #7, Wes "Doc" Robinson Flight Surgeon, Doug "Flounder" Hocking Public Affairs Officer and Mike Mullally as the new *Fat Albert* pilot. We newbies attended the end of season airshow briefs, debriefs and other functions in our khaki uniforms, remaining in the background, and saying nothing other than "Glad to Be Here," which was and still is the tradition.

Big Time was an RF-8 pilot and Leroy was an F-14 *Tomcat* fighter pilot. We newbies joined the veteran team members, which included #2 Mark "Birch" Bircher, who was an F-4 *Phantom II* pilot in the Marine Corps and an F-15 *Eagle* exchange pilot with the Air Force. Pat "Sponge" Walsh #4 came from the Light Attack community as an A-7E pilot and was the only *Blue Angel* with F/A-18 experience having been assigned to Air Test and Evaluation Squadron Five at China Lake. Curt "Griz" Watson #5 also came from the Light Attack world as an A-7E and A-4F driver. Dave "Hollywood" Anderson #6 completed representation of all naval aviation's major tactical jet carrier communities as an A-6E pilot. The Events Coordinator, #8, was Bill "Soupy" Campbell, who was a Marine Weapons Systems Officer with experience in both the F-4 *Phantom II* and the F/A-18 *Hornet*. The returning support officers included Jim "MO" Anderson Maintenance Officer, LCDR Vance Moore as the Supply Officer and Hans "Boba" Fett as the Administrative Officer. Filling out the team, Pete Donato and Stan Graham were both returning as C-130 *Fat Albert* pilots.

1986 Blue Angels, *from left to right: Bill "Soupy" Campbell, Dave "Hollywood" Anderson, Pat "Sponge" Walsh, Mark "Birch" Bircher, Gil "Boss" Rud, Donnie "Big Time" Cochran, Curt "Griz" Watson, Wayne "Leroy" Molnar.*

Unlike the *Hornet* turnover program, there really was no opportunity for any of us newbies to fly with our counterparts during practice airshows. As I mentioned earlier, the #7 jet was a TA-4J, with a J52 P6 engine, which was so underpowered that it was only used as an emergency spare. In other words, we started our training from scratch. I did fly with my predecessor, "Hoss" Pearson in the backseat of the #7 jet for one flight. It was just the two of us with no wingmen, but he did fly all the profiles to demonstrate what would eventually be expected of me as the flight leader. He basically flew profiles typical of a second year Boss that is finishing the season. The consistent 200-foot above ground level (AGL) bottoms that he easily performed to overhead maneuvers were incredibly intimidating to me. *I thought to myself for the first of many times during that first year of training, "Shit, I may have bitten off more than I can chew in this job."*

***

CHANGE OF COMMAND: Considering that Barb was teaching full time in a job she loved, the kids were all settled into their second year at their respective schools, and I would be deployed to El Centro, California for three months, we decided to delay the family move until the end of the school year. However, there was no way that we were not going to experience, as a family, the wonderful ceremony and celebration associated with a *Blue Angel* End of Season Airshow and Change of Command.

To make that happen, I found a beautiful condo on Pensacola Beach (it was out of season, so relatively affordable) that was large enough to not only house our family, but also to host a post-change-of-command party. It was a good thing I did because my buddies in the "Happy Hooligans" Air National Guard unit in Fargo decided that they were going to get in on the festivities. They did so by somehow finding and filling up, what I seem to recall was an old Convair C-131 with my relatives and friends from North Dakota. They parked that thing right on our flight line much to the horror of my Public Affairs Officer (PAO), Flounder Hocking. "Oh my God Boss, this can't be legal, and the press is going to have a field day with this obvious abuse of military funding."

I replied, "Flounder, this is the North Dakota Air National Guard. They work for the Governor through their state Adjutant General, and they can do whatever they damn well please, so let's embrace it as a great opportunity for a positive story." Since they also brought along a reporter who represented both the Fargo Forum newspaper and a local television station in Fargo that is exactly what happened. Kevin Wallevand, who at this point was just getting started as a reporter, followed us throughout the weekend, including the post change-of-command party at the beach condo. Kevin was a laid-back guy with a pleasant personality that was perfect to conduct interviews with many of my friends, both on and off camera. I can still clearly remember he and his crew, with help from my buddies, lugging their camera equipment along the beautiful sandy beaches of Pensacola conducting interviews on the go.

As far as the actual Change of Command, as is tradition, Hoss used the opportunity to recognize and thank the squadron for their support over his two-year tour as the commanding officer and flight leader. I kept my remarks short but did use the opportunity to thank Barb and the kids for putting up with all the moves and other challenges that gave me the opportunity to be *Blue Angel* #1. I then announced that when they did join me at the end of the school year, we would live wherever they chose (meaning very near the beach). Although certainly not a husband or father-of-the-year candidate, I did elicit some very broad smiles from my beloved family. *By the way, instead of living on the base, we did end up renting a house in Gulf Breeze, very near the beach.*

∗∗∗

INTERNATIONAL COUNCIL OF AIRSHOWS (ICAS): The yearly convention was held in Las Vegas during our November flying break. ICAS sets the standards for both excellence and safety for all airshows. This was my first opportunity to meet and hang out with other airshow performers, both civilian and military. This is the event that serves to solidify schedules for the upcoming year. The various airshow organizers have an opportunity to recruit performers, as well as to lobby the *Blues, Thunderbirds* and *Snowbirds* to headline their shows.

Some civilian performers prefer to headline shows that do not have a military demonstration team in attendance. Others prefer to show their stuff in front of the larger crowds that are associated with the military demonstration teams. The lobbying part was a heck of a lot of fun, as the various show sites competed for our attention in some rather unique ways. *I must admit that I was sorely tempted to ask for autographs from airshow icons like Bob Hoover, Leo Loudenslager, Jimmy Franklin, Wayne Handley, Gene Soucy and many others. I felt totally humbled and more than a little intimidated as I listened to their recollections of performing hundreds of airshows all over the world. I thought, "What in the hell am I doing here?"*

24. "Up We Go!"

WINTER TRAINING: We took a couple of weeks off prior to beginning to train the 1986 team. The previous year's accident at Niagara Falls during the New York Airshow on July 14, 1985, resulted in the tragic loss of #6, Mike "Dink" Gershon. This incident affected the last half of the 1985 season as well as the start of the 1986 season. The most efficient solution to the loss of the opposing solo would have been for #5, Andy Caputi to return for a third year as the lead solo in 1986. For many good reasons, he elected not to do that, which meant that we needed to find a solution or go with a five-plane airshow, which was not an option. There were no Ex-#5's available, so Curt Watson volunteered to move from #4 to #5. To my knowledge a move like this had never been done in the 40-year history of the *Blues*.

This also meant that Curt would end up with a four-year tour, which was not good for his career progression. Since we now had a situation where neither of our solos had any experience, Andy, very unselfishly, did stay with the team for an extra couple of months to help Griz and Hollywood Anderson get started on the solo program. We all agreed that the goal would be to get some extra practice time, and most importantly, to keep it safe, which they did. The loss of Mike was extremely hard on the team, and it also cast a shadow over the 1986 season. The media interview question that we were most often asked during the first months of the 1986 season usually regarded the accident at Niagara Falls.

We initially flew a few flights in the practice area just west of the Pensacola base, over the Gulf of Mexico. Taking into consideration the unusual situation of having no returning solos, we decided to make a special early deployment to El Centro during the month of December. As a first year Boss, this also gave me a head start on getting my act together, which I certainly needed.

The training is a building block approach starting with just two planes working together. #2, Birch and I worked on roll/loop combinations and other basics using a base altitude of 2,200 feet above the ground as our target bottom for all maneuvers. This means that we considered the ground to be 2,000 feet. The goal by the start of the actual show season was to make our bottoms 200 feet above the *actual* ground level.

While we were doing this, #4 was working with #3 and #5 with #6. After just a couple of these flights, #4 took over training the diamond formation. He did so by observing my profiles and voice calls, while working #3 into the formation with the lead and #2. Keeping in mind that the closeness of the *Blue Angel* diamond and delta formations are based on trust in the flight leader, it was imperative that I learned to be consistent in both my flying and my voice calls. This requires memorizing all the maneuvers for three different airshows: high, low and flat. *I had never studied so hard in my entire life.*

We flew at least two, sometimes three practices every day. During this add-on deployment to El Centro, we flew seven days a week. The A-4 had a "feel" system that required additional pull on the stick to get to neutral, so the grind was both mental and physical. All the newbies experienced some degree of wrist soreness. Also, since we were subject to high levels of acceleration force, and for good reasons do not wear anti-G suits, we all worked-out to stay in the best possible physical condition. As an avid distance runner, my resting heart rate was around 50 beats a minute, which is healthy, but not ideal for resisting G-forces. The good news was that as the flight leader, my performance was much more about smoothness and consistency than G-tolerance.

Because El Centro was a very popular winter training destination for both the training command and the fleet, there was very little housing available on the base. That meant that all the officers were billeted at the Ramada Inn located in the town of El Centro, which was about 10 miles from the base. Luckily, the Ramada had a "Denny's" restaurant attached to it, so we enjoyed especially good breakfasts. This was important because our days started very early. As I recall, our first brief was at 0715 with a launch at 0830. The early training flights would last as long as 1.5 hours, followed by a very intense two-hour debrief. We would take a short lunch break and do it all again, usually finishing our day at around 1430. Since I was not only the flight leader, but also the squadron commander, I would then accomplish those duties followed by a workout, early dinner, two hours or more of study and preparation for the next day's flights, then to bed. *There was absolutely no partying or really any leisure time, at least for the newbies during this initial training.*

Despite being an inexperienced newbie, a first-year Boss is still expected to lead all the flight briefings. Proper preparation is key to success. It is inevitable that you will make some mistakes, but there is an obvious difference between making a few errors and not being prepared. Every diamond and delta maneuver begins with a description from the Boss followed by the actual voice commands that are associated with the maneuver. These voice commands are what allows the team to fly so much closer together than anyone else. In other words, the Boss tells the rest of the formation exactly what he is going to do, then initiates the move with a cadence that allows all the aircraft in the formation to complete the movement together.

Let's use the diamond roll as an example. As the Boss, I would transmit, "We are behind the crowd setting up for the diamond roll. Coming left, coming further left a little pull, easing the pull, rolling out the diamond roll. Smoke on, up we go." (on the "g" of go I would pull back on the stick to begin the maneuver). "OK." (on the "k" I would initiate the left roll). As we completed the roll, "A little pull, a little more pull, smoke off." Then we would proceed to set up for the next maneuver. My biggest initial problem was that I tended to start maneuvering before—or simultaneously with—the voice command. Every time I did this, #2 would say, "simo." Eventually I got better, with a few less simo calls from Birch.

Every practice, whether in the desert or over the field was filmed. Also, every practice was critiqued by ground observers. Those observers were the maintenance officer and the flight surgeon. The filming included the initial march to the aircraft and the ingress into the cockpits. The closing of the cockpits in unison, smoke check in unison and finally the taxi evolution. The "shoulder" style intakes of the A-4 allowed for us to taxi all six jets extremely close together without fear of "fodding" (ingesting foreign objects into the engine). This meant that my voice calls began as soon as we started to taxi and continued until all of us were parked and shut down. The disembarking from the cockpit and the march back, including the final salute, were all part of the airshow and filmed every flight. *You might recall from my days as an Aviation Officer Candidate that my performance in the military bearing phase of training left much to be desired. Well, I still sucked, and had one heck of a time keeping my size 13 feet in step with the other guys, much to the delight of all the players during the debriefs.*

All six Blue Angel A-4s taxiing in close formation.

***

DEBRIEFS: These two-hour long sessions are what makes the *Blues* unique. We are always seeking perfection, and of course, never quite achieving it. During the early part of winter training, we were still seeking perfection, albeit at a very elementary level. Unfortunately, we didn't do that very well, either. It is a brutal experience for a first-year Boss, so you'd better have a thick skin. The debrief begins with a phase called "general-safe." This gives each of the pilots, starting with the Boss, an opportunity to share the mistakes that you remember making, especially the ones that could be considered to affect the safety of the flight. The more of these that you can recall the better, because it shows that you are paying attention and have a good chance of fixing the error prior to the next flight. In this phase, you do not question if other folks in the flight may have contributed to your performance issues. A good example would be me admitting to having a ratcheting (jerky) roll-rate in the diamond roll. Later in the brief #4 might say, "Boss, you had a ratcheting roll rate because both #2 and #3 were pumping through being flat and deep throughout the maneuver, which adversely affected your roll-rate."

Following the general-safe portion of the debrief, the real nitpicking got underway. As I mentioned, everything was filmed, starting with the march to the airplanes known as the "walk-down." In early training, we often had separate practice sessions for the Diamond and the Solos. Number 4 Pat "Sponge" Walsh, the slot pilot, headed up this portion for the Diamond, going through every maneuver that we practiced on that flight, also asking for comments from the ground observers. Number 5 Curt "Griz" Watson headed up this portion for the Solos, and as the Operations Officer he would head up this portion for all six of us when we practiced as a Delta. We all took notes so that we could attempt to fix our mistakes during the next flight. No one expected the newbies to fix all their numerous errors from one flight to the next, however, they did expect you to show progress. *I am certain that I speak for not only myself, but probably every first-year Boss, when I recall thinking, "Damn, I am not at all sure that I can ever get good enough at this to fly an actual airshow."*

***

Flyover of the 1986 Senior Bowl: After completion of the mini El Centro practice session, we had progressed to the point where we could fly in a Delta Formation. We were not performing aerobatics in that formation yet, but we decided that we could safely execute a flyover of the Senior Bowl in nearby Mobile, Alabama. This was a very special event because #5, Griz Watson, had been a standout running back at the University of Tennessee, and had played in the Senior Bowl prior to being drafted by the Green Bay Packers. Flyovers of most events are sequenced to occur on the last note of the *Star Spangled Banner*. This is not easy to do, especially if the event is on television and being held in a major metropolitan area. The good news was that although the Senior Bowl was being televised, one could hardly describe Mobile as a large metropolitan area, so we were able to pull it off. *This was the first of several flyovers that we would perform over the next three years. All of which would be more challenging than this one.*

***

Back to El Centro: In early January, we deployed the entire squadron for the real winter training. By this time, Griz and Hollywood were comfortable enough to proceed with the normal solo syllabus, and we quickly settled into our routine. Unlike the rest of the boys, I absolutely loved El Centro. The laid-back agricultural community reminded me of my time on the farm in North Dakota. I quickly made friends with folks in the local community. I was especially close to Donnie Lamb, a local crop duster, and Dick Wymore, a WWII Navy veteran and avid bird hunter. We usually flew once on Saturday morning and then took the rest of the weekend off. The boys would all head over the hill to San Diego while I stayed behind to hang out with the locals. This time away from the training grind was important for getting us ready to go on Monday.

Dick had a German shorthaired retriever that was especially talented when we hunted grouse. She was a great pointer as well as a retriever, so even a below-average shotgun shooter like me had some success. She did however have standards. We were on a dove hunt, and I missed a shot at a dove that flew near us. The dove then landed in a tree, so the great sportsman here shot it out of the tree. The German shorthaired refused to retrieve it. I swear she looked at me with contempt for pot shooting that dove.

We took our new flight surgeon, Wes "Doc" Robinson out on one of our hunts. This particular day we were shooting pigeons in a large feedlot, a couple of miles south of El Centro. The feedlot had a pen full of Brahma steers. The pen consisted of some open space and a tin sunroof. The sunroof was loaded with pigeons. Brahma cattle are a rather fierce breed; especially the bulls. Fortunately, this pen contained only steers, but Wes did not know that. The farmer was feeding these steers carrots, so I briefed Doc that we would climb into the pen, grab a handful of carrots, toss them onto the sunroof like a grenade, then shoot the pigeons when they took flight. This worked really well for Dick and I, but when I turned around to check on Doc, he was aiming his shotgun at the head of a Brahma steer that had been spooked by our shooting. Doc thought the steer was going to attack him. *This came very close to becoming a very expensive hunt. If Doc had shot it, we would have owned it.*

Donnie Lamb and I started what turned out to be a yearly tradition. We decided to celebrate reaching the halfway point of winter training with an "Imperial Valley Pub Crawl." We enlisted Donnie's daughter, Darlene, as the designated driver, and began our mission to hit as many bars as we could in Brawley and El Centro. We always ran into crop duster friends of Donnie, and rarely got to more than three bars before the "Tequila Shooters" got the best of us.

Donnie also hosted the annual hangar party at the Brawley airport. This was a particularly raucous event, well attended by Ex-*Blues* as well as our current

Blue Angel Team and local folks. About halfway through the event, Donnie would man up his crop duster and perform a night crop dusting routine right next to the hangar. *Donnie was one hell of a pilot and became my lifelong friend. When he passed away a few years ago, I attended his celebration of life. I had an opportunity to say a few words, "I just want all you crop duster guys to know that I always wanted to be a crop duster. Unfortunately, I wasn't a good enough pilot, so I had to settle for being a Blue Angel."* Both Donnie and Dick are honorary Blue Angels.

***

DEVELOPING A RELATIONSHIP WITH THE SAN DIEGO CHARGERS FOOTBALL TEAM: Part of our outreach to the various communities that we visited throughout the show season was to offer flights in the backseat of the #7 jet. These were usually reserved for local celebrities and community leaders to help raise awareness and increase attendance at that airshow. During winter training in El Centro, we flew some very high-profile folks including movie actors and professional sports stars. Since Griz was drafted by the Green Bay Packers and made their practice squad, he had credibility with the Chargers.

Hall of Fame Chargers quarterback Dan Fouts, following a flight with Griz Watson.

We developed a close relationship with Sid "Doc" Brooks, a 20-year Air Force veteran who was the beloved equipment manager for the Chargers for 27 years. In 1986, Sid talked Dan Fouts, the hall of fame quarterback, and the coach, Don Coryell, into flying with us in the #7 jet. After their flights, they stayed around to join the whole *Blue Angel* squadron for a barbecue. In return, they invited us to attend their pre-season camp, which was being held at the same time as the Miramar Airshow later that year. *During that visit to their pre-season camp in August of 1986, Griz and Hollywood gave a talk to the entire group of Chargers pre-season team members. As I recall they talked about the importance of teamwork. They showed film of their maneuvers both from the ground and from the cockpit with the theme: "If we don't work as a team, we will pay the ultimate price. We will die!" This got the attention of the 100+ players still competing for the final 53 slots.*

Although these are Hornets, *the solos briefing the Chargers players on "teamwork, the ultimate price," used a similar A-4 photo. It was very effective.*

***

SLOW LEARNER: As I have mentioned many times in this book, I was certainly not God's gift to naval aviation. In other words, I had to work very hard to become a good flight leader (just ask Sponge Walsh, who as the slot pilot, had to back me up on the various maneuvers). Some of the diamond maneuvers were more difficult than others to perform properly. One of these was the tuck under break, which we called the TUB. Our standard procedures called for any of the wingmen to "clear" the formation if they became uncomfortable or out of position.

We were out in the desert operating area on a rare third practice flight of the day to see if we could improve our performance of the TUB. About halfway through the 270-degree roll off required for the maneuver, I suddenly got a touch of vertigo, so I called, "Boss is clear!" Sponge quickly replied, "Boss you can't clear, you are the flight lead!" As it turned out, I was in perfect position although my brain was telling me otherwise. *Bottom line: Embarrassing, but also hilarious, and of course, something I never lived down, at least with Sponge.*

I also had some difficulty determining direction of flight from the inverted position. During early attempts at executing the double farvel (with me upside down, #2 and #3 right side up, and #4 upside down on top of me), I turned the wrong direction as I began to push out of the maneuver. To ensure it never happened again, I taped the letter "L" on the top of my left leg for a few days. I also carried a laminated "cheat" sheet in the map case. This cheat sheet had the exact order of all the maneuvers in all three of our potential airshow routines in case I would forget. *I never used it in three years of flying airshows, however, during one of my last airshows, I forgot to put it in the cockpit, and for a few seconds I was mortified. Silly, I know, but developing a habit pattern of preparation is important, and that cheat sheet was like a pacifier to me.*

***

First Real Airshow: Our first public airshow, as is the tradition, was held at El Centro. Although a great opportunity to get in another practice session, since it is the same venue that we have been using for three months, it does not feel like a real airshow. The first real show was held at Luke Air Force Base in Phoenix. Luke, being a premier Air Force Tactical Fighter Base, certainly presented a pressure packed opener for the 1986 team.

Just as we settled in to begin the brief for the airshow, the door to our briefing room burst open and a beautiful, truly exotic, belly-dancer leaped onto our briefing table and began an awesome performance. Poor Griz turned white as a sheet, obviously worried that me, as a first year, first-airshow flight leader would now totally lose all semblance of concentration, resulting in a disastrously awful airshow. Luckily, it had the opposite effect, totally relaxing me as I realized that the *Thunderbirds* had just pulled an awesome opening move on what was to be three years of punking one another. *The show was far from perfect, but it was a good one that set the tone for a great final year of performances flying the venerable A-4 Skyhawk.*

***

YELLOW FLIGHT SUITS: In my opinion, these things were incredibly ugly. However, since they were part of our allowance of show suits, we needed to use them. To prevent most people from seeing us in these clown suits, we only wore them while traveling to a show site, which was usually on a Thursday. The best and most humorous description of these suits can be attributed to Griz Watson: "Boss, I cannot stand still in this yellow flight suit because if I do, a dog will piss on me." Curt was a hell of an athlete but at 5 feet 10 inches and 230 lbs, in this flight suit, he did resemble a fire hydrant. *The good news is that we got rid of these things in 1987, although they have made a comeback from time-to-time over the last 35 years. It must have been particularly skinny teams that brought them back.*

***

PARKING SPOT AT THE NAS PENSACOLA OFFICERS' CLUB: Shortly after returning to Pensacola from El Centro, I decided to stop by the officers' club for a beer. Most of the boys drove some sort of sports car, but this good old farm boy was stuck with a 1978 Dodge Stretch-Cab, pickup truck. It had well over 100,000 miles on it and it also suffered from severe corrosion issues. In other words, it was one ugly truck. I was in civilian clothes when I pulled into the officers' club parking lot. It was "Happy Hour" so there were not a lot of available parking spaces. I was delighted to find that there was a spot with a nice big sign designating a parking spot for the Commanding Officer of the Flight Demonstration Team (the official title of *Blue Angel* #1).

I pulled my truck into the spot and got out when one of the civilian security police for the base walked up and said, "Hey! That spot is reserved for the Flight Leader of the *Blue Angels*. Get that piece of crap truck out of that spot right now!" I replied, " I am the Flight Leader of the *Blue Angels.*" He scoffed and said, "Oh sure, and I am President Reagan. Now get out of that parking spot." So, I got back in my truck and moved it out of the parking spot.

A few minutes later, I was having a beer when this same guy came up to me in the bar looking like somebody had shot his pet dog (apparently somebody who knew me had witnessed the eviction). He said, "Commander I am so sorry I kicked you out of your parking spot. Shit! I was sure that all of you guys drove Corvettes, and certainly not nasty old trucks." He was on duty so I could not buy him a beer, but we did have a good laugh and a discussion about not judging people by the vehicles that they drive.

***

40TH ANNIVERSARY OF THE *BLUE ANGELS*: We were especially honored to be the 40th *Blue Angel* Team and to have an opportunity to host the 40th Anniversary celebration in Pensacola. I must admit that there was a bit of pressure since instead of doing this at the end of the season, it was celebrated in May when we were still struggling a bit with our consistency. And of course, a huge number of Ex-*Blue Angels* were in attendance, including the first *Blue Angel* Flight Leader, my personal hero, Butch Voris.

The wonderful aspect of the reunion was that I got to meet so many Exes so early in my tenure as the Flight Leader. This event also featured Honorary *Blue Angel*, Bob Hope. He brought along a flight suit that we had given him several years earlier. He put it on, but it was a very tight fit, which elicited several awesome one-liners. He marched with us to start the airshow; cracking jokes the entire time. Bob Hope seemed to have the same wonderful persona whether he was on stage or just hanging with the boys.

We had an additional surprise guest that I knew nothing about until I saw this amazingly attractive young lady riding out in the back of a truck to fly in *Fat Albert* for their Jet Assisted Takeoff (JATO) and flyby. It was Brooke Shields. When I inquired as to why I had not been aware of Brooke flying in *Fat Albert*, I got the following answer from the "Bert" pilots, "Boss, we thought it would be better to ask forgiveness rather than permission." *I must admit that this is exactly what I would have done so I just smiled and asked if she had a good time.*

***

ANDREWS AFB AND THE U.S. NAVAL ACADEMY: Since my family had remained in Northern Virginia to complete the school year, this was a homecoming of sorts for me. They had not seen me fly an airshow and the kids were even dealing with questions from their friends regarding my whereabouts. When Ryan told his soccer teammates that I was gone because I was a *Blue Angel* they just smiled and said, "Oh sure he is."

With Barb at Andrews AFB show in 1986.

Whenever we are doing an airshow in a large city, we travel via an escorted caravan to the show site. This is always a blast for us as well as the motorcycle cops who get a chance to show off their skills in leading a dozen or more vehicles through some impressive traffic congestion. To keep this challenging evolution safe, the rules require that any vehicle driven in the caravan must be driven by a *Blue Angel*. The vehicle could belong to anyone that we approved for the caravan, but it had to be driven by a *Blue Angel*. I decided that it was time to show our kids' friends that I really was the *Blue Angel* Flight Leader. Ryan's soccer coach got a couple of parents with vans and loaded up the soccer team. I got two of the *Blues* to drive the vans, and they were off to be my guests in the caravan and at the Andrews AFB airshow. Since we did both Andrews and the U.S. Naval Academy, we spent a wonderful couple of weeks in the Washington, D.C. area, allowing me to reunite with my family, as well as to finalize plans for their move to Pensacola.

***

PREMIERE OF THE MOVIE *TOP GUN:* We flew the Andrews Show, the Naval Academy Show and a flyover of the Naval Academy graduation during our two weeks in the D.C. area. The Naval Academy is a remote show (airshow performed at a site other than the airport that we take off from) flown from Andrews. We had a very tight schedule, which included an invitation to the Kennedy Center to see the premiere of the movie *Top Gun.* Like most Kennedy Center events, it was a "coat and tie" affair. Our schedule indicated a 7 p.m. event, which gave us plenty of time to get back to the motel, change into commitment attire with tie, and head for the Kennedy Center.

It was a typical warm and humid day in the end of May, so with the limited air conditioning system in the A-4, we were some tired, sweaty and stinky boys when we landed at Andrews. To our surprise, a limo was waiting to take us directly to the Kennedy Center for a 5 p.m., not 7 p.m., event. No time to change, so off we went in our stinky flight suits. Due to the horrendous D.C. traffic, we arrived a few minutes late. No problem, we could just sneak in the back. Not so fast. They were holding the start of the movie until we arrived, and they had us sitting in the front row. Thank goodness the emcee explained our informal appearance, much to the delight of the audience who gave us a rousing welcome. *It may have been a special honor to be seated in the front row but watching a movie in the Kennedy Center from the front row is no better than doing so in any theater. We all had stiff necks the next day.*

***

THE ENFORCER: Ryan was completing 7th grade, and he was playing for a top-notch northern Virginia select soccer team. I had a chance to attend one of his games. His normal position was as a mid-fielder. He was a very aggressive player with the speed to create offensive opportunities as well as to play defense when required. They were playing another select team with one very large/imposing marking-back. He was manhandling the super-talented, but smallish forwards on Ryan's team. Following a vicious tackle on one of the forwards that sent Ryan's teammate limping to the bench, I noticed the coach whisper in Ryan's ear.

Ryan then moved from halfback to forward. It wasn't long before there was a contested ball between Ryan and the bully of a fullback. That kid ended up flat on his back writhing in pain. Ryan stood over him, then kneeled beside him and said something that made the kid just nod his head. I assumed Ryan had apologized for the legal but violent collision. Ryan then moved back to his normal position, and his team easily won the game. *After the game, I asked the coach about the switch of positions and he replied, "Ryan is my enforcer. If we run into an issue that requires physical retaliation, Ryan takes care of it. When Ryan leaned down to talk to the kid, he just warned him not to go after our little forwards again. Gil, you can teach kids how to play soccer, but you cannot teach them to do so with the courage and physicality that comes natural to Ryan."*

***

THE FAMILY MOVE: True to my promise at the Change of Command, we found a place to rent in Gulf Breeze at 1253 Ainsworth Drive. It was a long drive for me to work, but it was perfect for them, and the price was amazingly affordable. As I recall, it was $750 a month for a large 4-bedroom with a swimming pool, and it was only about a five-minute drive to Pensacola Beach. The public school system in Gulf Breeze was also excellent, which was especially important for Valerie, as she would be entering her third different high school in four years. Val had made both the basketball and the softball teams at Lake Braddock High School, which had an enrollment of over 5,000 students. Of course, she was initially reluctant to transfer again, but landing a job for the summer at a pizza shop on the beach was perfect to meet new friends from Gulf Breeze High School, which had a smaller enrollment than Lake Braddock. Ryan managed to make the West Florida Select Soccer team for his age group, and Sally enjoyed new friends from our family-oriented neighborhood. *Another great benefit of living in Gulf Breeze was that nobody in the neighborhood knew or cared what I was doing for a living, so I could get completely away from the pressure of the* Blue Angel *lifestyle.*

***

COMMUNITY OUTREACH: The mission of the *Blue Angels* is to showcase the pride and professionalism of the United States Navy and Marine Corps by inspiring a culture of excellence and service to country, through flight demonstrations and community outreach. The community outreach included visits to high schools located in the respective airshow area. We accomplished this on the morning prior to our Friday afternoon practice airshow, with two-member teams, consisting of a *Blue Angel* pilot and his crew chief. Both of us would address the entire assembly of students, which were usually gathered in the basketball gym. As one would expect, it was not easy to get, and hold, the attention of a bunch of rowdy teenagers. Luckily, we had the perfect solution.

The video *Dreams,* by Van Halen, was extremely popular and featured a collage of A-4 *Blue Angel* airshows. We began each session with that video, which was the perfect attention-getter. Often, the crew chief got more attention than the pilot because he was closer in age to the student body. He was also a great recruiter, providing an image that the kids who were not planning on attending college could relate to. We supported the Navy and Marine Corps recruiting efforts by inviting and hosting potential recruits at the practice airshow. The practice shows also provided an opportunity to host special guests, including Make-A-Wish kids, Special Olympics and developmentally challenged folks. All of us enjoyed these opportunities, almost as much as the flying.

***

OTHER MEMORABLE AIR SHOWS IN 1986: I would like to preface this section of the book by stating that all the airshows we flew were memorable in some manner. The following ones though, standout for many reasons:

***

FARGO, NORTH DAKOTA: This one was my homecoming, and as such, I will remember and cherish the experience forever. Airshow sites vary widely in their degree of difficulty to perform. All the over-water shows are challenging, as are those that take place at high-density altitudes. Ideal show sites include a runway show line with limited obstructions within the show area. Even more ideal are show sites that have a runway show line and surrounding show area, which consists of cardinal north-south and east-west streets and section lines. Hector airfield in Fargo is one of those ideal sites.

Also, we were hosted and supported by the North Dakota Air National Guard unit known as the "Happy Hooligans." They were an elite group of fighter pilots flying the F-4D *Phantom II*. In fact, even though they were a National Guard unit, they were the 1986, Air Force, "William Tell" aerial gunnery champions. In other words, they were shit-hot, and their members included many of my high school and college friends. Thank goodness this was such a squared-away show site, because I certainly had my share of wonderful distractions to deal with, including family and friends.

The standard procedure for all our airshows, began, with what we call "circle-and-arrival" maneuvers. If we had enough fuel on arrival at the show site (either it was a short distance from Pensacola or an en route fuel stop, or we were lucky enough to have an Air Force tanker accompanying us), we would do this before landing. In most cases, we would land, refuel and then commence the circle-and-arrival sequence.

We had aerial photographs of the show area taken by RF-8G photo recon aircraft; unfortunately, they were in black and white and were often quite dated. As the flight leader, I had over 20 check points on the ground that I needed to verify, to set up properly for the various maneuvers. All the wingmen needed to verify their individual flight lines and timing-check points for the "loop break cross" and "low break cross" maneuvers. The solos also worked their check points to ensure the impeccable timing needed to make their center point hits, which they accomplished with just under a 1,000-mph closure rate. All this rather random activity culminated in the final three delta formation maneuvers. The result was that if someone in the show site-city was not aware of a scheduled airshow, they certainly knew now.

Our Events Coordinator, #8, Bill "Soupy" Campbell and the Narrator, #7, Wayne "Leroy" Molnar, visited all the show sites via the #7 jet during winter training. In Fargo, they met with the show coordinators, Dick Walstad and Darryl Schroeder, plus representatives of the Happy Hooligans. They came back very enthusiastic about the site, especially regarding what was planned for our enlisted troops. So much so, that we decided to fly the jets in on Wednesday, instead of the usual Thursday arrival. The Wednesday arrival allowed for some family time, which included a reception at my sister Linda's and brother-in-law Dave's home near the airport. It was a great opportunity for me to spend time with relatives and close friends, prior to embarking on the rest of a very busy airshow flight and social schedule.

As a tribute to what a big deal this event was, Leroy flew Governor George Sinner as one of the VIPs. After the flight, Leroy remarked that the Governor enjoyed the flight, including a close, albeit high-speed look at his crops on the family farm near Casselton.

On Thursday evening, we all headed out to future Governor, Jack Dalrymple's farm, also located near Casselton, which was about 20 miles west of Fargo. Prior to arriving at the farm, we all gathered in a field a couple of miles from the farmstead. To my surprise and utter delight, they had positioned six giant Stieger tractors. The Steigers were the biggest and most powerful tractors in the world in 1986. Donnie Cochran and I, both being farm kids, were allowed to drive ours without a co-pilot, while the city boys had an observer ride along, just in case. We started them up and practiced driving them in formation. We then headed for the farmstead where a sizable crowd was on hand for the cookout-reception. We lined up the tractors just like we would our A-4s. They had gotten our crew chiefs to set up parking, just like at an airshow, so we simultaneously swung the tractors into parking, much to the delight of the crowd.

The whole event was really emotional for me. One of the duties of the Flight Leader, as I have mentioned previously, is to introduce the *Blue Angel* team and make a few remarks. Although I still had a long way to go to become proficient at flying airshows, I was, and still am, gifted with the ability to speak extemporaneously. In fact, I love doing it, especially at occasions like this one.

I noticed in the crowd, one of my personal heroes and a favorite mentor, coach Ron Pederson. Ron was my baseball and basketball coach at Portland High School. He was also a terrific science teacher. Shortly after I graduated, Ron had moved from teaching/coaching to join his father and brother at Pederson Implement, a very successful John Deere dealership located between Hatton and Northwood, North Dakota. I singled him out for how much he meant to me, and he replied, "Gil, I know you are a pretty decent pilot, but I bet that you cannot remember how to start the first tractor that you ever drove." He then pointed to a 1938 Model "A" John Deere that he had trailered down from his dealership.

I replied, "Coach, I am going to take you up on that challenge, but I am going to ask my sister, Linda, the worst tractor driver in the state of North Dakota, to help me out if I need it."

We headed over to the Model A, with the crowd quickly forming a circle around us and the tractor. I looked at Linda and in a very loud voice that the crowd could all hear, I said, "OK, I need to open both petcocks, one for each cylinder, turn on the gas, since it is a cold engine let's give it a little choke, open the throttle a bit, grab the flywheel and give it a turn." *Putt–Putt*, that old Model A fired right up. The crowd cheered as Linda and I drove that beautiful old tractor around the yard, parked and shut it down. *I consider this my finest hour as a combined farmer/Blue Angel. It is so important not to forget where you came from and the wonderful people that helped you achieve success. This was the perfect opportunity to prove, that although I was the flight leader of the Blue Angels, I was still the same "little old farm kid" from Portland, North Dakota.*

Me and my sister Linda taking a spin around the Dalrymple farm on the 1938 Model A John Deere.

***

SATURDAY AIRSHOW: We were all amazed at the number of people who showed up, and we were excited to get on with the airshow. We manned up and took off. We were about halfway through the show, when this nasty thunderstorm roared in, forcing us to land. We headed for the briefing room, noticing that despite 40-mph winds and torrential rain, most of the crowd remained on the show line. Griz said, "Boss, to my knowledge, we have never re-started an air-show that was interrupted by a storm. I recommend that these hardy folks deserve to see a full show. Let's fuel the jets up and try again!" We could hear the crowd cheer, as it was announced that we were going to give it another try. As is common in North Dakota, the storm was followed by a beautiful clear blue sky that afforded us the opportunity to fly a nearly perfect show.

***

SUNDAY AIRSHOW: We had decided at the beginning of the 1986 show season that we needed to establish a better relationship with the Federal Aviation Administration (FAA), and especially the local FAA representatives, tasked to monitor our compliance with established safety rules. To accomplish this, we invited the FAA representative to attend the first portion of our pre-airshow briefing. The idea being that he would then see us as dedicated professionals instead of wild-ass aerobatic cowboys.

Also attending this portion of the brief was LtCol "J.D." Nelson who was the flight leader of the Happy Hooligans flight demonstration team. They had flown a very impressive demonstration the previous day and were scheduled to fly again, just ahead of us. The FAA rep was a bit on the effeminate side, with a high-pitched voice. He said to J.D., "Colonel, your demonstration yesterday was unsafe. If you fly those same maneuvers today, I will give you a flight violation!" He then waved this piece of paper, representing a flight violation, in J.D.'s face. J.D. grabbed the piece of paper, glared at the FAA representative and said, "You go right ahead and give me a flight violation, and I will shove it up your ass!" *J.D. was a Vietnam veteran from a little town in North Dakota, and he really did not care what the FAA representative thought of his demonstration, as long as the crowd loved it, which they did.*

Upon completion of the Sunday show, Griz and Sponge suggested that I stay an extra day, while the rest of the team flew back to Pensacola. This was the second non-standard decision that was made on my behalf, and I deeply appreciated it.

The next day, my brother-in-law and good friend, Dave Kringlie, who was an avid photographer, asked if he could get a shot of me taking off. I suggested that he go to the north end of runway 36, and I would perform a low transition, followed by a high-performance climb. Prior to taking off, I asked for and received permission from the tower to perform a low transition followed by an unrestricted climb to altitude. I performed the low transition, gaining airspeed while about 30 feet above the runway, got about a mile past the end of the runway where I assumed Dave would be, and pulled the stick back into my lap. As it turned out, this put my exhaust pipe right on top of Dave, which destroyed any attempt at taking my picture, but was a thrilling experience for him.

***

NEW ORLEANS, LOUISIANA: We flew this show at NAS New Orleans. It was notable for being the only time in my three years as the Flight Leader that I did not march to-and-from my airplane. Why? Because, I had food poisoning from a delicious seafood dinner, served at a social commitment the night before the show. As previously mentioned, any of the five wingmen can be missing and we will still fly a five-plane show. The exception is the flight leader. If the Boss can't fly, the airshow is canceled.

It never happened. In other words, I flew every practice and some 240 airshows in three years, without missing a flight. How do you accomplish this? You do so with a damn good flight surgeon, Doc Robinson, and by behaving yourself on liberty. In this case I had thrown up a couple of times before the brief, which eliminated the worst of the symptoms, except for the diarrhea. That was still a huge threat, so we devised a diaper of sorts, just in case. We then set up next to our aircraft, instead of marching to them. On completion of the airshow, while the other five marched back, I hauled ass to the maintenance truck and then to the hangar toilets. The good news was that I made it, without the use of the diaper.

***

SAN FRANCISCO FLEET WEEK: This and probably Seattle are two of the more difficult over-water show sites. Of course, they also draw huge crowds and are an absolute blast to fly. In 1984, the mayor of San Francisco, Dianne Feinstein, had received complaints about the *Blues* flying over the city. Initially she reacted to these complaints by threatening to cancel the event. The *Blues* countered, by taking her for a flight in the #7 jet. She had an absolute blast, so by the time we came to town for the 1986 event, she was a strong supporter. So much so, that prior to the show, Dianne, her husband Richard Blum, Barbara and I had a wonderful private dinner at an Italian restaurant in the city.

With San Francisco Mayor Dianne Feinstein in 1986.

After the airshow, the Fleet Week activities included an event that involved formal introductions of all the commanding officers of the participating ships and other Navy units. Formal introductions like these, always follow the protocol of introducing the most junior officers first, culminating in the most senior Flag officer being the final introduction. I was certainly quite junior, so I took up my lineal position in the introduction line.

We were set up so that when your name was announced, you would walk down a staircase to be greeted by Mayor Feinstein. Just prior to the start of the introductions, an aide for the mayor grabs me out of the line. He said, "The Mayor insists that you be the last to be introduced." Not good. Fortunately, the senior admiral understood that I had nothing to do with this breech of protocol. Still, there was a lot of grumbling as I shrugged my shoulders and headed for the back of the line. As each officer was introduced with rank, name and command, the mayor greeted them with a firm handshake. That is until my name was called.

Since we had become pretty well acquainted at the private dinner, instead of shaking my hand, she gave me a big hug and a kiss! *Situations like this tend to paint the* Blue Angels *as prima donas. It makes it that much more important to handle oneself with humility, and even self-deprecation, when necessary.*

***

ABBOTSFORD, BRITISH COLUMBIA: We were fortunate to be invited to participate in the 1986, World Exposition on Transportation and Communications, better known as simply, Expo 86. The main event was centered in Vancouver, but the airshow was staged at Abbotsford, which is about 60 miles south of Vancouver, just across the border from the state of Washington. What was unique about this airshow was that there were five military teams from five different countries scheduled to perform. They included the *Snowbirds* from Canada, a team from Brazil, a team from France, a team from Italy and the *Blue Angels* representing the United States. It provided a wonderful opportunity for us to meet and hang out with all these other teams. The Canadians, French and Italians all flew jet-powered aircraft, originally designed as trainers. The Brazilians flew a turboprop trainer. All of them flew more aircraft in their show than we did, but the A-4 more than made up for the numbers with its superior operational performance.

On the first day of the show, the weather was awful. The Brazilians flew first. They had one type of show that they could accomplish in weather as low as a 2,500-foot ceiling. Unfortunately for them, the ceiling was much lower than that, so after attempting a couple of initial maneuvers, they gave up and landed. We were next. Their flight leader approached me and said, "Gil, the ceiling is only about a thousand feet and the visibility is, at-best, 3 miles, so you will not be able to fly a show." I replied, "We have three different airshow routines that we are prepared to fly. One of them, that we call a "flat show," only requires a thousand-foot ceiling and 3 miles of visibility." He was astounded and said, "You fly those tactical jets at over 400 knots in your airshow; how can you safely fly in these conditions?" While trying my best not to sound like a cocky asshole,

I replied, "Oh we can do it all right, probably because we are used to flying in shitty weather, while operating from aircraft carriers."

The weather was a little below a thousand feet, but the visibility was a bit better than 3 miles, so we launched and flew the flat show. The wingmen detested the flat show, because instead of me being able to use altitude to set up a smooth pattern for the maneuvers, we were committed to high-G, flat-turns and reversals, that were much more difficult. Also, we were now sharing airspace with the solos, so that coordination became much more critical. The good news was that we could fly the maneuvers at a distance of 500 feet from the crowd line, versus the normal 1,500 feet, so the crowd loved it. The weather did not improve so we were the only team to fly that first day. *We could not buy a beer at that evening's social event. The typical comment from show attendees was: "I can't believe that you guys were the only ones that had the balls to fly in that weather. The Blue Angels really are the best demonstration team in the world!"*

***

GRAND JUNCTION, COLORADO: It was a great show site, but on Saturday, we ended up flying in some really turbulent conditions. So much so, that for safety purposes, we needed to loosen up the formations a bit. It was probably the most challenging show that I had flown as a first-year Boss, but like all airshows, it was not over until we shut them down in the chocks.

I guess I was a bit distracted on the taxi back, because I inadvertently turned early into parking. Since, in the A-4, we all taxied very close together, then simultaneously turned into our parking spots, this error on my part was rather egregious. As soon as I gave the command, "Coming left, OK" I looked ahead and realized that I was looking at the #2 Crew Chief instead of mine. That meant that #6 had no crew chief to park him. "Dammit, I thought, how embarrassing." No problem though. The front man just directed all the crew chiefs into a right face movement, and they rapidly shifted down one slot to make everything look normal.

Of course, I got a ration of shit from the boys, as we all laughed about the mistake, but also discussed the potential seriousness of getting distracted into making such an error. The next day, when I cleared the runway for taxi back to parking, I discovered that the crew chiefs had devised a huge arrow out of duct tape, with the words: "This way Boss." That was only the first of many arrows leading directly to my crew chief. *I richly deserved this well-thought-out ruse and felt great that the team was comfortable in making fun of me, which they certainly did, more than once in my three-years on the team.*

***

200TH ANNIVERSARY OF THE STATUE OF LIBERTY: We were tasked to perform a flyover for the 200th anniversary of the Statue of Liberty. The *Thunderbirds* and the *Patrouille de France* aerobatic team were also scheduled to participate. Roger Riggs, the flight leader of the *Thunderbirds,* and I decided that we needed to do a joint fly-by of the Statue of Liberty, because the *Patrouille de France* had more aircraft than we did. We staged out of Republic Airport on Long Island and the *Thunderbirds* (they had their own KC-10 tanker) joined up with us over the Atlantic, south of Long Island.

I took the lead and proceeded down the Hudson River, past the Statue of Liberty and the main viewing area, which was aboard the USS *Intrepid* Museum. That gave us a 12-plane flyover versus nine for the French. We continued a few miles up the Hudson River, where, as briefed, we accomplished a 90/270 turn with a lead change. Following the lead change, I had one hell of a time keeping up with their brand-new F-16s, and we had to loosen up the formation until I finally got my act together for the reverse fly-by. *This is another example of us taking advantage of a great opportunity without asking permission. The Navy chain of command would have approved it, but the Air Force may not have, at least not in a timely manner.*

25. Preparing for the Transition to the F/A-18 Hornet

THE DECISION TO TRANSITION was made in 1985, because we had run out of A-4 *Super Fox* aircraft. Determining what type of aircraft that we were going to transition to was the critical decision that needed to be approved by the Secretary of the Navy (SECNAV), John Lehman. The T-45 advanced trainer was a contender, but if we flew that, we would need to fly at least a nine or 10 aircraft show to compete with the European teams. Also, the *Thunderbirds* had just transitioned to the F-16, so it was important to match their capability, as well as to demonstrate an operational fleet aircraft to the taxpayers. The F-14 *Tomcat* was considered, but quickly rejected as being too big and too costly to operate. The A-7 *Corsair II* also got a look, but it was dated, a bit underpowered, and all the birds were needed in the fleet.

Choosing the F/A-18 *Hornet* was the first step, but getting that decision approved by SECNAV required some creative thinking, to justify the use of a new fleet asset (it had only become operational three years prior to this decision) to be used for the *Blue Angels*. Since Secretary Lehman was still an active reserve A-6 BN (Bombardier Navigator), he understood the need to have a viable explanation for the transition decision. The explanation was that the *Blues* would be getting early model Low-Rate Initial Production (LRIP) aircraft, that were not carrier suitable, and therefore of limited use to the fleet. Also, if a war broke out, the *Blues* aircraft would be sent to VFA-106, the east coast *Hornet* RAG, and the *Blue Angel* pilots would then become instructors in the RAG. *This approach satisfied the decision makers throughout the chain of command. Now the real work got underway.*

About halfway through the 1986 A-4 season, one of the VFA-106 instructors, who also happened to be a *Blue Angel* applicant, flew a two-seat

Hornet into Pensacola. Sponge Walsh, who was our only qualified *Hornet* pilot, asked if the instructor would allow him to fly in the front seat. The Instructor Pilot (IP), (name omitted to protect the guilty) readily agreed, not realizing that Sponge intended to fly the *Hornet* in the slot position of our A-4 diamond. He did that, and while performing a diamond loop, Sponge was forced to clear the formation during the back half of the maneuver. He had been using full nose-down trim as the "feel system," which resulted in him not having enough stabilator (elevator in most aircraft) authority to safely recover from the loop. The good news was that we now realized the flight control system in the *Hornet* would need to be modified, to allow us to maintain the standard total overlap and 36 inch separation of a *Blue Angel* formation.

The required flight control modification was designed and tested by CDR Keith Crawford and his team of engineers and test pilots at the Naval Air Test Center in Patuxent River, Maryland. The key modification was to remove the Electronic Countermeasures panel (it was classified and needed to be removed for airshow flying anyway) and replace it with a blank-off plate. They mounted a spring on the blank-off plate that could then be attached to the control stick. We then trimmed to the amount of pull that was required to get the stick to neutral. This pull amounted to the equivalent of up to 30 pounds, which was similar to the feel system in the A-4.

As the flight leader, I adjusted mine so that when I rolled inverted, I did not have to push forward on the stick to maintain one negative-G required for level flight, but rather, I needed to pull slightly to stay level. This was important, in case of a bird strike or other incapacitating event, since the aircraft would climb, rather than nose down into the ground.

McDonnell Douglas and NAVAIR worked closely together to modify the selected Lot-4 *Hornets* designated for our use. They removed the gun and replaced it with a smoke-oil reservoir, and installed a temporary inverted fuel system that trapped enough usable fuel for about 45 seconds of inverted flight in basic engine (an inverted fuel pump was installed in 1987). The Automatic Carrier Landing System (ACLS) was removed, and the Canadian *Hornet* Instrument Landing System (ILS) was installed. Smoke oil lines were added from the reservoir in the nose, to the engine exhaust area. Software changes were also made to the flight control system.

All of this required a special flight clearance, and a designated safe-for-flight envelope that was slowly expanded by the test team. Unfortunately, during this flight test evolution, the inventor of the spring feel system, Keith Crawford, was killed in an accident, while he was flying wing during a diamond roll profile. The test team had another close call during testing of solo horizontal rolls. Unlike the A-4, which had a 700-degree per second rate-of-roll (awesome), the *Hornet's* rate-of-roll was only 260-degrees per second. We were able to safely complete four horizontal rolls in the A-4. We knew that would not be possible in the *Hornet*, but we were hoping for an envelope allowing at least two.

During the test, which was conducted at 5,000 feet versus airshow altitudes of 200 feet, about half way through the third roll, the *Hornet* coupled up and departed controlled flight. This event was recorded, as are all the test flights. The aircraft departed in an extremely violent manner. The pilot was narrating the maneuver for the test record, so when the aircraft departed, he exclaimed, "Holy Shit!" As it continued in uncontrolled flight heading rapidly for the water that he was flying over, he kept saying, "Holy Shit! Holy Shit! Holy Shit!" followed by, "I got it! I got it!" until finally recovering to controlled flight, about 500 feet above the water. *The really scary part of this event is that in the background you can hear the test engineers monitoring the flight saying, "Eject! Eject! Eject!" Had he ejected—or even worse—been killed, it might very well have been the end of the* Blue Angel *transition to the Hornet.*

At some point that summer, I was called in to brief the Chief of Naval Operations (CNO), head of naval aviation, (OP-05) on our transition plan. The admiral asked me if we could do the transition without standing down for a year. I replied, "Admiral, we can complete the transition with just a one-month extension of winter training if we freeze the team. If we must train new pilots to be *Blue Angels* while transitioning to a new aircraft, we will need much more time." He replied, "Rud, you do what you think will work best, and I will back up any decisions you and the Chief of naval aviation Training (CNATRA) make regarding keeping the team intact."

He then asked the following question, "Rud, this is not going to be easy, do you think you can do it without getting anybody killed?" I replied, "Yes sir, we are not going to kill anyone, although we will probably lose an airplane." He frowned and said, "That is not the answer I wanted to hear!" I went on to explain, "Even with all of the testing of the flight envelope, when we actually begin flying the *Blue Angel* maneuvers, we are certainly going to discover sur-

320

prises that will probably cost us an aircraft." "Alright Rud, good luck on the transition, and as is the case with the autonomy of command in our Navy, screw it up and I will fire you!" *He smiled when he said this last sentence, however, as all commanding officers in the Navy know, the price for the wonderful autonomy of command that we enjoy, is absolute responsibility for failure.*

***

FREEZING THE TEAM: We kept five of the six pilots to accomplish the transition. Griz Watson had already been on the team for four years, so he needed to get back to the fleet. Leroy Molnar moved from the narrator/#7 position to opposing solo. Since Leroy had been a superb *Blue Angel #7*, albeit not yet a member of the delta formation, we knew this would work. We then welcomed as the new #7 Cliff "Red" Skelton. Red was coming from being an instructor in the *Hornet*, so he brought a great deal of aircraft knowledge to the team. *I still vividly recall the day that freezing the team was officially approved. I addressed the ready room: "Boys, I have some bad news for you. Because you have as yet failed to perform a perfect airshow, you are all extended for a third-year on the team." Lots of smiling faces and high fives indicating that they were thrilled and excited to press on with the transition.*

***

END OF SEASON AIRSHOW AND *HORNET* **INTRODUCTION:** We flew our last airshow in the venerable A-4, *Super Fox* on November 8, 1986. It was very emotional for all of us and of course for the many Ex-*Blue Angels* in attendance. The A-4 had been, at least to that point, the best airshow aircraft in the *Blues* 40-year history. We had flown it for a wonderful 13 years. It was fun and relatively forgiving to fly, looked awesome, and was easy and relatively cheap to maintain. In other words, the A-4 was the perfect airshow aircraft.

Now, it was time to move on. We did this with an impressive ceremony. We parked our six A-4s and marched back for the last time in the 1986 season. We then had Red Skelton taxi in and park our first fully painted and ready-to-fly *Hornet*. An era was over, and a transition was officially underway.

26. An Adventure Filled First Year in the Hornet

INTRODUCING THE 1987 TEAM: As I mentioned previously, the only new pilot was Red. We did, however, add some very important officers, enlisted maintenance members and civilians to the rest of the transition team. LT Mike "Manny" Campbell, an A-6 BN, replaced MAJ Bill "Soupy" Campbell as #8, our Events Coordinator. Maj Frank Welborn and Capt Mark Mykityshyn joined Capt Mike Mullally to complete the *Fat Albert* pilot team. LT Jim Hawthorne replaced LT Hans "Boba" Fett as our Administrative Officer. LT Sam Walker joined the team midway through the 1986 season as an experienced fleet F/A-18 *Hornet* Supply Officer. He would play a crucial role in the all-important supply chain, required to support the transition. Alicia Petruska also joined the team as a Navy civilian support representative.

1987 Blue Angels, *from left to right: Mike "Manny" Campbell, Wayne "Leroy" Molnar, Pat "Sponge" Walsh, Mark "Birch" Bircher, Gil "Boss" Rud, Donnie "Big Time" Cochran, Dave "Hollywood" Anderson, Cliff "Red" Skelton.*

Jack Fallon replaced the McDonnell Douglas long-time A-4 technical support guru, Dale Specht, as the Senior Technical Representative for the F/A-18 *Hornet*. Bill Flynn assisted Jack as a second McDonnell Douglas Tech Rep, and Todd Perzanoski joined us as the General Electric Engine Technical Representative. We also added several Navy and Marine Corps enlisted maintenance personnel with *Hornet* experience. The *Blue Angel* 1987 Team was set and ready to go.

***

INITIAL *HORNET* PILOT TRAINING AT VFA-106: Following a brief end-of-season stand down, we discarded our *Blue Angel* flight suits and shiny flight boots, donned our regular fleet green bags and boots, and headed for NAS Cecil Field in Jacksonville, Florida. CDR "Swede" Peterson was the CO of the east coast RAG. He welcomed the challenge of getting a bunch of non-*Hornet* pilots qualified, via a specially designed 25-flight hour syllabus. The syllabus was structured to include most of the normal ground school and simulator training, so that we would become familiar with the aircraft systems. The flight syllabus also included exposure to air-to-air, air-to-ground and aerial-refueling training. Although none of us had actual fleet operational experience in the *Hornet*, we wanted to be exposed to all the capabilities, so that we could intelligently discuss them with the public.

***

USE OF THE HEADS-UP DISPLAY (HUD): Sponge and I, having A-7E experience, immediately fell in love with the much-improved HUD in the *Hornet*, as did Birch from his F-15 *Eagle* days, during an Air Force exchange tour. Although the F-14 did have a HUD, it was not ground-stabilized so Leroy and Big Time, along with Hollywood (no HUD in the A-6) had to learn to use this amazing device. I caught Hollywood and Leroy initially flying the simulator with the HUD in the "off" mode. "No, boys! Turn it on and use it." The good news was that they soon realized the added situational awareness that it provided, and within a few days, they also were hooked.

***

ADAPTING TO FLEET RULES: We discussed at length the importance of acting like normal flight students regarding how we flew the aircraft, adhering to altitudes, airspeeds, etc. On one of my early dual flights with instructor pilot, LT Steve "Cec" Dallaire in my backseat, we departed Cecil Field via a max performance climb so that I could experience full afterburner capability. While enjoying this experience, Steve reminded me that we had a 10,000-foot ceiling to adhere to. Instead of slowly reducing power and lowering the nose to a smooth level off, I reverted to *Blue Angel* mode. Passing about 9,000 feet, I reduced power, rolled inverted, pulled the nose to the horizon and rolled upright, level at 10,000 feet. Steve said, "Wow! That is the first time I have had a student use that level off technique." "Dammit!" I thought, "Shame on me for not following my own guidance." The good news was that as I was profusely apologizing, Steve said, "Boss, I absolutely loved it!" "OK, but don't tell anybody, especially any of my boys."

***

BACK TO PENSACOLA AND SOME FAMILY TIME: While we were at Cecil Field and a short time at Key West, completing 25-hours of flight time in just 10 days, our families were preparing for the holidays. I mentioned earlier that my daughter Valerie was contemplating trying out for the Gulf Breeze basketball team. She did that back in October while I was off on the west coast swing of airshows. When I got back from that trip, I asked her if she tried out. She handed me the sports page from a local weekly newspaper. The title of the article was "Gift from Northern Virginia fills out Dolphin Roster."

Val not only made the team, but she also added her larger school experience, especially on defense, which helped eventually propel the team to a 29-2 record and a spot in the Division Three Florida "Final Four." I did not get to see many games, but I can tell you that they were fun to watch. They pressed on defense and ran the floor from the opening tip to the final buzzer, scoring over 100 points a couple of times. They lost to the eventual state champions, Miami Hallandale 65-55. I flew a *Hornet* from El Centro to Cecil Field, then drove to Orlando to watch that game. Gulf Breeze rotated 10 players including two very talented shorter players. However, at one point, to try match Hallandale's exceptionally tall lineup, Valerie, who at 5 foot 10 inches tall, was the smallest player from either team on the court. *Although she had to endure attending three different high schools in four years, this last move provided a wonderful opportunity that she took full advantage of.*

Barbara was back to teaching math full-time at Gulf Breeze High School. Ryan was enjoying playing highly competitive soccer as the only freshman starter for Gulf Breeze High School and also for the "West Florida Under 14 Select team." All three levels: grade school for Sally, middle school for Ryan and high school for Valerie, were in the same general area of town, so even working full-time, Barb could keep a close eye on all three kids. *The Ruds had settled into life in Gulf Breeze, and we loved it.*

***

START OF THE BLUE ANGEL HORNET TRAINING PROGRAM: Our goal was to work our way through the basic maneuvers that we had flown with the A-4 *Skyhawk*, then work in some new stuff made possible by the increased performance of the *Hornet*. That increased performance involved the use of afterburner. The great news is that instead of a staged afterburner, the *Hornet* had a fully modulated afterburner, which allowed us to use it in formation. We quickly found out that this was easier said than done. If any of the wingmen were late in selecting their burners, they would quickly fall out of formation. The same was true of deselecting burner. The key to success lay in the consistency of my voice commands. We settled on this command: "Burners, ready, now!" I then selected and deselected the burners on the "n'" of now.

It might sound a little silly, but we literally ran up and down the north coast of the Gulf of Mexico in straight and level flight, running these burner drills until we were all in sync. The next issue with the burners was how much could I, as the flight leader, use, and still leave some room for the wingmen to make corrections. Initially, I just ran the throttle to full burner and quickly throttled back to mid-burner range. This was not very effective, so we finally settled on installing a burner-stop for my throttles that allowed me to go to about 70% of full burner. This was enough to safely make all the required maneuvers and still give the wingmen enough extra power to make corrections while in formation.

***

1987 WINTER TRAINING ADVENTURES: We had seen a video of four MIG-29s completing a Diamond Loop on Takeoff. They were going the same way on the same day, but nowhere close to a *Blue Angel* diamond formation. Initially, we started working the profile with just two aircraft. I distinctly remember staying in burner all the way over the top of the loop, which resulted in a nasty, high-G recovery process on the backside. We kept coming out of burner earlier in the profile, until we settled on about 90 degrees nose up as the right spot to come out of burner, with enough energy to comfortably make it over the top and keep the recovery airspeed under control. It was not long before we got comfortable enough to perform the loop with the entire diamond. We had discovered a great way to begin the airshow demonstrating the incredible performance of the F/A-18 *Hornet*.

***

DIAMOND DIRTY LOOP: Hollywood and Leroy came back from a solo practice where they had been experimenting with a combination clean loop/dirty loop. Hollywood said, "Boss, this aircraft dirty loops so nicely that I think you can do it in diamond." We gave it a try the next day with mixed results. Similar to the loop on takeoff, the real challenge was controlling the recovery airspeed on the backside of the loop. I don't recall the maximum gear down airspeed allowed in the *Hornet*, but I do recall that it was based on the gear doors. And yes, we lost a few gear doors prior to beefing them up a bit. The good news was that we were then—and as far as I know, still are—the only military flight demonstration team in the world to perform this awesome maneuver.

***

SOLO FLIGHT ENVELOPE CHALLENGES: The very nature of the solo flight demonstration, features maneuvers flown at the edges of the aircraft flight envelope. The problem for Hollywood and Leroy was that the Naval Air Test Center was still working on flight clearances to expand that envelope, while the solos were putting together their airshow package. The solos experienced flight control caution lights and some flight control anomalies that needed to be resolved throughout winter training. One of the unintended consequences was that the diamond airshow came together quicker, putting pressure on the solos to catch up.

***

DREADED BLACK SMOKE CLOUD IN THE DESERT: A serious deficiency that we had to work with that first year was the lack of an inverted-flight fuel pump. Instead, we had a baffle system that trapped enough fuel for 45 seconds of inverted flight at military power. That was just enough capability for both the diamond and the solos to fly their inverted routines. As I mentioned previously, the solo flight envelope was still being tested while they were developing their airshow maneuvers. Hollywood was under some real (and some self-imposed) pressure to catch up to the diamond, and to initiate some new *Hornet*-specific maneuvers.

One of the maneuvers he was working on involved an outside half Cuban eight. One of the surprises he encountered was that the *Hornet* lost airspeed when it was rolled inverted. This meant that the inverted half Cuban eight that he was working on would be difficult to complete in basic engine. Although the envelope was not yet cleared to do so, he began experimenting with the use of a few seconds of afterburner to complete the maneuver. Without getting into detail, he found out the hard way that the warning light system associated with flying inverted was not sufficient to warn of and prevent engine flameout when the afterburners were initiated. Oops! The baffle fuel was depleted in seconds and the *Hornet* engines quit, with no possibility of a re-light. The good news was that Hollywood ejected successfully. The bad news was that his aircraft crashed right next to a major railroad track.

The diamond had just landed, and we were walking back to the ready room when this ominous black smoke cloud appeared in our desert practice area. It is difficult to explain the awful feeling of dread one feels when you know it is one of your guys. "Oh shit! I hope whoever it was ejected successfully and is OK."

We soon learned that Hollywood had ejected successfully and had sustained no injuries. *The Blues use a special harness that essentially glues you to the ejection seat. The purpose is to keep the same sight picture whether right side up or inverted. In a normal fighter aircraft, when you roll inverted, at one negative G, you will experience some separation from the seat. That is not acceptable in the* Blues. *A side benefit is that you will most likely avoid injury in an ejection.*

Our maintenance officer, Jim Anderson, witnessed both the crash and the maneuver leading up to the crash. Hollywood soon confirmed the cause of the crash, so we proceeded to remove the aircraft debris from the crash site. We did this to avoid a shutdown of a major railroad line that would result in unwanted press interest. The Naval Air Safety Center, however, was apoplectic when I told them that we had removed the debris from the crash site before they had an opportunity to conduct an onsite investigation. My boss, RADM Dave

Morris, the Chief of Naval Air Training, strongly supported my decision.

To ensure minimum disruption to our transition efforts RADM Morris then sent his Chief of Staff, Captain Ray Alcorn (a former long-term POW and all-around great person) to El Centro to conduct a required Field Naval Aviator Review Board (FNAB) on Hollywood. Hollywood had flown the aircraft beyond the cleared flight envelope, so he was placed on double-secret probation and we pressed on with the training. *Thank you RADM Morris and Captain Alcorn for your unwavering support and common-sense approach to what, could very likely, have been a show stopper.*

A special hats off to both Hollywood and Wayne "Leroy" Molnar. As the opposing solo, and the only new guy, Leroy did a hell of a job keeping up with the rest of us, who all had at least one-year of experience flying airshows. Hollywood had the added responsibility of training Leroy while he was still dealing with unknowns in the flight envelope. Leroy was a graduate of Florida State and had taught Physical Education for a year prior to joining the Navy. He had also been a "plowback" (assigned as an instructor immediately after earning his wings) instructor in the training command before becoming an F-14 *Tomcat* pilot. The added flight time and maturity achieved from being an instructor, combined to make him an outstanding, albeit inexperienced opposing solo.

We found to our delight that we could safely perform most of the maneuvers flown in the A-4 Airshow. We did have to overcome some airframe fatigue issues caused by the close formation flying (#4 having his vertical tails buffeted by my exhaust), and the violent maneuvering involved in the solo timing pattern and the high speed/high closure rate rendezvous performed by all the wingmen during the airshow routine. NAVAIR added the Leading-Edge Extension (LEX) fence that solved the vertical tail problem, but the solo program and the aggressive rendezvous continued to beat the shit out of the aircraft. *To fly a tight, entertaining and timely airshow, we decided to move many of our rendezvous from behind to in front of the crowd. In other words, they became part of the airshow. During most of these events, I would use a base speed of 350 knots. When the solos joined the diamond for the delta portion of the show, they would often arrive with a 250-knot closure rate. They mitigated this with speed brakes and a vicious +7.5 G/-3.5 G nose-pumping maneuver, which rapidly bled off their airspeed. It was very effective and entertaining, especially to other naval aviators (it also put several fleet back seat riders to sleep).*

***

FLYING THE FLEET: We discovered that the two-seat F/A-18B performance was very close to that of the single-seat birds. We took full advantage of this option to invite operational fleet pilots (selected by the Team) to fly during our practice airshows. Usually, they flew in the slot with Pat Walsh, but later in the year, a few also flew with Hollywood in the lead solo position. This interaction with the fleet included those riders attending our entire brief and debrief, which resulted in a better appreciation of the professional manner that we conducted our flights. *Our debrief method of each pilot recalling his own miscues and potential safety violations at the beginning of the debrief, eventually was successfully adopted as a best practice by many of the fleet squadrons.*

***

DEVELOPING A MORE EFFECTIVE *BLUE ANGEL* **STANDARD OPERATING PROCEDURES MANUAL (SOP):** Pat Walsh took the lead on putting together a detailed SOP that covered every maneuver in the airshow. He did this with his state-of-the-art (for 1987), 386-Word Processor. The SOP was especially geared toward flying a safe airshow. *In the first 40-years, the* Blue Angels *lost 20 pilots to fatal accidents. Following the introduction of the* Hornet *and the new SOP, we went 14 years before losing a pilot. Thank you Sponge!*

***

AIRSHOW TRANSIT ADVANTAGES OF THE *HORNET:* Getting from Pensacola to the various airshow sites was really a challenge in the A-4 with its very limited navigation system. In fact, it only had a Tactical Air Navigation System (TACAN), which was not widely available, especially at civilian airfields. We did not have the most common form of navigation found throughout the country, which was the VOR. If the weather required Instrument Flight Rules (IFR) and there was no Ground Controlled Approach System (GCA), I would simply declare (lie) that my VOR was malfunctioning, then ask for vectors to the field. We also flew most of our transits with all six aircraft together in the delta formation. Our *Hornets* were modified with the Canadian VOR/ILS package, and they also had a very effective Inertial Navigation System (INS), which made the transits much easier.

In the A-4, the fuel burn rate at lower altitudes allowed us to transit at altitudes where we did not even need to use oxygen masks (we had specially designed lip mikes that were used in the airshow and for low-altitude transits). The *Hornet*, however, was a different story. To make use of the high-altitude fuel efficiency, we often transited as high as flight level (FL) 430 (43,000 feet). At this altitude, we were above the normal commercial traffic, which is usually FL 320 to 370, so we could proceed "direct" to our destination. The problem we encountered, involved how to efficiently get from sea level to these transit altitudes.

The Pensacola training area included our home base of Sherman Field, Whiting Field and several outlying fields. The huge number of aircraft, and the restricted operating areas associated with these fields, made our departures a nightmare of an inefficient stair-stepping climb to altitude. Pat Walsh got together with the local air traffic controllers and came up with an approved departure that took us directly from sea level to FL290. This departure was predicated on a 25,000 feet-per-minute rate of climb remaining within 5 miles of Sherman Field. To accomplish this, I would take off with the diamond on one of the parallel runways while the solos would use the other. Once the solos were joined (just a few seconds after launch), they would check in with, "You got six, Boss." I would call for the after burners; pull the nose up 30+ degrees and start a circling 250-knot climb to altitude. Imagine 12 after burners in delta formation (the no shit sound of freedom) roaring overhead Sherman Field as we climbed to altitude. Once we got to transit altitude, we got terrific performance with numbers of about .94 Mach at around 1,800 pounds-per-hour of fuel flow per engine. This would allow us to transit nearly 1,000 miles without refueling (our aircraft were clean with no tanks or ordnance).

Since we normally returned late Sunday afternoon, we encountered far less traffic, which allowed us to use an idle descent for the last 100 miles. On one of our first returns from an airshow site, we ran into some weather in the Pensacola area. Since we would not be able to come into the "carrier break" as a six plane, we broke up into two-triads, with me leading number two and number three, and Sponge leading number five and number six. Keeping in mind that the local controllers were still getting used to the *Hornet* performance and high fuel usage at low altitude in the landing configuration, we got the following instructions from the air traffic controller: "*Blue Angel* One, Sherman Field is below circling minimums so I cannot approve multiple plane approaches. You will have to break up for individual GCAs.

I answered, "Negative, our fuel states require that we land in three plane elements." The controller—now agitated, "I repeat, we are below circling minimums so you must breakup for single aircraft approaches!" I replied, "We are going to land in two three-plane elements, and you are going to give us those approaches right now!" Controller, "No sir, I cannot do that." Me, "Oh yes, you will do it, or you will have six parachutes instead of six aircraft landing at your field. Now, give us the God damn approaches!" The boys were getting a huge kick out of this conversation of course, and we landed in triads, uneventfully.

The controller filed a flight violation against me, and came over to maintenance control at our hangar, to meet me and present me with the flight violation. Our Maintenance Master Chief, Jack Cragen advised the controller that it would be smarter to leave the flight violation with him. "Have you ever met Boss Rud? He just might not be in the mood to accept any flight violations, if you know what I mean." The Controller reconsidered, left the flight violation with the Master Chief, and in the end, the controller and I met in person, resolved the issue, and set up some flexibility in the rules for our future returns to Pensacola. *This incident was another example of common sense triumphing over rigid rule interpretation. Sometimes you must be a bit of an asshole to make things work.*

***

1987 Airshow Season Highlights: We completed Winter Training on April 23, exactly a month-longer session than we had in 1986. Our first show was at MCAS Yuma, then we headed for Pensacola. 1987 was a great year for airshows. The unprecedented success and popularity of *Top Gun*, combined with our powerful new *Hornet* show birds resulted in huge crowds at all the shows. A couple of the larger ones estimated attendance at close to one million people. We had a few maintenance challenges, but still managed to fly six jets for at least a portion of all the shows. The fleet and the civilian crowds loved the new aircraft. The exception might have been some of the Ex-*Blues*, who felt like the show was too much like the A-4, with not enough new stuff highlighting the added capability of the *Hornet*. That was certainly understandable, but we achieved our goal of flying a challenging, but safe routine.

A safe routine is one in which you can repeatedly fly a maneuver with significant margin for error. An example being the "dirty roll-on takeoff" where about half of the available roll rate is used to complete the maneuver, with the excess roll rate available, if needed. The solos were initially flying a maneuver called the "opposing square loop." It was very impressive, but the profile was too dependent on density altitude, and after some unpleasant surprises, Hollywood decided to stop flying the maneuver. *His decision was exactly right since soon after, an overly aggressive Marine* Hornet *fleet demonstration pilot crashed at the airshow at MCAS El Toro in southern California, when he started the recovery from a square loop profile without sufficient altitude. He miraculously survived the crash, albeit with serious injuries.*

***

NAS PATUXENT RIVER, MARYLAND: We dedicated this show to Keith Crawford, who was killed testing the "feel" system and flight control modifications for the *Blue Angel Hornets.* It was very emotional for his family and his fellow test pilots, as it was for us. I think we all felt a little pressure to put on the best possible show, so that these guys could see what their hard work and personal risk had achieved.

***

HAMILTON, ONTARIO, CANADA WITH A STOP AT THE MCDONNELL DOUGLAS FACTORY: During the transit to this show, we stopped in Saint Louis for fuel. We parked the aircraft outside of the main factory where McDonnell Douglas was building *Hornets.* The *Hornet* Leadership Team set up a platform so that I could address all the workers. It was a great opportunity for us to thank these incredible folks for building such a great aircraft. *I also credit the decision to make this stop as a factor in me eventually landing a job with the company.*

Located near Toronto, Hamilton was a well-attended airshow. We were briefing in an older hangar that had plumbing problems. Just before we were heading to man up for the show, the toilets quit working altogether. Call it superstition, nervousness, or just a habit; in any case, I needed a rest room before getting into my aircraft. The problem was that there were incredibly long lines for the port-a-potties on the flight line. As a solution, I walked to the front of the line and stated in a very loud voice, "Folks, if you do not let me cut to the front of this line, there is not going to be an airshow." The line parted like the Red Sea for Moses, and I got into the port-a-potty with the added pressure of several airshow fans loudly discussing the *Blue Angel* in Potty #7.

***

NAS MOFFETT FIELD: We flew this show over the Fourth of July weekend. It was a great show site and always drew huge crowds from around the Bay Area. Although we pilots generally behaved ourselves when it came to the party scene, it was common practice for our much younger and more socially active crew chiefs to enjoy the local area's entertainment options. Although they certainly did not openly pretend to be pilots, if they were mistakenly identified as such, let's just say that they might not immediately correct that label.

Sometimes a seemingly innocuous situation can rapidly turn into real trouble. This is what happened to five of our best maintenance technicians and crew chiefs. They met a very attractive and boisterous young lady at a local San Jose nightclub. She invited them back to her apartment building to join with other revelers for a midnight swim, sans clothing.

Realizing that this behavior was a bit on the risky side, they declined the naked part, at first, but were soon coerced (called chicken) by this young lady into joining her in the pool. It was not long before someone called the cops. Still should have been no problem except that the "Devil" woman physically assaulted one of the policemen who answered the noise disturbance call. Shit! At that point, everyone was arrested and carted off to jail. To add to the bad timing of this incident, there had been a large gang-related disturbance that evening that brought reporters from several bay area newspapers to the San Jose precinct. So, the arrival of these five *Blue Angel* swimmers now got the total attention of the press.

I got a phone call from the Maintenance Officer at about midnight informing me of the incident. The PAO was also now on the scene bailing out the infamous five. Once he related the press interest, I decided to quickly send out what I seem to remember was called an "OPREP 3, Navy Blue Pinnacle" report to my boss, and up the chain, all the way to the Pacific Fleet Commander. This action insured that nobody in the Navy chain of command would be surprised by the newspaper articles that were sure to be on the front page the next morning. It was a good thing that I did that because the headlines on the front page of the San Francisco Examiner blared out in huge print: "*Blue Angels* Busted!"

Of course, the information in the article was sketchy and inaccurate, however, it referred to the swimmers as *Blue Angel* pilots. The good news was that the article also encouraged the public to attend the airshow, so they could see, in person, these heavy-partying pilots perform their death-defying aerobatic maneuvers. The result was a record-setting crowd of close to a million people attending the show, creating a monstrous traffic jam on the heavily traveled 101 freeway. *I remember flying the "Double Farvel" maneuver, inverted, right over the freeway. I couldn't believe the number of people who were outside of their cars, in the traffic jam, watching the show.*

At the completion of the show, we were literally accosted by reporters. I was selected for an interview with an associated press reporter, whose article we knew would get nationwide attention. I explained the entire incident, including the denial of any pilot involvement. After the interview was complete, he looked me in the eye and said, "Off the record, how in the hell do you guys party as hard as you do and still fly these airshows?" I replied, "Off the record, I can tell you that it is not easy, especially at my age."

The tough part of this incident was the need to discipline the swimmers. Although they did not get officially charged with anything, their actions had damaged the reputation of the *Blue Angels* and had broken all our rules regarding proper behavior. They knew it, and they knew what was coming. I fired all five of them. However, I also contacted the enlisted detailer and informed him that he was getting five awesome sailors available for assignment, and that wherever they went, they would be great additions to their unit.

A week or so later during the Redding, California airshow, I got a late-evening call from a radio talk show host. That was not altogether unusual, however, this guy really went after me for firing the swimmers. He had several callers who agreed with him that I was an asshole. I went on to share why the firings occurred, and how the five would still get the opportunity to be productive sailors. It was a fun interview, but in the end, the host and his listeners still considered me to be an asshole.

***

BROOMFIELD, COLORADO: Located on the front range of the Rocky Mountains, this show site, at approximately 5,000 feet of elevation, was our first real test of the *Hornet's* performance in a high-density altitude environment. I distinctly remember performing the "Diamond Dirty Loop" at the Friday practice. It was a very warm day with probably a density altitude of 8,000 feet. It did not go well. I had to stay in burner, and we made it over the top maybe five knots above stall speed. As we recovered on the backside Sponge said, "Hey Boss, I suggest that we do a clean loop tomorrow." And that is exactly what we did.

***

NAS OCEANA, VIRGINIA: The shows we flew at the Navy and Marine Corps tactical jet bases, offered an opportunity to get expert feedback from the fleet on the quality of our airshow. It also afforded us the opportunity to socialize with former squadron mates and shipmates that we may not have seen in years. At this show, my daughter Valerie, now a member of the Virginia Tech Corps of Cadets and the Navy ROTC unit, somehow talked the Professor of Naval Science into allowing a bus full of midshipmen to attend the airshow. They were all in uniform, and we had them hanging out with us as we gathered near *Fat Albert* prior to the start of the show. As they were about to leave the area to join the rest of the audience, Valerie gave me a big hug and a kiss. A few minutes later, I got a shout from the crowd from an old buddy, Sergei Kowalchek, "Hey Gil, is that pretty lady your new girlfriend?" I replied, "No you old fart, that is my daughter Valerie." Now appearing mortified, he said, "Oh my God Gil, we are getting old!"

***

MCAS BEAUFORT, SOUTH CAROLINA: Like Oceana, Beaufort was the Marine east coast base for most of their tactical jets. Normally at the social functions, as the Boss, I would do the speaking and introduce the team. Since Mark Bircher was a Marine, we decided that he would have that honor at this event. He did a great job, including a ruse where he described, in some detail, the physical fitness requirements necessary to be able to handle the high-G environment without a G-suit (the Marines are loving this with several shouting "Oorahs.") He then further described the opposing solos high-sustained G, 360-turning maneuver as the most challenging. "Now, let me introduce the man behind the maneuver, Wayne 'Leroy' Molnar!" Leroy, at 6 feet 4 inches, 165 pounds, walked onto the stage in a purposely-stooped posture with his chest sunken and his little belly extended. The Marine audience went nuts; they absolutely loved it.

***

CARSWELL AIR FORCE BASE, FORT WORTH, TEXAS: This one stands out because it was a homecoming for Sponge Walsh. "Wally World," as the large Walsh clan was called, consisted of Sponge's parents plus five brothers and sisters. Sponge was the oldest, and all of them had attended the Jesuit high school in Dallas where Sponge's dad had been the basketball coach, and I believe, eventually, the principal. Also, except for Sponge, all the rest had attended Texas Tech University. As we have covered earlier discussing the professionalism of the team, we always wore the appropriate *Blue Angel* flight suits, which included a khaki-fore and aft cap, commonly referred to as a "piss cutter." At the end of this airshow, however, we exchanged our piss cutters for giant 10-gallon white cowboy hats that we wore for the walk back at the end of the show. *This situation was another prime example of asking forgiveness rather than permission. Texas loved us, so it was well worth the risk of some potential leadership criticism for a uniform violation.*

***

San Francisco: As usual, we arrived on a Wednesday. Since it was too late to do circle and arrival maneuvers, we scheduled them for Thursday. Unknown to us, a film crew was in the process of filming the movie *The Presidio* with a scene from the movie actively taking place right under our flight paths. You can imagine how disruptive six noisy blue jets were to this very expensive process.

Chatting with Sean Connery on the set of The Presidio.

336

The cast for the movie included Sean Connery, Mark Harmon and Meg Ryan, who were all being paid huge salaries, which added to the explosive reaction we got from the movie's producer. Our events coordinator, Manny Campbell, worked his magic and managed to defuse the situation as well as potential bad press. He did such a good job that the movie producer invited us to attend the filming of a scene in downtown San Francisco where we had a chance to mix and mingle with the actors. Mark Harmon and Sean Connery both turned out to be great guys, swapping their many acting experiences/stories for our description of the challenges involved in flying airshows.

27. President Reagan's Boys

Selecting the 1988 Team: As I recall, we had over 50 applicants for the three pilot slots that were available. Since most of these applicants were more than qualified, the selection was a difficult process. The final selectees and newest pilots to join the *Blue Angels* were Marine Captain Kevin "Tow Bar'" Lauver, LCDR Mark "Ziggy'" Ziegler and LT Doug "Hound Dog" McClain. Tow Bar came from the *Harrier* community, Ziggy was an A-7E pilot and Hound Dog was an A-6E *Intruder* driver. Following selection, all three were sent to the F/A-18 RAG for training. Tow Bar would replace Mark Bircher as #2, the right wingman. Ziggy would become #3, the left wingman, and Hound Dog, as the most junior in rank, would replace Red as #7, the narrator. Big Time Cochran would now move from #3 to become #4, the slot pilot. Leroy would move from #6, the opposing solo, to #5 the lead solo, and Red would move from #7, the narrator to the opposing solo position. The other new officers selected included LT Fred "Derf" Cleveland as the Maintenance Officer, LT Rusty Holmes as the Public Affairs Officer, and Captain Ken Hopper as the third C-130 pilot.

***

Highlights from 1988 Winter Training and Airshow Season: I was now into my third year as the commanding officer and flight leader. With the two years of experience came self-confidence, to the point where I was relaxed enough to thoroughly enjoy the entire *Blue Angel* experience. Derf Cleveland had the maintenance department squared away. Sam Walker and the McDonnell Douglas Technical Support folks had streamlined the parts supply system so that we had six jets available to fly for every practice and airshow for the entire year.

1988 Blue Angels, *from left to right: Mike "Manny" Campbell, Cliff "Red" Skelton, Donnie "Big Time" Cochran, Kevin "Tow Bar" Lauver, Gil "Boss" Rud, Mark "Ziggy" Ziegler, Wayne "Leroy" Molnar, Doug "Hound Dog" McClain.*

We did have a few misadventures including a little paint-swapping incident very early in the winter training syllabus. I don't recall what maneuver it was, but #2 was in the process of clearing the formation when his vertical stabilizer made contact with my wing tip. I did not even feel the impact, however the cap on his vertical stab came loose and fell into the Gulf of Mexico. We thoroughly checked out both aircraft before deciding that precautionary straight-in landings were the best option. Keeping in mind that everything we do is highly scrutinized by both our leadership and the press, I called the tower and said, "Instead of our usual entry into the break, today, the Diamond would like to practice individual straight-in approaches." The tower responded, "Roger *Blue Angel* One, you are cleared for four individual straight-in approaches." We had alerted maintenance control so instead of our normal parking, they opened the hangar bay doors so that we taxied the #1 and #2 jets directly into the hangar. *The damage was minor and both aircraft were flyable the next day. I elected not to report the incident, a decision that would probably get me fired in today's Navy.*

Winter Training gave us the opportunity to introduce a couple of new maneuvers. One of them was what we called the "Diamond afterburner turn." We brought the Diamond in from the left at 500 feet of altitude, just 500 feet from the crowd line, at the relatively slow airspeed of 220 knots. I then called for afterburners, as I rolled the Diamond into about a 60-degree angle-of-bank turn, which placed all eight afterburners directly on the crowd line. I continued a level

500-foot, accelerating turn until I reached 4-Gs and 400-knots. At that point, I called, "Burners, ready now," and we deselected the afterburners, continuing the tight turn with a "legal" recovery ending 1,500 feet from the crowd line.

Because of the dynamics involved, we flew this maneuver just a little looser than the normal diamond formation. It was a ground-shaking, crowd-pleasing maneuver, but unfortunately, after just one season, following the Italian demonstration team's catastrophic midair at a show in Germany, the FAA banned all maneuvers that they considered might endanger the crowd. According to them, the "Diamond afterburner turn" fell into that category. *This was also a maneuver that, due to its relative difficulty, would have been appropriate only for a second-year flight leader. Although short-lived, it certainly was fun to fly, and the current team is now using a similar FAA-approved maneuver.*

***

FLYOVER OF SUPER BOWL XXII: The Super Bowl was scheduled for late January in San Diego. Even though it was early in the training cycle, being invited to perform that flyover was something we were not going to turn down. In those days, the football stadium, located in Mission Valley was open on the east end. This afforded a perfect approach for us, which gave the audience a chance to see us coming from several miles away. Timing is absolutely crucial, of course, with the goal being to be over the field on the last note of the "Star Spangled Banner," which was played by Herb Alpert.

We listened to a recording of the practice to determine the exact length and then we placed our recently departed #2, Mark Bircher, in the stands with a hand-held radio so I could listen to the music and adjust my timing. Although the FAA had supposedly cleared the area of all traffic, it was disconcerting to see all the blimps and news helicopters very close to our route of flight. As I rolled the formation out on our final approach course east of San Diego State University, Birch came on the radio, "Boss we can see you, turn on the smoke." I turned on the smoke and Birch said, "Add it up Boss and bring it down lower." I did just that and we hit the center of the stadium at 400 feet of altitude and almost 400 knots of airspeed. It was the only time in my three years as the flight leader that I had the opportunity to hear the roar of the crowd during a flyover. What a thrill. *If the team today used the altitude and airspeed that we used for this event, they most certainly would get a flight violation, and more than likely the Boss would face disciplinary action of some sort.*

***

VISIT TO THE WHITE HOUSE: In 1988, we once again headed to Andrews AFB to fly the U.S. Naval Academy airshow and graduation ceremony. Due to a close relationship that Pat Walsh had established for the team with Press Secretary, James Baker, who was badly wounded during the assassination attempt on the President, we were fortunate enough to be invited to the White House to meet with President Reagan. That is an experience that I will never forget. He was every bit as charismatic in person, as he was on television. The President spent almost 30 minutes chatting with us and entertaining with both jokes and true stories.

Just before heading to his next event, he turned to his Chief of Staff, former Senator Howard Baker, "Howard, wouldn't it be great if the boys here flew over the Rose Garden in the morning, as part of the press event prior to my departure to Russia for talks with Gorbachev?" The Chief of Staff looked directly at me, and before he could ask, I said, "Mr. President, we would be honored to fly over the Rose Garden in the morning." #8, our events coordinator, Manny Campbell, was already out the door to start the massive coordination required to carry out the flyover.

Somehow, Manny performed another miracle, managing to coordinate the timing of the Rose Garden with the flyover of the Naval Academy graduation at Annapolis. The coordination included a total shutdown of National Airport. So how could this event get even more complicated? Shitty weather. It was just barely VFR with a 1,000-foot ceiling and 3 miles of visibility in rain. Fortunately, we had the airspace in Washington, D.C. all to ourselves. During our preflight planning, we found that if I flew the Delta straight down runway 01 at National (ironically later to become Reagan National), it would lead us on a flight path directly over the White House. We also decided that 400 feet and 325 knots would be an appropriate flyover altitude and airspeed (#6 said after the event that he was worried that I would scrape him off on the top of the Washington monument. This would never happen today).

As I rolled the formation out over the Potomac a couple of miles south of National Airport, I became concerned, that with the limited visibility, I would not be able to see the White House. We placed our smoke on before we

341

even got to the runway at National, and by the time we were 3 miles from the Rose Garden, the flashes from the numerous press cameras guided me right to the target.

After completing the flyover, I started a climb into the overcast, picking up an IFR clearance that we had worked out prior to takeoff. The weather in Annapolis was a little better, but I still needed to use my radar to pick up the water tower at the football stadium that guided me right over the graduation ceremony. *Prior to the Rose Garden flyover, Senator Baker warned me not to fly directly over the President. How do you respond to such an order? You ignore it and proceed with the best flyover you can perform (how in the hell am I supposed to pick out the President among all those trees in the Rose Garden at 325 knots.) I was prepared to ask forgiveness, but nobody complained.*

Meeitng President Reagan with Donnie Cochran. The President spent 30 minutes visiting with the team before requesting that we do a flyover the next morning.

***

GRAND FORKS AFB: This venue would serve as my second chance for a homecoming opportunity in my beloved North Dakota. We did not have a tanker for this trip, so we stopped for gas at Offutt AFB in Omaha, Nebraska. There was a significant weather disturbance between Omaha and Grand Forks, so we climbed up to 41,000 feet to get over it. I checked in with Minneapolis Center, "Minneapolis Center this is *Blue Angel* One with a flight of six, level at Flight Level 410." The reply from Minneapolis Center in a very sweet female voice, "Roger *Blue Angel* One, Gilman, is that you?" Now you can just imagine the reaction of the five boys on my wing to this transmission. I reply, "Yes this is Gilman, and who do I have the pleasure of speaking with?" This is Marchelle Amb, the little neighbor girl that grew up a few miles up the river from your farm." I remembered her of course, and especially her brother, Brad "Mad Dog" Amb, who was an outstanding high school and college wrestler. This encounter really set the stage for another week immersed in typical North Dakota hospitality.

The weather improved as we neared Fargo, so I canceled our IFR clearance and proceeded low level toward my hometown. As we neared the town, I called for the formation to tighten up for a flyover. I then transmitted, "Smoke on!" A few seconds later I transmitted, "Smoke off!" Big Time was surprised so he said, "What happened Boss; did we miss it?" I said, "Nope, that was it." We then proceeded toward Grand Forks AFB. Our flight path happened to pass directly over the Pederson Implement dealership. I had given my former coach (now part owner of the dealership) a "heads up" that we were going to perform a flyover at a certain time. For this one, I took the Delta down to 200 feet, as we were now close enough to the base to get by with that altitude. It was an amazing sight to see many people, some on top of John Deere combines to get a better look. And they certainly got a close-up look. What a thrill for both them and me.

Following our arrival maneuvers and debrief, I headed for my hometown to attend a function at the Portland American Legion Hall. I don't recall for sure, but I think it was Big Time and Leroy that asked if they could come, so the three of us drove the 40 miles or so to the metropolis of Portland. We had a blast. I got a chance to see folks that I had not seen in 20 years. I am pretty good at remembering names, so I managed to recall who most of them were, by name. Of course, Leroy and Big Time were deluged with autograph, photo and hug seekers. *This event, much like the one in 1986 in Fargo, made me realize how incredibly lucky I was to have grown up in this wonderful community, and I vowed never to forget that, and to give back every chance that I got. I am still to this day, an active and proud member of the Portland American Legion Post 93.*

***

A Conversation With the Shuttle Discovery: On October 1, 1988 we were flying an airshow at Ellington AFB in Houston. During this same time frame, the shuttle Discovery was in orbit led by the Mission Commander, my personal hero, Rick Hauck. I got an invitation from NASA to visit the space center, where they connected me with the shuttle and a conversation with Rick. What a thrill that was.

***

"Engine Left! Engine Left!": After completion of the airshow, we departed for Pensacola in delta formation. As we were climbing out over downtown Houston, I felt and heard an ominous pop and deceleration followed by this warning: "Engine Left! Engine Left!" Having been a single-engine flier for so many years, I immediately began looking for a place to steer my aircraft away from the city prior to initiating a possible ejection. In short order, I remembered that I had two engines on this machine, thank God. Although all five of my wingmen cleared the formation as soon as I lost the engine, four of them continued on to Pensacola as #2 accompanied me back to Ellington for an uneventful emergency landing. The *Hornet* flies incredibly well on a single-engine. I shiver at the thought of this occurring in an A-4.

***

Flying Hurt: My relief, Pat Moneymaker, had been selected, and had just joined the team for the last month or so of the 1988 airshow season. Pat was a former squadron mate in VA-147, a great stick, and a good friend. On the Monday of the week that we were going to begin a late season, extended west coast trip, I was working in my yard in Gulf Breeze. I noticed a bent nail on the gate latch leading to my backyard. Being too lazy to get my toolbox, I picked up a brick from the garden and began using it to straighten the nail. Yes, I know, incredibly stupid, but most accidents are the result of a stupid decision. As I thrust upward with the brick, the brick broke, and my right index finger went right through the head of the nail. The cut was right through the middle knuckle. It was deep and hurt like hell, but did not bleed much, so I wrapped it in a bandage and elected to see our flight surgeon, Wes Robinson, the next morning.

I walked into Wes' office on Tuesday morning prior to our brief for the practice and showed him the finger. As we were unwrapping the bandage, he suddenly blurts out, "Stop! Do not move that finger. You have cut your flexor tendon, and we need to perform surgery as soon as possible." "Wait," I responded, "I have to fly a practice today." "Not with that finger, you won't. We need to get it fixed so that hopefully you can fly to San Francisco, tomorrow."

Thank goodness that Wes had great connections that included a talented surgeon, who was stationed right there in Pensacola. Of course, we had to keep all of this secret, because there was no way in hell that I was going to end up with a medical "down" chit. (Remember, anyone can be missing from an airshow except the Boss; if he doesn't fly, the airshow is canceled.) With this in mind, we checked into the hospital as a Master Chief or some other assumed identity. I underwent surgery with the usual medication required and spent one night in the hospital. The next day, I reported to work, but Wes told me not to fly until late that afternoon. To ensure secrecy to this whole mess, maintenance painted a number one on a spare jet, and I think Pat, or one of the other newbies flew it to San Francisco, so that everything would appear normal. My finger was placed in a metal splint that still allowed me to manipulate the control stick.

The plan was for me to fly a solo practice flight in my #1 jet to see if I could do basic roll/loop combinations with this device on my finger. I took off a few hours after the other folks had departed for the west coast and headed to an operating area over the Gulf of Mexico. I successfully performed the basic aerobatic maneuvers, albeit, in a bit of discomfort, and headed back to the base. Feeling pretty good about the whole situation, I came in for my normal 500-knot break. The duty runway was 07R, so it was a right break. I rolled into a 90-degree bank and pulled. "Shit! I can't see!" Apparently, the medication had not worn off yet, and the high-G pull literally blacked me out.

I was awake but could not see anything but a sort of flashing light. I called the tower with some bullshit lie like, "Tower, I am having some problems with my attitude instruments and would like you to confirm that I am now in level flight." Tower response, "Yes sir, you appear to be level, but heading south versus downwind." "OK," I responded, "I am just going to continue south for a few minutes, let me know if you detect a change in altitude or angle of bank." I finally started to regain the ability to see the HUD instruments and luckily determined that I was, indeed, wings-level with a steady altitude. I then requested a full stop from my present position and landed uneventfully.

As I taxied to the flight line, I thought about how close I had come to dying. Imagine the accident report when they discovered that I had undergone major surgery, less than 24 hours prior to flying into the water. And if anything were left of me to perform an autopsy on, they would have found a good deal of some type of drug in my system. Well, the good news was that I could fly normal airshow maneuvers, just fine. They refueled my aircraft and I took off for San Francisco.

I never said a word to Wes or anyone else about my near-death experience. Now, it was time for a practice show. The show was flown as a "remote." In other words, we took off from NAS Alameda, flew the show with the center point just off Fisherman's Wharf, then returned to NAS Alameda. To make use of the opportunity to practice the maneuver, we decided to do a "loop on take-off" from Alameda. As I began pulling the G required on the backside of the loop, my finger began to throb with pain. I thought, "Holy shit! This is going to be one long airshow practice." Fortunately, the pain soon subsided, and outside of having to be very careful when I saluted or shook hands, I was back to a full-up round.

***

EARLY PROMOTION AND SELECTION TO SHIP DRIVING SCHOOL: It must have been a "Norwegian quota," or some other miracle, but I was shocked to receive notice that I had been promoted to Captain, a year early. I could not help but think, "How in the hell did I even make it to Lieutenant? Now I am actually a Captain (06) in the United States Navy?" Unfortunately, this early promotion adversely affected my goal of being selected to be a CAG. I was now too senior to become a Deputy CAG, so instead of getting an opportunity to fly for another three years, I was now on a path to drive ships. I distinctly remember getting the call from my boss, RADM "Snake" Morris congratulating me on selection for "deep draft" command. My response was not exactly enthusiastic, "A deep draft? I don't know anything about ships. Why a deep draft?" His typical "no-nonsense" reply, "Well, frankly, in my opinion, you should not have gotten anything." After a short pause, I said, "OK, Admiral, I guess I'd better break out the books and figure out how to drive a ship."

***

THE LAST AIRSHOW: Flying an airshow, I assume, is much like competing in a NASCAR race. It takes total concentration for the entire event. If you let anything interfere with that concentration, you endanger yourself and everyone else involved. The last show is certainly a very emotional event. The brief includes attendance by any Exes and Honorary *Blue Angels* who desire to attend. Families of all the *Blues* who will be leaving the team are in a special area of the crowd, so that they can get a close-up look at their respective loved one, whether that person is flying, or in a vital support role that keeps the aircraft in the air.

As we had done for three wonderful years, just before we started the show, we spent time with the crew chiefs and maintenance troops. These are the finest the Navy and Marine Corps have to offer. My crew chiefs including John Cordova, Tom Cooper, Mike Belisle and others will always have a special place in my heart. The front man, Terry Hynds, and I had a ritual that we performed before every show. Terry was "bald as a billiard ball." I would make a great show of rubbing his head, and then look very closely at it, as though it were a crystal ball. I would then step back and announce, "I see the sun reflecting off Terry's head, so let's do a 'high show!'"

Not that they needed it, but I gave the boys this simple advice prior to walking down for the show, "This is just another airshow that needs to be performed to the best of our ability, like all of the others we have done. We will fly it no lower, no faster and no closer to the crowd than any other show. Now let's go out there and have a good one."

We did just that, and when it was over, I had a sense of both sadness and relief. Sad, that I would never get to fly another *Blue* jet, and relief, that we had been able to complete three years and hundreds of practices/airshows without anybody getting hurt.

28. "Selected for What?"

SHIP DRIVING SCHOOL: I was still reeling from the shock of my selection to drive a ship. In preparation for the upcoming curriculum, the school, located in Newport, Rhode Island, sent to me a "Math for Dummies" book. Unlike most of my fellow students, I completed every problem in this book from simple trigonometry to the polar form of complex numbers. Luckily, my math teacher wife was a very patient and skilled tutor. The result was that when we attended our first math class, the instructor, a nuclear power "egghead," opened the session with a description of the various college majors that were in attendance. When he got to me, he literally snorted, "Agricultural economics?" He then began this irritating "nerd" laugh. It really pissed me off, so I said, "Just give us the God damn test and we will see how an agricultural economist does." Well, since I was probably the only one in the class that had completed the prep course, I got a perfect score. The "egghead" was flabbergasted. Of course, as the math progressed in complexity, I quickly fell into the mediocrity range on the test scores.

Although we had not met personally, Captain Bob Houser and I decided to team up and get a house in the city of Newport. Bob was originally an A-4 pilot. He transitioned to A-6s and had risen to be the Commodore heading that community in Virginia Beach. Like me, he had a large family, so we rented a house big enough to accommodate potential visits from our families.

In addition to the math, the classes taught at the school were very technical in nature. Thermodynamics was the nastiest one for this struggling student. The exams were blue book type, with lots of long formula-centric answers. The exams were also timed, so it was a brutal scenario. I decided that there was no way that I could solve the problems, so I memorized the appropriate formulas to use, wrote the necessary numbers into the formula and then went on to the next problem. In this manner, I got enough points to pass the course, albeit with a 61 (60 was passing). At the end of the class, they gave me the "Thermo-dynamics Award" for being the most innovative test-taker. A "slam" on my lack of technical intellect of course, but I loved it.

Bob, on the other hand, was a brilliant student. With what I seem to remember was a master's degree in psychology, he absolutely nailed every test, with minimal study time, I might add. Oh yes, we had study sessions at our house, which pretty much always ended in a trip to the nearest sports bar.

After completion of the ship systems and engineering stuff, we got into some more interesting things, which included driving ships in a simulator. I really enjoyed this and decided that I just might be pretty good at it. I also decided that whatever ship they assigned me to, I just wanted to make sure that during my tenure, it would never go near a shipyard. I certainly got my wish as they assigned me to the Replenishment Oiler, USS *Wabash*, (AOR 5), based out of Naval Station Long Beach, California. It had just come out of the yard and was scheduled to be physically at sea for 13 of the 18 months that I would be in command.

The last few weeks we were in Newport, we attended the Command-ing Officers course. The weather had warmed up, so we began to participate in the slow-pitch softball league. Despite our relatively advanced ages, we had a very good team. The best team by far, though, was made up of a bunch of En-signs fresh out of being commissioned at the U.S. Naval Academy, OCS or an NROTC Unit. They were attending Surface Warfare Officers School (SWOS). Both teams were undefeated, although they had been scoring many more runs than we had.

There were several fields that we played on, but only one of them was fenceless. We decided that it was critical to not play with a fence, since those youngsters would just slam one home run after another over the fence. They were young and in great shape, but we had some very experienced former base-ball players. This included a former starting catcher and a shortstop from the Naval Academy, plus one of our instructors who had made it all the way to "Double A" at the professional level. The rest of us all had played in high school and in my case, a year of college. Another advantage we had was that these youngsters were arrogant as hell.

We purposely took our time warming up, showing off none of our significant talent. Although maintaining a certain amount of military respect, they began to make fun of us. Big mistake! They were up first. Those of us playing in the outfield played extremely deep, which turned their normal home runs into routine fly balls. When we came to bat, we played "small ball," which resulted in a couple of runs. A few innings later, adjusting for our singles falling in, they moved their outfielders in. Bam! One of our guys hit it over their heads for a "grand slam" home run, which put us up something like six to four.

In the last inning, they got a couple of runners on, one of whom tried to score on a fly ball to me. I still had a hell of an arm and I nailed him at the plate. As the inning progressed, they got a runner on first base with two outs. That is when their hitter connected with what should have been a base hit between shortstop and third base. However, our former Naval Academy shortstop backhanded it deep in the hole, fired it to the second baseman for the force out, and the game was over. Those kids were in shock!

We all headed for the officers' club where one of the Ensigns, who knew my daughter from Virginia Tech, came over to talk to me. He said, "I can't believe we lost to you guys." I replied, "Maybe this will make you feel a little better," and I introduced him to all our players, with each one of them sharing their personal background as baseball players. He literally ran back to his teammates, who were sulking in their beers with this enlightening news. We bought them a few pitchers of beer and soon the "sea stories" began to flow. *It was a great lesson for these youngsters not to be so quick to judge an adversary, and damn, it was fun for us old guys too.*

***

Back to California: Since my training was only six months long, Barbara, Ryan and Sally stayed in Gulf Breeze until I finished. During this time period, Barbara got a call from the principal at Lemoore High School. He seemed to know more about my orders than even I did. The point of the call was to convince Barbara to move back to Lemoore. Why? There were two reasons: The first one was that she was one hell of a good math teacher and they wanted her to teach advanced math classes at the high school. The second was that Ryan was a very good soccer player, and the coach felt that with Ryan, he could possibly win the league championship.

Armed with the knowledge that I would be gone to sea most of the time anyway, I let the family make the decision. It was unanimous—let's move back to Lemoore—and that is what we did. It was a four-hour drive for me, but since many of the *Wabash* crew were geographic bachelors from San Diego, when in port, the ship had been working four 10-hour days, so they could enjoy a three-day weekend. My plan was to live on the ship Monday through Thursday, and then drive to Lemoore for the weekends. It worked quite well, and certainly made life easier for Barb and the kids. They were back in a familiar environment with great friends and support when I was deployed.

29. Underway on USS Wabash (AOR 5)

ENTERING A NEW WORLD: I checked aboard *Wabash* as the prospective new commanding officer in the middle of August 1989. I was welcomed aboard ship by the captain, a friend and fellow light attack aviator that I had known for years, D.J. Wright. The ship was at the pier in Naval Station, Long Beach, so they had reserved a parking space for me. I was still driving the same old rusty, 1978 Dodge truck, which just barely fit into the parking spot. After boarding the ship, as D.J. and I were walking to his in-port cabin, a salty old First Class Boatswains Mate said to me, "Sir, is that your truck parked down there?" I answer, "It sure is." He smiled and said, "The crew has been worried that you, as the new skipper, might be an up-tight snob. Now that I have seen what you drive, I think we can relax." *At least this time, they did not ask me to move it out of the parking space.*

***

CHANGE-OF-COMMAND CEREMONY ON *WABASH*: While I was in the *Blue Angels*, I had the opportunity to meet and become friends with the famous actor, Ernie Borgnine. Ernie was a Navy veteran of WWII, and of course, starred in the television series *McHale's Navy*. Since my change of command on *Wabash* was scheduled to take place while the ship was tied up in Long Beach, I gave his agent a call to see if Ernie might be interested in attending. As busy as his schedule was, he graciously accepted the invitation, showed up early for the ceremony and stayed afterward to hang out with the crew.

We decided to stretch normal Navy protocol a bit by playing the *McHale's Navy* theme song for him when he came aboard the ship. My boss, the Logistics Group Commander (COMLOGGRUONE), RADM Bitoff, gave me a rather stern look when he heard this music being played. I smiled and said, "Admiral, would you like to join me in welcoming aboard Ernie Borgnine?" The Admiral immediately brightened and said, "Is Ernie really joining us for the ceremony?" *Ernie was a genuinely nice guy, an Honorary* Blue Angel *and my friend. Fifteen years later when Barbara passed away, Ernie was one of the first people to call with condolences. I miss him.*

Ernie Borgnine aboard USS Wabash.

***

INTRODUCTION TO THE SHIP AND CREW: USS *Wabash* (AOR 5), also known by the crew as the "Cannonball Express," was a fleet replenishment oiler. Her primary mission was to replenish operating ships at sea, with a variety of provisions including fuel, food, parts and munitions. She was 659 feet long, displaced 41,360 tons with a draft of 36 feet. Her power plant consisted of three boilers that fed two steam-turbine engines with shafts capable of 32,000 total horsepower, ahead. Unfortunately, her astern thrust was limited to 3,000 horsepower. In other words, it took one hell of a long-time to back down from a top speed of 22 knots.

The crew numbered 420. This included 22 officers and 398 enlisted personnel. There were many different ratings of sailors aboard, but because of the mission, the crew was heavy with Boatswain's Mates and Supply-rated sailors. We also had ordnance-rated sailors responsible for our self-defense capability, which consisted of a NATO *Sea Sparrow* launcher and two *Phalanx* Close in Weapons System (CIWS), 20 mm cannon, each capable of firing 4,000 rounds per minute. Other armament included 25 mm chain guns that we could set up to ward off small-boat attacks.

In addition to our shipboard underway-replenishment stations, we also had a CH-46 *Sea Knight* helicopter detachment aboard. The detachment included two CH-46s, eight pilots and 20 enlisted crewmen and maintenance technicians. The helicopters, acting in concert with our multiple replenish/refueling stations, gave us significant capability. LCDR Ken Bitar was the Officer in Charge of the detachment, which included LT Joy Smith, the first and only female on *Wabash*.

I was incredibly lucky to have Ed Lapating, initially, then Bill Kirkland, who were both great Executive Officers. Kirkland was a Naval Academy graduate and a perfect mentor for the JOs, and for that manner, me, teaching, instead of scolding the watch team. Bill always wore his engineers cap in honor of the Cannon Ball Express.

Our first underway period was the start of a massive exercise entitled Pacific Fleet Exercise, 1989 (PACEX-89). The exercise involved three carrier battle groups, *Constellation* (CV 64), *Enterprise* (CVN 65) and *Carl Vinson* (CVN 70). Each of these battle groups included an AOR with *Wabash* being assigned to *Carl Vinson*. The battleship USS *Missouri* (BB 63) was also participating with an accompanying surface action group.

My boss, RADM Bitoff decided to begin the exercise aboard *Wabash*. Of course, this added a bit of stress to an already nervous aviator, as we cast-off and headed for "Queen's Gate" to exit Long Beach Harbor. As part of the exercise, we were told to expect a simulated attack from small boats, while still in the harbor. Since we were full of supplies and fuel, and drawing over 36 feet, it was critical that we stay in deep water on the way to the mouth of the harbor. With this restriction in mind, I briefed the OOD (Officer of the Deck) and the Conning Officer (officer giving the orders for the course and speed of the ship), not to over-react to any simulated attacks. We devised a silent hand signal from me to them, indicating when they should ignore the attack and continue straight ahead, keeping enough speed to safely maneuver the ship.

USS Wabash *(AOR 5)*.

Shortly after getting underway, still very close to dangerously shallow water, two rigid hull inflatable boats (RHIBs) simulated a very realistic attack, which included firing of blanks from various weapons. The admiral was on the bridge in the XO's chair, watching closely, as I gave the hand signal to ignore the attack and proceed straight ahead. The incredulous admiral finally came over to my chair and said, "Captain, why are you not reacting to the attack?"

I got up from my chair, grabbed a bullhorn, went out on the bridge wing and literally shouted at the attackers, "Get the hell away from my ship or I will run over your little butts!" I then explained my actions to the admiral saying, "This part of the exercise was poorly conceived, as it puts deep-draft ships like *Wabash* in jeopardy of running aground. *It was my first day at sea, I had been underway about 10 minutes, and already I was in danger of being fired.* Luckily, RADM Bitoff was a veteran ship driver, so he understood why I had ignored the attack, and we proceeded to rendezvous with the rest of the battle group.

One of the goals of PACEX-89 was to practice operating in the challenging weather environment of the North Pacific and the Bering Sea. The environmental rules involved in at-sea refueling required that we accomplish the event no closer than 50 nm from land. The three carriers were operating in a weather-protected Aleutian Island Bay, while we proceeded past Adak, Alaska into the Bering Sea, just in time to encounter a storm with 70 knots of wind and 35-foot waves.

A view from the stern of Wabash *during a storm with 35-foot waves in the Bering Sea.*

Even though we were still quite heavily loaded with fuel and provisions, we were tossed about like a large cork.

The weather conditions were such that all we could do was steer the ship, to limit the effect of the enormous waves. Oh by the way, it was also snowing and—unbelievably—it was foggy! (How in the hell does fog occur in 70-knot winds?). At one point during the first night, and at the height of the storm, the rather inexperienced, but well-intended OOD, decided to make a turn, to stay in a certain operational area.

I had just gotten up from a couple hours of sleep and had stepped into the shower. Suddenly, I was tossed against the shower door and looked up to witness a chair in my office overturn (this is not good). I quickly realized what must be happening, grabbed the sound powered phone and called the bridge. The OOD answered, "Yes Sir." I said, "Right full rudder, all ahead flank! I will be on the bridge in a few seconds." We had been steaming at a minimum steerage speed of five knots into the 35-foot seas. This was the best way to minimize the effect of the sea on the ship. I had correctly deduced, that upon reaching a predetermined location, where we would normally turn the ship around, the OOD had ordered a port turn, which basically put the ship abeam of these giant seas.

My stateroom was only a few feet from the bridge, so I pulled on a pair of pants with no shirt or shoes, soaking wet from the shower and bleeding from a cut over my eye caused by slamming into the shower door. As serious as the situation was, we later laughed about the reaction of the bridge team to the arrival of the disheveled Captain announcing, "… this is the Captain, I have the conn!" (I was now giving the direct commands to the helmsman to steer the ship, and the lee helmsman to control power to the engines.)

Since this incident occurred when the crew, other than those who were on watch, was in their bunks, no one was seriously injured and there was no significant damage to the ship or our cargo. Still, it was a reminder that no matter how big your vessel is Mother Nature can quickly humble you.

The following day, with the weather a bit better, but still with a challenging sea state, we were scheduled to attempt a refueling from a civilian tanker. We really did not need any fuel; however, it was important to demonstrate that we could go alongside and hook up to a non-Navy ship. We shot the approach and arrived at the required 140-foot of vessel separation. At that point, things began to go bad! We could manage the 140-foot separation; however, we were struggling to stay in position, fore and aft. I finally called that vessel's captain and said, "Captain, have you set your engine power to maintain 13 knots?" He replied, "I am doing everything I can to maintain 13 knots, which is not easy in this sea state."

No wonder we were having such an awful time trying to stay in position. As the "guide" vessel, it was critical that he not touch his power setting once it was set for 13 knots. I finally got the point across, but by that time, we had enough of this exercise and got the hell away from him.

We proceeded from the Bering Sea to Okinawa, where we joined with the other two battle groups and some Japanese vessels. Once we were all in the area, we proceeded to create a giant 47-ship formation with the three aircraft carriers in the center. This made for an awesome photo opportunity but was also fraught with risk.

Shortly after we formed up, one of the single-screw frigates lost power, including the ability to steer his ship. This resulted in slow-motion chaos. The frigate began drifting out of position toward *Enterprise*. *Enterprise's* Captain, my old *Oriskany* roommate, and my commanding officer in VA-192, H.T. Rittenour, reacted by backing down to avoid the frigate. Captain Denny McGinn on USS *Wichita* (AOR 1) was 500 yards astern of *Enterprise*, but like my ship, he had limited ability to back down, so he turned to starboard to avoid *Enterprise*. Eventually, everyone got back into position for the photo, followed by a welcome scattering of vessels to safer waters.

From the Okinawa area, we proceeded into the Sea of Japan. Since *Wabash* was scheduled to refuel/resupply in Sasebo, we were about 100 miles ahead of *Carl Vinson*, and the first ship through the Tsushima Straits. The weather included a low cloud deck, with pretty good visibility underneath. Our NATO *Sea Sparrow* weapons system included some limited surface-to-air radar capability that picked up a contact at about 10 miles, heading straight for us. I went out on the bridge wing just in time to witness this massive Russian Tupolev Tu-95, *Bear* bomber, pass overhead at an extremely low altitude.

Since he was so low, it was going to be difficult for the *Carl Vinson* to pick him up in time to launch interceptors. I immediately got on the radio to alert them and get the *Tomcats* launched. "Alpha Bravo (Battle Group Commander), this is *Wabash*, we were just overflown by a Russian *Bear* at very low altitude, and he is heading directly for you!" The reply, "Negative *Wabash*; that aircraft is a Navy P-3 *Orion*." I was astounded at this reply, and wasted no time in rather unprofessionally responding, "This is Captain Rud, I am a naval aviator, and the aircraft inbound to you is a God damn Russian *Bear*, so, launch the alert now!" They did and managed to keep from being embarrassed by an over-flight of an unescorted Russian. I heard later that the rest of the Alpha Bravo watch team got a kick out of the watch officer getting schooled on aircraft recognition.

***

SASEBO, JAPAN: Located on the west coast of Japan, Sasebo is a beautiful and very historic port. Like all large and busy seaports, all ships entering the harbor are required to enter with a local harbor pilot aboard. The Sasebo Harbor pilot arrived by pilot boat, outside of the harbor entrance. As he arrived on the bridge, I was surprised by his small stature and his limited command of the

English language. He smiled, bowed, then produced two gloves, one green and one red. He then put the red one on his left hand, and the green one on his right announcing, "Red is port, and green is starboard."

The good news about Sasebo Harbor and the approach was that the water was deep, much like a Norwegian fjord. You just needed to stay clear of the obvious rocks. I noticed though, that the area where we were going to have to make a sharp 90-degree port turn was full of small fishing vessels. I pointed at these boats and asked the pilot for a recommendation on avoiding them. He said, "Captain, what is your gross tonnage?" I replied, "41,000 tons." He then said with a big smile, "We shall invoke the '41,000-ton' rule. The fishing vessels will get out of your way." And they did. We docked inside of what was once a massive-dry dock area used to support battleships in WWII. There was limited space between vessels, but the pilot and his super-capable, vectored-thrust tug-boats easily and safely got us to the pier.

Later, near the end of the exercise, and just prior to heading back to the States, the entire battle group visited Sasebo. I was fortunate to get together with one of the F/A-18 pilots, Chuck Wright, who was one of my junior officers in VA-192. Chuck had been an Olmsted Scholar (awarded to outstanding junior officers for foreign language graduate studies) and had attended the Japanese War College. He was conversational in Japanese, so I really had a wonderful tour guide for the visit. We had dinner with some local folks, including one of the last Japanese soldiers to surrender on Guam (years after the war was over). He was a real character, with an amazing story of survival. He was short in stature and joked that his survival was mostly based on the Americans aiming too high.

***

INCIDENT WITH CHINESE-FLAGGED FREIGHTER: Following our pier-side refueling in Sasebo, we headed back into the Sea of Japan. By this time, there were a lot of thirsty ships in the battle group, so we set up a course and speed as the station/guide ship, and they began lining up for fuel. As the station ship, we set a steady course by gyro, and speed by engine RPM. The other ships then came to station alongside of us and maintained that position throughout the evolution. The course and speed is known as the *Romeo Corpen*. The *Romeo Corpen* depends on operational needs, as well as sea state and vessel traffic. *Wabash* could refuel two ships simultaneously. Aircraft carriers including the smaller LHDs and LPHs were only capable of refueling from the port side of *Wabash*. This restriction was because of the location of the bridge and the fuel receiving stations on their starboard side. *Carl Vinson* came along our port side and USS *Vincennes* (CG 49), a Ticonderoga-class cruiser, hooked up on our starboard side.

As the captain of the station ship, it was my responsibility to keep an eye on other ships that might be a collision hazard. The rules of the road allow for an at-sea refueling evolution to have the right-of-way over all other traffic. That being said, the oceans are full of idiots that either don't know the rules of the road or choose to ignore them. In this case, a Chinese-flagged freighter, which was already the burdened vessel by the normal rules-of-the-road, continued to steam a course in direct conflict with our refueling evolution.

I tried to raise him on the bridge-to-bridge radio several times, to no avail. As the situation would soon cause us to perform emergency breakaways if he did not respond, I called LCDR Ken Bitar, who was flying one of our airborne CH-46s. I said, "Ken, go over and get that Chinese freighter's attention, and get him to turn around." His answer, "No problem, Skipper, consider it done." The next thing I saw was this very large and intimidating helicopter, literally hovering in front of the bridge of the freighter, with one of the crewmen leaning out the door gesturing to the freighter's captain to turn his ship around—now! And by God, he sure did. The last we saw of that ship it was still heading back toward China.

***

Transit to Long Beach with a Typhoon Problem: We departed Sasebo with a real sense of urgency. A typhoon had formed and was threatening our track across the Pacific. Also, shortly after beginning the transit, the oiler servicing *Missouri* and her surface action group experienced a major engineering casualty, requiring her to return to Yokosuka, Japan for repairs. *Wabash* was now supporting two battle groups. To stay ahead of the typhoon, all the ships were steaming at a higher speed than normal, which, of course, resulted in more fuel used. By the time we reached a spot approximately 1,000 miles north of Hawaii, our fuel state was so low that the bow of *Wabash* was much higher out of the water than our stern. At that point, we rendezvoused with a USNS oiler (non-commissioned ships that are the property of the U.S. Navy and manned with a mostly civilian crew) that had sailed from Hawaii. Due to the low fuel state and resulting ballast issues, *Wabash* was a bit unstable, so XO Kirkland and I worked extremely hard for almost 14 hours (one of us was always on the bridge wing), to take on about four million gallons of fuel. Imagine maintaining a position of 140 feet from another vessel for 14 hours. The positive side of this ordeal was that the junior officers really gained confidence in ship handling.

***

WABASH **EARNS A NICKNAME:** During the short turnaround between PACEX-89 and our next full western Pacific deployment, we had an opportunity to be involved in a training exercise for our weapons systems. This involved the live firing of our NATO *Sea Sparrow* and our Close in Weapons System (CIWS). The CIWS is a point-defense weapon system for detecting and destroying short-range incoming missiles and enemy aircraft. The two CIWS mounts were located on the bow. Although intended for use against airborne targets, XO Kirkland and I could also set them up in a manual mode to be used against a potential small boat attack. We figured that a 4,000 round-per-minute barrage of 20 mm, plus our 25 mm chain guns, would discourage any boarding attempt by a potential enemy.

The exercise for the CIWS consisted of firing at a target missile that was towed behind an A-6E *Intruder*. Normally, they would tow the missile almost parallel to *Wabash*, with enough of an angle to make the target appear as a threat to the ship. This would result in acquisition by the CIWS and subsequent automatic activation of the guns. I managed to convince the A-6 crew to instead pull the target so that it was coming directly at the ship. I promised not to activate the system until the A-6 had passed over the ship and out of danger. We did exactly that, and the results were spectacular. The CIWS automatically acquired the target and fired, breaking the missile into pieces. Alarmingly, the largest remaining piece of the missile continued toward the ship, appearing as though it would hit us. No problem, the CIWS picked that up too, and blasted it to bits just before it hit the ship. Most of the crew were on deck to witness this event, so it served as a real confidence builder, and even got us the nickname: "The Destroyler."

During the relatively quiet time between PACEX 89 and our western Pacific/Indian Ocean deployment in 1990, we decided to get the crew into better physical condition. We had a small gym on board, and we encouraged all sorts of physical activity, especially running. The Command Master Chief, RMCM Mike Ellington, and I decided to initiate a physical fitness contest consisting of a 1.5-mile run from the pier we were docked at. The course would go out .75 miles, then return to the pier.

Since I was a runner, I challenged the crew with the following announcement: "Any crew member who beats me in this run will get a 96-hour liberty chit." Initially, the XO was a bit nervous, "Are you sure you want to do this Skipper? We may end up with half the crew on liberty for 96 hours!" I assured him that would not happen, however, we did start to include comments in the ship's Plan of the Day to encourage participation in the race. The comments included hints like this: "The Captain is 46 years old, which is older than most of your fathers. He weighs 185 lbs., which is kind of porky. You have three weeks to get in shape, beat him in the run and earn a 96-hour liberty."

The reaction was amazing. Sailors were seen running everywhere on the streets of Naval Station Long Beach. The day of the run arrived, and I would estimate that we had about 200 runners on the pier, ready to kick the captain's ass. Of course, there were probably about 10 real runners in the group that I knew would beat me, but the rest were a bit of a mystery. Since the out-and-back course offered a great opportunity for me to keep track of how many were ahead of me, I stayed in a pack, a short distance behind the leaders. The leaders included some good runners, but also many who thought they were good runners and had charged off at an impossible-to-sustain pace.

As we made the turn, there were many shocked looks at how close behind the "old man" was. RMCM Elliot and I had decided that we could afford to have 15–20 sailors awarded 96-hour liberty chits, so as we neared the end of the run, I started to pass most of the "pretend" runners. Just before we reached the finish line, where the rest of the crew were cheering wildly for their buddies to stay in front of the fast-closing "old man," I started to pass this obviously-struggling young Seaman. He was a small kid, wildly popular with his shipmates, now throwing up and about to pass out just a few steps from the finish line. I picked him up, carried him the few steps to the finish line and deposited him ahead of me. The crew went nuts! *This small action on my part turned out to be a real morale booster and encouraged improved physical fitness for the arduous upcoming deployment.*

***

SUPERSONIC *TOMCAT:* We were tasked as the duty oiler for a short at-sea period, prior to the start of our next extended deployment. Since we were operating under the restricted air space that the fighters from NAS Miramar used to train, it was not unusual for them to request a flyby. The *Wabash* crew loved these, so when an F-14 *Tomcat* crew requested a supersonic flyby, I quickly approved and made an announcement so that as many crew members as possible could get on deck to observe the flyby. Normally, the flyby was parallel to the ship, but in this case, they asked to fly directly over us.

I did not really think this one through before saying, "Sure, bring it right over the ship." They did, and it was spectacular! So spectacular that it literally blew off our hangar bay doors. Now what? I reacted by sending a message questioning the viability of hangar bay doors that could not even withstand a supersonic flyover. The reaction was quick, in that a new and better set of doors was waiting for us at the pier when we got back to Long Beach. *This is one of those actions that managed to avert some sort of investigation; that would have been a huge pain in the ass, both for me and the* Tomcat *crew.*

***

WESTERN PACIFIC/INDIAN OCEAN DEPLOYMENT, 1990: On February 1, 1990, we departed for a six-month deployment with the *Carl Vinson* Battle Group. Having the experience of PACEX-89, our crew was ready to go. Typically, a battle group that is transiting to the western Pacific (WESTPAC) will maintain a speed of advance (SOA) of about 14 knots. This may not seem to be very fast, however, this includes flight operations and other events that, depending on the wind, may not take place on the transit course. With a maximum speed of 22 knots, *Wabash* was by far the slowest of all the vessels in the group.

Bottom line: Our chief engineer, LT Jim Calabrese, and his boys were tasked with maintaining what amounted to a "full-bell" throttle setting for most of the transit. The typical settings for the Engine Order Telegraph (EOT) (this sends a signal from the navigation bridge to the engine room to set certain power/rpm for the engines) are Stop, Ahead 1/3, Ahead 2/3, Ahead Standard, Ahead Full, Ahead Flank. At the full and flank settings the temperatures and noise level in the engineering spaces are nasty. Congratulations to Jim, CWO-3 Ron Nelson and the entire engineering department, as they would complete the entire six-month deployment without ever going "cold iron" (shutting down the plant).

***

PATTAYA BEACH, THAILAND: The battle group stopped for a port call in the Philippines with the next one scheduled to be Singapore. Although a beautiful and exotic port, Singapore was also very expensive and not exactly sailor friendly. Keeping this in mind, I asked RADM Dan March's Chief of Staff (RADM March was the battle group commander) if *Wabash* could make a port call in Pattaya Beach instead of Singapore. The answer, "Yes you may, but you will need to have one of the destroyers accompany you." That was certainly not a problem, as Tim LaFleur, the skipper of Destroyer USS *Elliot* (DD 967), immediately volunteered to be our escort (yes, the same young officer who eventually became a three-star Admiral and the head of the entire Pacific Surface Force).

Since there were only two ships anchored in the bay at Pattaya, versus a whole battle group, our crew was treated like gold and had a wonderful time in Pattaya Beach. We were now about a day behind the rest of the battle group, so we cut short our visit and headed for the Straits of Malacca. The Straits of Malacca is a narrow band of water, 580 miles in length, located between the Malay Peninsula and the island of Sumatra. It serves as the main shipping channel between the Pacific and Indian Oceans. The plan was to enter the straits at sunrise so that we could transit the incredibly busy shipping lanes—and especially Singapore itself—in daylight.

That plan quickly fell apart when *Elliot* came across a disabled boat loaded with Vietnamese refugees. Following the rescue of 30 people, *Elliot* came alongside *Wabash* to get some fuel. Ironically, one of the newest members of our officer wardroom was Ensign Tran. He was a recent graduate of the Navy ROTC program at Northwestern University in Chicago, and he was also a refugee who had escaped from the communist regime via the same method that these folks were attempting. Ensign Tran was fluent in Vietnamese, so he went back to our helicopter platform, which was now 140 feet from the *Elliot's* helicopter platform, where all the refugees were gathered. Using a bullhorn microphone, he shared his story with these people. Every so often, you would hear a huge cheer erupt from those folks, as they listened to his story of obvious success. *This incident brought back memories of "Operation Frequent Wind," the evacuation of Saigon. The fact that 15 years later, people were still willing to risk their lives to escape the communist regime helped those of us who were veterans of that war to feel a bit better about our efforts to keep that country free.*

***

TRANSITING THE STRAITS OF MALACCA: Because of the delay involved with the rescue, we now entered the straits at sunset versus sunrise. Not good. I had never seen so many ships in such close proximity. It would have been a challenge even in daylight, but it was a "wild and crazy" evolution at night. We had the most experienced folks on the bridge as we neared Singapore. LCDR Ira Stokes was the Officer of the Deck (OOD) and our most talented JOs filled out the rest of the watch team. The XO and I were both on the bridge the entire time. You might think that with so much traffic, it would be a good idea to proceed at a slower speed. However, that would put your vessel in more peril, as faster traffic would be constantly passing you. As I recall, we steamed at a "flank bell" making about 20 knots through the water. At that speed, we were comparable to the massive 100,000-ton super tankers and able to hold our position in the queue of transiting ships.

As we neared the especially challenging waters of Singapore Harbor, I noticed a super tanker displaying all-around white anchor lights. Although this indicated that she was anchored, there was a great deal of activity on board, which led me to believe that she was about to get underway. The problem was that if she got underway, she would be coming into conflict with us—and she would have the right-of-way. Suddenly, the anchor lights extinguished, and she was underway with her massive bow swinging to port directly in our path. It was too late to explain what I was seeing to the watch team, so I said, "On the bridge, this is the Captain, I have the conn. Right 10 degrees rudder, all stop."

This move cleared her stern comfortably, and we were then able to pass her to starboard and resume our course (the all stop of the engines probably resulted in only a couple of knots decrease of speed, but it was enough). *It takes an entire team to safely navigate these waters. Because we trusted the bridge team and did not micro-manage them, the XO and I were able to concentrate solely on unexpected issues, like the no-notice getting underway of that super tanker.*

***

INDIAN OCEAN OPERATIONS: The weather in our operating area was incredibly hot and humid. Since so much of our mission involved manual labor, we relaxed the uniform rules to dungaree pants and tee shirts. We also decided to initiate a beard-growing contest. With this combination, the crew of *Wabash* looked more like an eighteenth-century pirate vessel, than a normal U.S. Navy ship. About a month into the beard-growing contest, the battle group commander, and my boss, RADM March, announced that he was going to personally visit *Wabash*. Our dilemma now was whether to continue the beard-growing contest or clean up our act for the admiral's visit. We decided on a compromise solution that required all the officers to shave their beards, while the enlisted crew members continued the contest.

During the admiral's visit, as he and I were meeting in my cabin, the traditional eight o'clock report was delivered by one of our most squared away first class petty officers. He reported in his immaculately cleaned and pressed dress-white uniform. He was also sporting a full beard that would eventually win the contest. This young man had combed his chest hair up into his beard. The reaction of the admiral was priceless! He did not say a word until the sailor left. He then looked at me, smiled and said, "CAPT Rud, I think I now know why you have such great morale on this ship. Now, how about declaring a winner in this contest and getting your crew cleaned up and prepared for our upcoming visit to Perth." *Once again, I was pretty sure that I had sealed my fate regarding any chance of a potential positive fitness report.*

***

PORT VISIT TO PERTH/FREEMANTLE, AUSTRALIA: Although *Carl Vinson* anchored out, which required shuttling the crew to the wonderful liberty associated with Perth, *Wabash* tied up to a pier in Kwinana. This allowed our crew easy walk-off access to the local train system, so that they could conveniently enjoy the best liberty of the entire deployment.

Shortly after we tied up to the pier, I got a call from the battle group chief of staff. He said that he had been totally surprised by the arrival of seven female midshipmen with orders to ride *Carl Vinson* for their summer cruise. He went on to explain that the carrier had no women on board and no facilities to berth them. He said, "Rud, you have a female aviator on board, is there any chance that you might be able to host these women?" I replied, "Let me talk to

LT Smith (the female CH-46 pilot) and I will get back to you in an hour."

I arranged for a quick meeting of the officers and Master Chief Elliot. Just before the all officers meeting, I asked LT Smith if she would be willing to basically chaperone seven female midshipmen? She readily agreed, suggesting that the junior officer bunkroom would be a perfect solution to provide acceptable accommodations. I opened up the all officers meeting with the following comment, "Boys, I have some bad news. We are going to have to move you out of the bunkroom for the rest of the cruise. The good news is that we are going to have seven female midshipmen moving in tomorrow." Since most of the *Wabash* junior officers were young bachelors, they eagerly agreed to the now "small" inconvenience of a move.

The next afternoon, as the XO and I were standing on the bridge wing, a van pulled up with our seven midshipmen. The XO made a praying motion with his hands, looked up to the heavens and said, "Please God, make these ladies all butt-ugly!" His prayer certainly was not answered, as seven attractive young women emerged from the van to join our crew.

In 1990, the qualifications required for a female to get an NROTC scholarship or Naval Academy appointment was higher than that of their male counterparts. In other words, they all turned out to be quick learners, and in some cases intimidated our junior officers with their performance on the various watch teams.

The next day, I got another call from the chief of staff. He said, "Rud, I understand that you come from a farming background. We have been asked to host a rancher VIP who is the son of the former Premier of Western Australia. Would you mind hosting him for a lunch aboard *Wabash*?" I replied, "Sounds like a great opportunity to get to know more about this awesome country. Send him over." As I have often found to be the case, this opportunity turned out to be quite an adventure. Following lunch aboard *Wabash*, our guest invited me and one of the CH-46 pilots, who was also a farm kid, to visit his station (ranch), which was about 100 kilometers from Perth. We did just that and had a wonderful day learning about Merino sheep and the wool industry.

At the completion of our day in the country, we were invited to dinner at our VIP's beautiful home in Perth. He had a large family, and while we were getting a tour of the residence, his twelve-year-old daughter asked me the following question, "Gil, would you like to see my mother's trophies?" Quite curious at this point, I responded, "I sure would." She then showed me this giant trophy case filled with dozens of impressive awards including what turned out to be 24 major tennis championships.

Yes, my VIP guest was Barry Court and his wife was Margaret Smith-Court, the first woman to win the singles Grand Slam. I returned to the kitchen where Margaret was preparing dinner and asked, "Margaret, do you still play tennis?" She beckoned me to join her at a large picture window overlooking their property. She then flipped a switch, which lit up a beautiful tennis court. "Gil, would you like to take me on for a set after dinner?" "No thanks Margaret, maybe next time I am in Perth."

***

RUD VERSUS RUD: Following a brief stop in the Philippines, and another in Pearl Harbor, we began the final phase of our transit back to Long Beach. Prior to pulling into port, we needed to offload some ordnance to an ammunition ship. USS *Shasta* (AE 33) was scheduled to rendezvous with us for this underway evolution. The morning of the rendezvous, I was on the bridge as *Shasta* appeared on the horizon. They were flashing a message to us, so I queried the signal bridge as to what the message said. I got the following reply, "Midshipman Rud has the conn."

This meant that my daughter was giving the orders to drive the *Shasta*. My XO said, "Holy cow, Rud versus Rud for an underway replenishment. This has got to be a first." I grabbed some binoculars and by God, there she was, blonde hair flying in the wind, smiling from ear to ear (of course the captain, CDR Dan Gabe, was standing just out of sight, giving her advice on what orders to give). Still, it was a wonderful ruse, and certainly one of the proudest moments in my life. By the way, since I was familiar with her early attempts at driving a car, I had the signal bridge send the following message back to the *Shasta*: "I sure hope it is not a straight-stick."

We hooked up and began both a ship-to-ship and vertical transfer of our ordnance. Since Valerie had almost completed her summer cruise requirements, the captain of *Shasta* suggested that my daughter should jump in one of the CH-46s and ride *Wabash* into port. It just so happened that LT Joy Smith was the pilot that picked up Val and transferred her to *Wabash*. A couple of years later Joy would be one of Val's instructors in the training command. It really is a small world.

It is tough enough for a father-son situation to follow military protocol. For a father-daughter, it is literally impossible. Val showed up on the bridge and gave me a huge hug and kiss, followed by a "What's up Dad?" The bridge team loved seeing the "old man" light up and relax a bit from his usual "alongside" personality. Following completion of the ammunition offload, the XO, Val and I headed to my inport cabin for a cup of coffee. As we entered this rather impressive space, Val said, "Wow Dad, this is perfect. Where are *you* going to sleep?" Did I mention that my daughter has a wonderful sense of humor?

Father and daughter aboard USS Wabash.

***

BACK IN THE GOOD OLD USA: Unlike my previous returns from cruises, there was no family to greet Val and me. The reason was that Ryan had managed to make the roster of a soccer team that was training and traveling in Europe. Despite a lack of financial resources, Barbara decided that this was a once-in-a-lifetime opportunity, so both she and Sally were also in Europe. Although it took me about three years to pay off the credit card they used, it was worth it. They had a blast, and Ryan got the exposure that eventually led to him being recruited to the University of California, Berkeley soccer team.

***

LEADERSHIP FAILURE: *Wabash* had just completed a second deployment, and we were a shit-hot crew and a great operational ship that had safely completed the refueling and replenishment of over 277 ships. I could not have been prouder of the performance of everyone, especially Jim Calabrese and the engineers, who had kept the plant going for six consecutive months. Unfortunately, now it was time for the inevitable round of "assist" visits by the various inspection teams. Assist turned out to be a huge misnomer, as these folks simply came aboard to inspect and condemn the condition of the engineering plant, as well as the engineering training program.

Of course, after six months of continuous steaming, the plant needed some tender loving care. Also, since we had been operating all this time, training had all been on the job, and not always according to the rigid rules of the surface Navy. During one of these assist visits, the senior inspector physically grabbed one of our young engineers and started screaming at him about how incompetent he was. I quietly asked this inspector to accompany me out of the space into an adjacent passageway. I then looked him in the eyes and said, "If you ever touch one of my sailors again, I will kick your ass!" I then said, "Get your assist team the hell off my ship, right now!" They left, and the eventual result was that we got some folks to actually help us get ready for the inspection cycle. *I was a success as an operational leader, but I failed to provide the leadership required to do well on these inspections. I would like to take this opportunity to apologize for my shortcomings to LT Jim Calabrese, CWO-3 Nelson and the entire engineering crew. You guys were awesome.*

30. Chasing Drug Traffickers

ASSIGNMENT TO JOINT DUTY: The next step on what the Navy considered to be an appropriate career path, needed to be an assignment to a joint service job. The one that caught my attention was the Chief of Staff and Deputy Commander position at Joint Task Force Five (JTF-5), located on Coast Guard Island in Alameda, California. This was a fairly new command, formed in 1989, when the U.S. military was given statutory responsibility to detect and monitor aerial and maritime illicit drug shipments to the United States. The unit consisted of representatives from the Army, Navy, Marine Corps, Air Force and Coast Guard. Since the purpose was to bring military and law enforcement agencies capabilities together to combat drug-related transnational crime, we also had representatives from the Drug Enforcement Agency (DEA), the Federal Bureau of Investigation (FBI), the U.S. Customs Service and the National Security Agency (NSA).

The Commander of JTF-5 was a Coast Guard one-star admiral, RADM Leahy. His deputy and chief of staff was Navy Captain Jim Ellis, who was leaving that position to become the commanding officer of the nuclear-powered aircraft carrier USS *Abraham Lincoln* (CVN 72). Jim eventually became a four-star admiral, so as the new deputy, I had some big shoes to fill. We were tasked to direct and coordinate the detection, monitoring and sorting of suspect drug-trafficking vessels and aircraft. Targets were then turned over to U.S. law enforcement authorities for apprehension. It was an exciting and rewarding tour of duty. Unfortunately, due to the still-sensitive nature of the mission, it is not appropriate for me to share the details of the many successful interdictions and arrests that were made possible by the actions of the talented folks in this command.

Ryan was now a freshman at Berkeley, so at the close of the school year in Lemoore, Barb and Sally joined me in government housing in Novato, California. Our house was located on the old Hamilton AFB complex. It was built in the 1930s, but it was very nice, and the location was convenient for both my work and school for Sally. As usual, Barb soon secured a job in the Marin County school system, and we settled into life in northern California.

The Rud kids, Sally, Ryan and Valerie at our Bay Area home on Hamilton AFB.

It did not take me long to realize that I was basically inept as a staff officer. Word processors were in all the offices and were now the standard method of communication. I had never taken a typing class (big mistake) so I was, at best, a "hunt and peck" typist. Fortunately, I was blessed with Connie Kozlusky as my awesome secretary, so she was able to interpret my awful, longhand penmanship, and turn my scribble into readable documents.

Shortly after I took over the chief of staff job, RADM Jack Linnon replaced RADM Leahy. RADM Linnon was a no-nonsense Coast Guard warrior who had come up through the enlisted ranks. Being accustomed to real action daily, he was bored stiff by this staff job and had little patience with my incompetence. Part of my responsibility was to prepare speeches for him. I had done some of this in the past, using a bullet/outline format, which was OK for a common luncheon speech, but not for congressional testimony. Congressional testimony was "for the record" and therefore needed to be read verbatim from a script. This task was way over my head. I realized this and turned to the staff intelligence officers, Howie Ehret and my good friend and running partner, Geno Spatafore for help. They were both awesome writers, so once again I managed to avoid getting fired.

To survive in this job, I purchased a "Mavis Beacon" typing tutor disc that I could use with my word processor. I spent hours with that silly music in the background, insects smashing against a motorcycle face shield every time I made a mistake (which was quite often), and other demeaning typing lessons that made my 14-year-old daughter, Sally, giggle at her cussing and struggling father. I eventually graduated from the course with the ability to type about 40 words a minute without looking at the keyboard. *You are never too old to teach yourself a new skill, which I certainly put to good use, both for the rest of my Navy career—and later for 14 years in the aerospace marketing business.*

Although I cannot get into specifics, there was a considerable amount of activity in our spaces that included undercover DEA and Customs agents. Being undercover, these folks were some of the wildest looking characters you might find on the streets. My son, Ryan, who was a student at UC Berkeley, wore his hair long and pretty much resembled the typical Cal student. He was not only a member of the Cal soccer team; he was also an excellent skate boarder, with the ability to skate fast enough to safely travel through some of the tougher areas of Oakland. One day, he skateboarded to Coast Guard Island to pay me a visit. The staff all thought that he was an undercover agent, so no one paid any attention as he entered my office for a family visit.

Gus Fassler, the DEA agent on the staff, was a real character and shared many hair-raising stories of his adventures. Gus drove a large, four-door sedan with a huge engine and what I seem to remember was a sawed-off shotgun slipped between the bucket seats. He also had a police flasher and siren that he could stick on the roof if needed. One day, he invited me to a Federal Law Enforcement Officers Association (FLEA) luncheon. The event was being held in San Francisco, so we had to cross the Bay Bridge, which was always a traffic nightmare. I noticed that on that day, there seemed to be an extraordinary amount of police activity on the bridge. Suddenly, Gus placed his flasher on top of the car and turned on the siren. Curious, I said, "What is going on Gus? Is there a police emergency?" He calmly replied, "Nope, just a bunch of cops not wanting to be late for lunch."

31. Command of Aircraft Carrier USS Constellation (CV 64)

TOWARD THE END of my two-year assignment in JTF-5, I got the unbelievably great news that I had been selected to be the captain of an aircraft carrier. Even better, the orders put me back into a flying status. Most carrier commanding officers, did not take the flying status seriously, as they expected to be too busy to get in any real flying. I, on the other hand, looked at it as a great opportunity to get back into the air.

I soon learned that my assignment was to USS *Constellation* (CV 64). She was just completing a three-year service life extension at the Philadelphia Naval Shipyard, under the leadership of Captain Mike Nordeen. I had previously served on *Connie* as a member of the VA-147 *Argonauts*, so I had great memories, and felt extremely fortunate to get an opportunity to command "America's Flagship."

As part of my training track prior to taking command, I went back to the *Hornet* RAG in Lemoore for a CAT 4 (shortened training cycle for experienced pilots) refresher syllabus. It felt wonderful to get back in a jet, once again. Kind of like they say about riding a bike, I quickly became comfortable enough to embark on a cross-country flight with Marine Col Dave Percy. He was the NAS Lemoore Marine Detachment Commander and a *Hornet* flight instructor. Dave was married to Sue Creed, the widow of Bart Creed, my next-door neighbor who was shot down and killed in 1971. Both of Dave's stepchildren were flight students in the training command, as was my daughter Valerie, so we decided to pay them all a visit. This involved stops in Pensacola and Corpus Christi, Texas. During the various legs of the flight, we performed all sorts of training so that I finished the syllabus in just a couple of weeks. We did disrupt the normal traffic pattern at Corpus Christi, just a little, with a 500-knot carrier break. Boy did that feel great! *Even a couple of old guys can still enjoy giving the training command students a thrill.*

***

CHANGE OF COMMAND AND THE START OF A GREAT ADVENTURE: *Constellation* had only been out of the shipyard for a couple of months, during which time they had been conducting sea trials and initial operations with *Carrier Air Wing Two*. Captain Nordeen was thoroughly enjoying an opportunity to finally take the ship to sea, after spending most of his tenure in the shipyard. During this initial operating period, the ship was berthed at Naval Station Mayport, which was located in Jacksonville, Florida. On May 7, 1993, I relieved Mike Nordeen as the Captain of USS *Constellation*.

Much like taking over as the commanding officer and flight leader of the *Blue Angels*, this event made big news in my home state of North Dakota. Since Don Weiss, a native of Jamestown, North Dakota was the Captain of USS *Saratoga* (CV 60), also berthed at Mayport, the headlines in North Dakota newspapers read something like this: "As of today, twenty percent of the nation's carrier force is under the command of North Dakota natives." This statement was slightly embellished, but close enough, and a great source of pride for my beloved North Dakota.

***

USS *CONSTELLATION* (CV 64): To better grasp what this little old farm boy was now in charge of, I would like to share a unique description of the *Connie*. Clarence Stilwill, a writer for *Heartland USA* magazine visited the ship, interviewed me and several members of the crew, and produced a piece he called "A City at Sea." I am going to take the liberty now of repeating, word for word, his description of the *Connie: Imagine living in a small town (about 5,000 people). Then imagine one day you were asked to turn your town into an aircraft carrier. No problem. In my town it would go something like this: First we'd take over two blocks of Main Street, dig down about 35 feet (3 ½ stories) and throw the town's power station, sewer plant and water works into the hole. After covering that at street level, we'd have to gather all the rooms in town — living rooms, bedrooms, kitchens, bathrooms, stores, closets, etc. — and stack everything about four stories high on top of our two Main Street blocks. Now we'd bring in all the town's spaces — garage spaces, storage spaces, shops, gas stations, empty lots, and build a shell yet another four stories* *high to contain this empty space. Around this shell, we might scatter whatever is left in town — the radio and TV stations, the Planning and Zoning offices, the mayor's office and council chambers, the post office, and maybe the Street Department office. After all this, we would have to somehow drag in about 1/3 of the county airport and set it up on top of everything we've put together. We'd need all the airport equipment, too — plows, fire trucks, forklifts and cranes. As a final project, we'd take the tower, the terminal, and the offices of the people who run those places; throw in a radar dish and a couple of radio antennas and stack them all up another seven stories high on one side of the airport. One more thing, to do its intended job, all 5,000 people in town should be on it. Add 80 aircraft of different types, send it off into the middle of nowhere for months on end with all the fuel, food and fighting spirit it will hold and you might have something similar to USS* Constellation — *a real American Navy aircraft carrier.*

Some of the actual statistics of *Connie* include — She was commissioned in October 1961. The propulsion plant consists of four steam turbine engines, each rated at 70,000 shaft horsepower for a total of 280,000 horsepower, with a speed through the water of over 30 knots. Each of the four propellers is 21 feet in diameter and weighs 44,000 pounds. The two anchors each weigh 30 tons. The flight deck area covers 4.1 acres. The distilling plants make over 400,000 gallons of water a day and the crew is served over 18,000 meals a day. In other words, this is one big vessel!

Either naval aviators or Naval Flight Officers command aircraft carriers. The reason for that is because it really is a floating airport. Of course, prior to commanding an aircraft carrier, we have all had the experience of being in command of a deep-draft ship, which in my case was USS *Wabash*. I felt that I was ready, but of course, I was also intimidated by the sheer immensity of the challenge we now faced — getting the ship around South America's Cape Horn, and back to our home-port of San Diego. Unlike smaller ships, *Connie* would not fit through the Panama Canal, so like the seafarers prior to the canal, we would be taking the much longer and treacherous route. And we would be doing it at the worst possible time of year: June. June in the southern hemisphere is December in the northern hemisphere—in other words, the dead of winter. As if that were not enough of a challenge, the Cape area is surrounded by wild seas, created both by weather and the meeting of the Pacific and Atlantic oceans.

USS Constellation *(CV 64).*

***

LOSS OF MY EXECUTIVE OFFICER: As we began preparations for the voyage, I got a call from Washington, D.C. Although this was now the spring of 1993, the fallout from the 1991 Tailhook event was still affecting naval aviation. I won't go into detail here, other than to say that the "witch hunt" to fry anyone in attendance was still underway. The sad thing is that it seemed to take precedence over everything, including operational capability and safety. My executive officer was a shit-hot E-2 *Hawkeye* NFO, Mark Milliken. He had been the XO of *Connie* for over a year and brought the perfect experience and continuity that would make us the ideal leadership team.

Not so fast. The Navy and civilian leadership running the Tailhook investigation now claimed that he was needed to testify regarding some part of the scandal, and would be taken off the ship before our upcoming voyage around the Horn. I was infuriated, and once again placed my career in jeopardy when I said to my boss, VADM Rudy Kohn, "No problem, sir; I will just delay our transit until Mark is back aboard *Connie*." I then went on to explain what I considered to be the ridiculous priorities that endangered the operational Navy. VADM Kohn agreed with me, but cautioned that threatening to delay getting a carrier underway was a risky decision on my part.

I am not sure how close I came to getting fired over my stance, but sometimes you just must stand up for what is right. Finally, the Bureau of Personnel agreed to send me a temporary-replacement XO. Captain Gary Stubbs joined the crew, and although not as experienced as Captain Milliken, he did a great job until Mark returned to the ship near the end of the transit to San Diego.

***

UNDERWAY FOR SAN DIEGO: Thank goodness that I was blessed with having what I consider to be the best navigator in the Navy. Gene Smith was, like me, a farm kid. He was born and raised in Iowa where he was still participating in the operation of a large family farm. Gene was a superb A-7E pilot but had interrupted his career to join the airlines. Bored by that job, he returned to the Navy with what is referred to as "broken service." In my opinion, if he did not have broken service, he would more than likely have been in command of his own carrier. *I had total confidence in my "Gator," his assistant Darryl Barrickman and his bridge watch teams led by folks like LT Kevin Keilty. They allowed me to concentrate on other areas of the ship that were struggling due to underfunding/scheduling problems in the shipyard.*

We did not have a full air wing on board, however, the CAG, Captain Dan Hacker did have representation of all the aircraft types totaling about 50 aircraft versus the normal 80. This number was perfect for getting the flight deck crew back up to speed and to allow most of the aircraft to be placed in the hangar bay for the anticipated nasty weather around Cape Horn.

***

INTERACTION WITH OUR SOUTH AMERICAN ALLIES: A big part of our mission during the transit was to interact with the various countries in South America. The largest of these were Brazil and Argentina. Both countries had the capability to fly from aircraft carriers, so we had some S-2 Anti-Submarine Warfare (ASW) aircraft practice landings when we passed by Brazil. The CAG had a foreign exchange pilot from Argentina, who was qualified in the *Hornet*. He also happened to be an LSO, was qualified to wave the air wing aircraft, and served as the LSO for Argentina to fly their Dassault-Breguet *Super Etendards* aboard *Connie*. Because we no longer had bridle launch capability, they only made touch-and-go landings.

An admiral from the Argentine Navy came aboard *Connie* and was on the bridge watching our *Hornets* operating. He was listening intently to the radio chatter including the LSOs talking to the *Hornet* pilots. Suddenly, he tapped me on the shoulder and said, "The controlling LSO sounds like he has a Spanish accent." I replied, "Of course he does, because that is Carlos, your LSO from Argentina." He then said, "But he is controlling your aircraft." "Yes, he sure is, and he is damn good at it too!" The admiral then puffed his chest out so far that I thought all his gold buttons would pop off.

Later, this same admiral was getting a tour of our recently upgraded and very impressive Combat Direction Center (CDC). The space had some very modern electronic displays, including maps of the area. CDR Nick Petriccione, the CDC officer, was briefing the capability, when the admiral suddenly interrupted the brief and pointed to the map. He said in an angry tone, "What is this?" He was pointing to the Falkland Islands, which Argentina called the *Malvinas*. Without any hesitation Nick replied, "Sir, that is a hazard to navigation." The admiral smiled, looked at me, pointed at Nick and said, "This young man will go very far in your Navy."

***

AROUND THE HORN: The next step in the transit was to complete the furthest south portion of our voyage. There is a less exposed route, the Strait of Magellan, but it is fraught with risk for a large vessel like *Connie*, should we have an engineering issue causing loss of power and no tug to keep us off the rocks. At this latitude, the fronts and associated weather systems move much quicker than what one might normally encounter. With that in mind, we held in the lee of the Horn area for a few hours, until the weather looked the most favorable, and then we started around. The seas were rough and turbulent, but not bad enough to cause damage. Since the flight deck was clear of aircraft, except for alert helicopters staged in case of a man overboard, we allowed the crew to complete a 5K run during this portion of the transit.

***

LOSS OF OILER SUPPORT: The good news was that we made it around safely. The bad news was that shortly after starting northwest along the coast of Chile, we got hit by a nasty blizzard, with huge waves and lots of snow. The sea state was severe enough to cause some damage to our catwalks and other exposed areas along the deck edges. About this time, our assigned oiler experienced an engineering casualty that forced her to go into Valparaiso, Chile for repairs. *Connie's* engineering department had been short-changed regarding some scheduled maintenance in the shipyard, which resulted in a salted-up condenser and the precautionary shutdown of one engine room. We were now in a situation where we had only three screws and no oiler support.

Now, it was time for a lesson in geography. Although we were in the Pacific Ocean, we needed to keep in mind that we were so far east that the closest replacement oiler would sail from Norfolk, Virginia to transit the Panama Canal and meet us for a much needed, at-sea refueling. *The Panama Canal is almost straight south of Miami, and contrary to what most geographically-challenged folks think, it runs in a north-south direction.*

We did have an opportunity to do some joint operations with the Chilean military. This included having one of their destroyers come alongside *Connie* where we hooked up for a simulated replenishment evolution. Well, it was not all simulated, as the captain of the Chilean ship sent over a bottle of wine and I sent him a *Connie* baseball hat. *Although I could not drink it until we got to San Diego, I certainly got the better of that exchange.*

We met up with our replacement oiler as we neared the Panama Canal area. Since we were pretty low on fuel, both for the ships boilers as well as jet fuel for the air wing, we spent several hours alongside. I was very comfortable, as a veteran of some 277 evolutions on the *Wabash*, and it was a great opportunity for the watch teams to get folks exposed to the standards that the Gator and I set. That standard was 140 feet of separation, with no more than 160 feet between vessels. *Conventional carriers like* Connie, *refuel often and are much more proficient than their nuclear-powered counterparts, which only refuel when they get low on jet fuel.*

***

PORT CALL IN ACAPULCO, MEXICO: For the entire transit, we only had two port calls; both were accomplished from anchorage, versus being tied up to a pier. The first was Trinidad and Tobago, which was just a few days into the journey, and now, just a few days from San Diego, we anchored in Acapulco Bay. Described as the Mexican Riviera, it was a beautiful area with lots of entertainment for a well deserving crew. In addition to our liberty boats, the local marina used glass-bottomed tourist boats and converted shrimp boats to transport supplies and people on and off the ship, via the stern dock. This normally worked well when we were at anchor if the sea state was relatively calm.

The glass bottom boats were pretty cool, however, they were certainly not the most seaworthy vessels for this sort of activity. On our last morning in port, the sea state had turned a bit rough. Rough enough that one of these boats banged into the side of the ship while attempting to tie up to our stern, and quickly sank. There were several Mexican nationals on board, all of which were pulled from the water within a few minutes. Since we were already in the process of weighing anchor (pulling up the anchors), we decided to get underway and then return the Mexican crew back to the marina via helicopter. We gave them some dry coveralls and the XO brought them to the bridge, where I presented them with *Connie* ball caps. Rather than being upset over the loss of their boat, they were super excited to be getting their first ride in a helicopter. *I fully expected that we would be asked to compensate them for their boat, however, apparently, the VIP treatment we gave them was compensation enough.*

***

ARRIVAL IN SAN DIEGO: Following a three-year absence, *Connie* was finally back at our home port, which was the carrier pier at Naval Air Station, North Island, located next to the beautiful city of Coronado, California. During our transit from the east coast, Barb and Sally had moved from our home in Novato to our new digs at Quarters "Q," in senior officer housing on North Island, near the famous I-Bar. Our neighbors included Bill "Bear" and Nancy Pickavance. Bear was the CO of *Connie's* sister ship, USS *Kitty Hawk* (CV 63), which was berthed right next to *Connie*.

The Deputy Commander of CAG Two, the carrier air wing aboard *Connie*, Dave Nichols, and his wife Denise, also lived in the same area. We were all good friends and of course, Bear and I shared all sorts of information and supported one another, as we prepared these two great ships for deployment. All moves for teenagers are tough, but at least Sally was reunited with one of her best friends from Lemoore, Kelly Wallace, who was also starting her sophomore year at Coronado High School. Sally made the soccer team so that helped create a circle of friends too. As usual, Barb, the brains in our family, soon landed a job as an advanced math teacher at Coronado High. The Rud clan was settled in for another Navy adventure.

***

BLESSED WITH AN AMAZING LEADERSHIP TEAM AND CREW: I have previously mentioned that carrier aviators love to fly from the carrier, but disassociated sea tours (non-flying jobs) are usually avoided, if possible. But, if you must endure a non-flying job on a carrier, then getting assigned to one stationed in San Diego is the first choice, at least from the aviation communities that are based there. I was the beneficiary of some top-notch folks electing to serve aboard *Connie*. The Air Boss, Jim Destafney, and the Assistant (mini) Air Boss, Dan Cain, were both from the F-14 community as was our Air Operations Officer, Art Gratus and the incoming Navigator, Brian "Beef" Flannery. The new XO, Al Haefner, was from the HS community in San Diego. The Operations Officer, Nick Petriccione, came from the local S-3 community, and the Supply Officer, Bill Bickert, had been San Diego-based and had a home just outside the gate of NAS North Island. Dave Beck, who led over 400 personnel as the head of the Aircraft Intermediate Maintenance Department (AIMD) was San Diego-based too. Master Chief John Martin also joined our team as the Command Master Chief with a background, which included assignments in the San Diego area. CDR Gar Wright, who eventually became a two-star admiral, led our designated Naval Reserve augmentation unit. The bottom line: The entire *Connie* leadership team and most of the crew were from the San Diego community and very happy to have the opportunity for their families to stay in the area.

***

ENGINEERING CHALLENGES: The 600 officers and crew assigned to *Connie's* engineering department were the hardest working group on the ship. Unfortunately, they were dealt a bad hand from the Philadelphia Naval Shipyard. I don't recall the exact issues, but *Connie* did not get the same number, or quality, of engineering upgrades that our sister ship, *Kitty Hawk* received. The result was that we struggled to keep four screws online for most of my tenure as the commanding officer. During this period of time, we were also slated to relieve USS *Independence* (CV 62) as the forward-deployed aircraft carrier. This meant that we would eventually be stationed in Yokosuka, Japan, which adversely affected volunteers asking to be assigned to our engineering department. These factors combined to result in our engineering department being low on manning. The good news was that we were able to overcome these issues with a tremendous performance from the rest of the ship, and we maintained a reputation as a top-notch operational carrier.

***

AWESOME SUPPLY DEPARTMENT: The morale of a ship's crew is directly tied to the quality of meals they are served and the condition of their living spaces. Bill Bickert and his Supply team were absolutely superb in providing both quality and quantity. They also kept the cleanest and most efficient mess decks in the fleet. I must admit that I have great memories of the sumptuous meals and decadent desserts. Luckily, I still possessed a furnace for a digestive system.

***

HIGHLIGHTS OF *CONNIE'S* PREPARATION FOR A WESTERN PACIFIC CRUISE: Everyone was chomping at the bit to get going with our primary mission of being the best floating airfield in the fleet. After a couple of months of trying to catch up on repairs not completed in Philadelphia, we finally were ready for sea. Our air wing was the most modern in the Navy with three squadrons of brand-new F/A-18C *Hornets*. These squadrons were VFA-151, VFA-137 and VMFA-323 from the Marine Corps. We also had the first F-14D squadron, VF-2, with new *Tomcats*. These were the first *Tomcats* that included a ground-stabilized Heads-Up Display (HUD), so their performance around the ship, especially boarding rate, was much improved. Our E-2 squadron was VAW-116. VAQ-131 provided the EA-6Bs, VS-38 the S-3 contingent, and HS-2 the much-appreciated and multi-tasked SH-60 Helicopters. Two detachments, VRC-30 and VQ-5 provided the C-2 CODs and the spooky (can't tell you what they did) ES-3s. We were one of the first, if not the first, ship/air wing team to deploy without the A-6E *Intruder*.

Because we were now the most modern and capable air wing/ship team in the fleet, we needed to take a close look at how we operated. There had always been a tendency to stick with proven methods of carrier operations, but in this case the CAG, our new boss, RADM Bordy and I all agreed that we would test out new concepts of operation. What we found was that all the new and easily maintained aircraft, provided us with the flexibility to move from the standard seven event air plan, to one of continuous operation, where we could react to a threat in a much more effective manner.

The CAG, the COS and I (the navigator/operations officer often represented me) formed a team that planned and executed together. No surprises and very little "head-butting" occurred. And we had fun!

***

On the bridge of Connie *with CAG, Dave Nichols. We made a great ship/carrier air wing team. Dave went on to be a three-star admiral.*

PERFORMING A PRIVATE AIRSHOW: During one of our many at-sea periods during the workups, we were tasked with providing a deck for the various RAGs to qualify their students. This is always interesting, especially when these folks are getting their first exposure to night landings. On this evening, the prevailing wind was such that we were operating only about 20 miles off the coast, near San Diego. The bridge watch team brought to my attention a small radar return very close to the course we were using for launching and recovering aircraft. I asked the signal bridge to light up the area with their powerful searchlight. The light revealed a small sailboat bobbing about in the relatively calm ocean.

I called him on the common frequency of the bridge-to-bridge radio.

"Sailboat located 20 miles west of San Diego harbor, this is U.S. Navy warship *Constellation*, over." Pretty soon, a sleepy voice replied, "*Constellation*, I think that may be me." I then said, "Well, if you come topside, we will provide you with your own private airshow as we pass by." All this time, the searchlight was lighting up his boat, and soon we saw a man and his wife (we surmised may have been girlfriend), come out of the hatch, just as we launched a *Hornet* less than 1,000 yards from him. They were observed clapping and cheering as we steamed by at about 25 knots. I apologized for the giant wake we created, but he was so excited he replied, "Oh my God, thanks so much for defending our country, and for the airshow. What a story we now have to tell!"

***

RIM OF THE PACIFIC EXERCISE (RIMPAC): This is the world's largest international maritime warfare exercise. It took place in July of 1994 and was a great opportunity for the *Constellation/Carrier Air Wing Two* team to evaluate our new approach to war fighting from a carrier deck. The exercise pitted two battle groups, *Constellation* and *Independence* against each other. If you are looking for results, each side claimed to have kicked the other's ass. To be honest, the *Independence* group was constantly at a wartime readiness, so they could have done a bit better, at least according to my good friends Denny Irelan, skipper of *Independence*, and Hugh Butt, Commander of *Destroyer Squadron Fifteen* of that battle group.

At the end of the exercise both carriers pulled into Pearl Harbor. The entrance into Pearl is a relatively narrow channel with a significant current running perpendicular to the course required to enter. Since we were over 1,000 feet long, and one knot of current has the same effect as 30 knots of wind, we entered this channel at a relatively high rate of speed. My boss, RADM Bordy, was on the bridge as we entered, and asked me what the bridge team was using to stay in the middle of the channel. I pointed to some channel marker boards on a distant hill and said, "We just line those two up and it keeps us in the middle." Although he was a very experienced ship driver having had command of the nuclear cruiser, *South Carolina* (CGN 37), he was not accustomed to being on a ship so large that all you saw was dirt on both sides of the channel.

Both carriers were berthed across from each other at Hotel Pier. Since *Connie* was a bit larger, and in my opinion, had the best supply officer in the

Navy, I volunteered to host the RIMPAC gala on my flight deck. Admiral Barney Kelly was the Commander in Chief of the Pacific Fleet (CINPACFLT). He was very pleased with the condition of *Constellation*, and especially the wonderful array of international food that Bill Bickert and his team somehow procured and flawlessly prepared. Toward the end of the event, the admiral pulled me aside and said, "Gil, let's retire to your in-port cabin for a little chat. This will not be admiral to captain, but rather a Barney to Gil conversation." Oh sure, I thought. Over a cup of coffee in my cabin he began the conversation, "Gil, don't hold anything back, and tell me what you think about my leadership."

I replied, "OK, overall, you are supportive, and I appreciate that. What I don't appreciate is your order that no carrier commanding officer be allowed to fly off his ship! He frowned (not good) and said, "Goddammit Rud, your job is to drive this carrier, not to fly aircraft from it." I should have shut up and replied with a "yes sir," however, I pushed it further. "The crew and the air wing would absolutely love to see the carrier CO flying, so that they would know that we are personally aware of any issues one might face flying from our carrier." He put his coffee cup down and said, "I was hoping that you would have a more mature approach to this opportunity for a conversation, let's get back to the flight deck." *Although I really had no chance at any more promotions, and I knew it, this event certainly did not enhance that extremely remote possibility.*

***

FLYING WHILE IN PORT: I was not allowed to fly from the *Connie*; however, I used every opportunity to fly both the *Hornet* and the S-3 *Viking* ashore. Since the S-3 was now considered to be a single-piloted aircraft, I could fly from the right seat without being legally qualified. I would fly with VS-38 to Lemoore. I would then do all the exams/simulators to be NATOPS current in the *Hornet*. VFA-151s LCDR Tom Gurney, an outstanding aviator, test pilot, and eventually a key contributor to the successful evaluation of the *Super Hornet*, drew the short straw and was designated to be my wingman/instructor.

I had so much fun. We flew all sorts of training flights with interesting results. Keeping in mind that although I had 1,100 hours in the *Hornet*, most of it was airshow related with very little tactical expertise. On one flight, I was involved in a four versus four air-to-air engagements near China Lake. During the debrief of the event, I was asked to relate what my radar picture was just prior to the merge. I replied, "I knew exactly how much gas I had in every tank." The other pilots laughed heartily at this because they realized that I had pressed one too many tiles on my display, which showed me the status of all my fuel tanks, instead of the radar.

At times, one of the VFA-151 pilots would bring an aircraft to North Island for me to get a flight. On one of these occasions, I was lining up to depart from runway 29 at North Island. The approach/takeoff end of this runway is located right next to senior officer housing and very close to the residence of Commander Third Fleet (COMTHIRDFLT). Normally we would do a basic engine departure to hold down the noise a bit. I saw that something was going on at his home, so I purposely lit the afterburners and roared off into the skies. The next day I got a "personal for" message from the admiral's chief of staff: "The Admiral was just wondering if you were flying yesterday at about 1600?" *Ha, he knew full well it had to be that smartass "farm kid."*

***

OPERATIONAL SUCCESSES: The *Connie* and *CAG Two* team absolutely nailed all the operational exercises leading up to the deployment. Dave Nichols had now taken over as the CAG and Steve Kunkle was the new Deputy. Our experimentation with the most capable carrier air wing in the fleet resulted in a record number of sorties generated during one of our final readiness exercises. As we approached 72 hours of continuous operations, one of the inspectors approached me on the bridge. He said, "Captain Rud, do you know that you just broke the record for most sorties in a 72-hour period?" I replied, "Nope, I had no idea. We have just been reacting to your virtually continuous threat scenarios by launching a response."

***

CONTINUING ENGINEERING ISSUES: Fortunately, we were able to make our commitments, but not without a struggle. What we were unable to do was to satisfactorily pass the Operational Propulsion Plant Examination (OPPE). My engineer at the time was a very smart, hard-working officer with experience working on a nuclear-powered ship. Unfortunately, he was also a micro manager, who drove himself too hard. During the "in" briefing, at the very start of the inspection, he fell asleep. And I am not talking nodding off here. It did not get any better from that point, and with the combination of material issues and manning challenges, the inspection team gave us a failing grade. Normally, a failure like this results in the firing of the captain—me. I was totally prepared for that ignominious possibility. Fortunately, VADM Rocky Spane, who was Commander, Naval Air Forces, Pacific (COMNAVAIRPAC), took into consideration the existing material and manning issues that we were dealing with. Instead of firing me, he provided us CDR Kevin Gannon, a talented engineer from his staff, who helped "right the ship," so to speak.

***

NAVAL AVIATION, A FAMILY AFFAIR: The *Kitty Hawk* was still equipped with a bridle arrestor on one of its catapults, so they were assigned to qualify the training command T-2C students. We were operating just a few miles away with our normal carrier air wing. I was in a meeting with RADM Bordy and his staff when I kept getting calls from the bridge. This was quite normal, however, the frequency of the interruptions got the admiral's attention, so he said, "Captain Rud, is there something going on that I should be aware of?" I replied a bit sheepishly, "No sir, I am just getting calls from my friend, Captain "Bear" Pickavance, on *Kitty Hawk*, each time my daughter completes a landing on his ship. The admiral (a great guy) said, "Gil get back up on the bridge so you can keep personal tabs on this great event."

Val enjoying some downtime while a student at Virginia Tech. She would earn a Navy commission and follow in her father's footsteps.

Val completed the carrier qualifications and went on to get her wings. Like me, she was a slow starter, but eventually, became the first woman to command an operational E-2 Hawkeye squadron, and the first woman to be the commodore in charge of all the E-2s and C-2s in the Navy. She also won the Top Hook Award (best landing grades out of 100+ pilots) for an eight-month combat cruise on USS Nimitz (CVN 68). Val and I are the only father-daughter combination to share centurion status (100 arrested landings each) on the same aircraft carrier, USS Enterprise. I got my 100th during Operation Frequent Wind (the evacuation of Saigon) in 1975, and she got her 100th during Operation Desert Fox in Iraq in 1998. She claims to be the better pilot though, since she completed all her landings without the use of a Heads-Up Display (HUD). Val is a veteran of the Bosnia conflict, along with two Iraq, and two Afghanistan combat cruises. I am an incredibly proud father to say the least.

***

UNDERWAY REPLENISHMENT FROM AN AVIATOR'S PERSPECTIVE: As a conventionally powered aircraft carrier, we burned over 100,000 gallons of fuel a day, just for normal operations. To always be prepared for potential orders to steam at "best speed" to wherever we were needed, we always kept our fuel state as close to full as possible. We did this by utilizing the San Diego area "duty" tanker or our battle group tanker, whenever the opportunity arose. The navigator, Beef Flannery, an F-14 *Tomcat* pilot, oversaw preparations for these underway replenishment events. It was our responsibility to execute a rendezvous with the tanker. Beef did this by asking the tanker to set a certain course and speed. We would then execute a rendezvous that would set us up 500 yards astern of the tanker. From that point, we would start our approach to end up 140-160 feet abeam of the tanker at co-speed.

Just prior to one of these evolutions, the chief of staff called me and said, "The admiral would like to attend the navigator's pre-rendezvous briefing and observe the event from your bridge." Beef overheard this conversation and inquired, "Captain, can I do my aviation style briefing, or do I have to go standard "black shoe" for the admiral?" I replied, "The admiral wants to see how we normally do it, because we are damn good at this, so you are cleared in hot with your normal aviator's-style brief!" It just so happened that the conning officer for this approach was one of our junior watch team members, who, although a

LTJG, looked like he was fresh out of high school.

The admiral arrived on the bridge, the navigator gathered his team around the "Gator" chair on the starboard side of the bridge and started the brief, "Gentlemen, our oiler is on a course of 270 at 13 knots. We are on a course of 090 at 24 knots." He then pointed out the bridge window at the oiler and continued the brief, "What we have here is a classic beak-to-beak situation. I want you to take a little separation to port, slow down a bit, and then turn the ship to point the nose of *Connie* just slightly behind the stern of the oiler. Remember, if you are not pointed directly at him, you are not going to hit him."

At this point the admiral excused himself from the brief, looked at me and said, "I am going to observe the rest of this evolution from my own bridge." The kid then executed a perfect rendezvous and within a few minutes we were taking on fuel at the rate of 400,000 gallons an hour. *I later asked the chief of staff what the admiral thought of the evolution? He said, "Initially, the admiral was appalled at the informal, aviation-laced terms used in the brief. However, as he observed the incredibly efficient and effective rendezvous, he remarked that he thought it was amazing how aviators handled relative motion and drove the ship into position, without consulting maneuvering boards, radar, etc. I should not have said anything more, however, I just could not help but make the comment: "It really boils down to 500-knot minds vs. 30-knot minds." I am sure the chief of staff did not share that comment with the admiral.*

***

NEW LEADERSHIP AT CINCPACFLT: A month or so prior to the start of our Western Pacific/Indian Ocean deployment, Admiral Zlatoper replaced Admiral Kelly as CINCPACFLT. The staff had given us a heads up that "Zap" intended to allow the carrier commanding officers to, once again, fly from their own carrier decks. Of course, one could only do that when fully flight qualified, which nobody was, except for the "farm kid." I flew in an S-3 to Lemoore, jumped into a *Hornet*, got the required FCLPs and was ready to go as we went to sea for the last time prior to our extended deployment. That same morning that we went to sea, Admiral Zlatoper sent a "personal for" message, officially lifting the ban on us flying from the carrier. As we cleared the channel leaving San Diego, I looked at Beef and said, "Gator, you got it. I am off to brief and get my ten-day landings."

Although I had been in the cockpit much of the time, it had been 10 years since my last trap. Walt Stammer was the CAG LSO. He was a typical shit-hot aviator and had been an A-7E pilot prior to flying *Hornets*. In the brief he said to me, "Captain, I know that in the A-7 you preferred to fly without the auto throttle in the day pattern, because it lagged too much on throttle inputs. The *Hornet* has a much better auto throttle, so I strongly suggest that you use it." I followed his advice and had an absolute blast flying this awesome machine, on and off the carrier, at the age of 50. I am sure that my landings were less than stellar, but to me, they were all perfect. I could not believe how much I had missed the thrill of both the cat shots and the landings. The Air Boss, now Dan "Darth Vader" Cain, asked if I would like to keep bagging following my tenth landing. I replied, "Nope, lets reserve the bagging for the nuggets." I said this because I had observed too many senior folks using their position to achieve some trap number milestone, like the coveted 1,000. In my opinion, this type of leadership behavior certainly did not promote the morale and retention of our first-tour pilots.

***

Thick Skin a Requirement: I mentioned in an earlier portion of this book, that the carrier air wing had a tradition of holding an event to celebrate the end of a line period, and to pass out awards for the top tail hooker, both individual and squadron. At sea, they were held on the ship's Forecastle (Foc's'le) deck, which is a large space that holds the ships anchors and associated equipment. All the squadrons were expected to put on some sort of entertainment, and of course, this was usually done at the expense of the leadership. And yes, the captain of the ship was certainly a popular target. VMFA-323, the Marine *Hornet* squadron, had among its pilots a young captain by the name of Pat "Kato" Cooke. Pat had an uncanny ability to perfectly imitate my voice, especially the Norwegian accent. During the follies, he would do this in the form of hilarious 1 MC (the sound system that is in every living and working space on the ship) announcements that he made from behind a curtain. As part of the fun, I would fake an angry reaction and threaten to have him put on bread and water. *Pat would go on to become* Blue Angel #2, *and we have remained life-long friends. Whenever we get together, I ask him to demonstrate his special talent, and of course, he does.*

***

UNDERWAY FOR THE WESTERN PACIFIC: Instead of the normal transit, which would take us to Hawaii, we took the northern route. The northern route was a bit more of a challenge due to the weather patterns that involved a generally rougher sea state, and more unpredictable storms. We were conducting flight operations approximately 1,000 miles north of Hawaii. The weather was quite nice as we launched the first event. I then started to notice a threatening sky on the horizon that set off alarm bells in this commonsense farm boy (I had always wanted to be a meteorologist but could not handle the physics courses involved). I called the CAG and the admiral and said, "I am going to pull forward the second launch and get the first launch back on deck as soon as possible." The admiral was a bit incredulous, because the weather report did not include any threat of storms, but the CAG readily agreed to my suggestion, so the admiral did too. Thank goodness they did, because we managed to get everyone back aboard, except for the E-2 *Hawkeye*, when this vicious squall hit us with 50-knot winds and heavy, blinding rain. The E-2 had lots of gas, so we just kept him airborne until we got a break in the weather, which he found with his own radar.

***

LOSS OF A *HORNET:* I made a good call with the unexpected storm; however, I was about to make a bad one. We had become comfortable with sea states that were marginal, especially considering that we were conducting "Blue Water" operations (no emergency bingo fields). On this fateful day, the sea state was such that the boarding rate of the carrier air wing was affected, but it was certainly not unsafe. That is, until the sun went down. During the first night recovery, the deck was pitching and rolling to the point where the Fresnel lens was unusable (it is stabilized to handle 20 feet of deck movement). The LSOs set up the Manually Operated Visual Landing Aid System (MOVLAS), which is a series of lights in a vertical row installed in front of the now inoperable Fresnel lens system. As I had mentioned earlier, the MOVLAS is manually controlled by the LSO, who uses it more to initiate proper pilot control inputs than to show him where he really is on the glide slope. For example, if the pilot is high, but has too much of a sink rate going, the LSO will lower the MOVLAS presentation so that the pilot applies power even if he is still above the normal glide path. The air wing practiced with the MOVLAS, usually one event out of the normal seven each day, so that both the LSOs and the pilots were comfortable using it. But we did not practice very often at night, because if the deck is moving that much, you should not be flying.

LT Paul Horan was flying a VFA-151, F/A-18C *Hornet*. As fate would have it, my name was on the aircraft (I was very flattered to have my name on one of their aircraft). There was so much natural wind that we were steaming at minimum steerage, which was about five knots. He was a bit high on his approach, so he took off some power just as the flight deck began a wild upward fluctuation. The LSO waved him off and even called for afterburner. Despite this correct action by both the LSO and the pilot, his aircraft touched down, with the main wheels *only*, at the end of the flight deck landing area.

The nose of the aircraft fell through and now pointed directly at the ocean, which was only 60 feet below him. Paul pulled the control stick into his lap, stopping the descent of his aircraft so close to the water that his afterburners were throwing up geysers of salt water. At this point, he made a decision that any sane pilot would do. He ejected from the aircraft.

On the bridge the team reacted properly with a right full rudder and engines stop command. We did this to avoid running over the pilot. I can still visualize the next few seconds as if it was yesterday. Instead of immediately crashing, the now pilotless *Hornet* began a slow climb. I told the OOD to sound "General Quarters," as it appeared that the aircraft might now crash into our crowded flight deck. Thankfully, it continued to climb, slowly turning toward the destroyer that was acting as our plane guard. I quickly got on the bridge-to-bridge radio and said to the destroyer captain, "Do you see that *Hornet* in full afterburner coming toward you?" He replied, "Yes sir, pretty impressive." I then quickly added, "There isn't anybody in it!" The reply, "Oh shit!" Fortunately, it turned away from the destroyer before crashing well clear of any vessels.

Our HS-2, H-60 helicopter picked Paul out of the water in just a few minutes, and he was immediately sent to medical. Our chief medical officer soon sent me word that other than being wet and cold, he was in perfect health. Great news! We successfully recovered the rest of the carrier air wing and then knocked off flying for the night.

The next morning, LT Horan reported to me on the bridge with the following comment, "Captain, I am so sorry that I lost your aircraft." I replied,

"Paul, we can easily replace aircraft." I then gave him a big hug and said, "Welcome back from a very close call. I am looking forward to seeing your name on the flight schedule again, in the very near future." *As is the case with any Class Alpha aircraft accident, a thorough investigation is conducted as to the probable cause, as well as recommendations for assignment of responsibility (blame), and most importantly, recommended actions to prevent another accident. I was in the endorsement chain, so when it came to me, I accepted blame, because, I should not have allowed flying from the* Connie *in that weather. When the endorsers above me saw my acceptance of responsibility, the LSOs and the pilot were pretty much absolved, which was appropriate. I also knew that this was going to be my last job in the Navy, so no six-checking was required to end up doing the right thing.*

The Battle Group proceeded directly to an operating area near mainland Japan. We were not scheduled for a port visit until after my change of command. Captain Mark Ostertag, a good friend and legendary F-14 *Tomcat* pilot, was already aboard the ship and would soon become the new Captain of *Connie.* However, a couple more adventures were still in my future before that event was to occur.

We had another mishap where an F-14 *Tomcat* lost his brakes on the flight deck and then ran into a parked E-2 *Hawkeye.* The damage was severe enough that we needed to offload the *Hawkeye* onto a barge located at Naval Station, Yokosuka, in Tokyo Bay. My good friend, Captain Denny Irelan, recommended a Senior Chief Quartermaster from Commander, Fleet Activities Yokosuka to act as a harbor pilot to help us get *Connie* into Yokosuka for the offload of the E-2. The offload was set up to occur at an anchorage versus alongside a pier. The approach to the anchorage required us to steam up the extremely busy entrance to Tokyo Bay fast enough to keep our position in the queue of inbound ships, then make a 90-degree port turn across outbound traffic, followed by a "back full," backing down bell, to drop the anchor for the offload. With the experience and guidance of the Senior Chief Quartermaster and some great work from our Gator and his bridge team, we safely accomplished this task. Whew!

At this point in the deployment, CDR Dan Cain was the Air Boss and CDR Mike Wertz was the mini boss. Our best cat and arresting gear officer was a P-3 pilot by the name of Joe Rixey. The mini boss was on emergency leave due to the loss of an immediate family member, so Joe was filling in as the mini boss. It was not uncommon for us to be having some fun-banter between the bridge and primary flight control (the aircraft control tower). We knew that Joe was a bit nervous in his new position of responsibility (I don't know of any other P-3 pilots that had ever served as a mini boss on a carrier), so we decided to punk him a bit.

It was his responsibility to arrange for the barge to be brought alongside *Connie*. Before the barge could be tied up, tugs needed to tow out a fender camel to be placed between the barge and *Connie*. From the bridge, we sighted a tug towing out one of these fender camels. I called Joe in the tower and said, "Goddammit Joe, that barge (actually a fender camel) is too damn small to fit an E-2 on!" He now sighted the camel, and said, "I am so sorry captain, I will call them right now and get a bigger barge."

Since the Air Boss was in on the ruse, I continued to badger the now quite-flustered P-3 pilot, with all sorts of ridiculous questions like, "Christ Joe, you flew P-3s, which are about the size of an E-2. How could you have let them bring that shitty little raft (the camel) out here and waste our precious time?" Finally, I said, "Joe, get your butt over to the bridge, right now!" He shows up in just a couple of minutes, white as a sheet, and of course we all lose it in laughter. I gave him a big hug, and said, "Joe, that is only a camel. The barge is just behind him, coming around the pier now. I only make fun of folks on this ship that I respect and that I know can take it." *Joe went on to command a special operations P-3 Squadron and eventually became a three-star admiral.*

***

LAST FLIGHT AS A NAVAL AVIATOR: On December 3, 1994, *Connie* was in the middle of a maritime exercise just off the coast of Okinawa. As usual, I was on the bridge as we conducted flight operations. Suddenly, it dawned on me that the next day I would be giving up command of this amazing aircraft carrier. About the same time, I realized that this would be my last chance to fly an F/A-18 *Hornet* from the deck of an aircraft carrier. Normal protocol would have been for me to ask permission from RADM Bordy, prior to going flying. But, due to the busy schedule and the large number of ships steaming near our battle group, I pretty well knew that my request would be turned down. So, I decided to do what I had done throughout my 28-year Navy career. I decided to ask forgiveness instead of permission.

I called down to VFA-151's ready room and asked to speak to the skipper, Tom Trotter. Before I could even request to get on the flight schedule he said, "Captain, how about you and I for an air combat maneuvering (ACM) flight?" I replied, "Perfect, however, you will need to lead the flight, and we will have to brief up here on the bridge." Tom came to the bridge and brought with him all my flight gear that they kept in their pilot flight gear locker. The brief was short and simple. Since I had not had a chance to fly for a couple of weeks, we would launch and do some basic maneuvering prior to setting up for some one-on-one ACM. I then added, "Trots, since I am the Captain, I am going to ask the Air Boss to clear us for a high-speed pass just before the recovery. I want you to lead, and I want you to make the pass from bow to stern at 400 knots. You are a hell of a good pilot, and I know you are interested in possibly applying to be the *Blue Angels* Flight Leader. This is an opportunity to do a little audition. I am going to be welded to your wing, closer than normal parade formation. I can only do this if you keep a positive-G pull on during the entire flyby. "No problem, Captain, I can do that."

I turned to the Gator and said, "Gator, you got it." I am going flying!" Of course, the Air Boss let everyone on the flight deck know that Captain Rud was launching on his last flight as the commanding officer of USS *Constellation*. It is difficult, even now, for me to describe the emotion that I was feeling as the flight deck crew welcomed me to my aircraft. The cat shot, like all daytime cat shots, was another amazing thrill that I will remember and cherish forever. The flight included all the fun things that one gets to do with a high-performance fighter aircraft. So much so, that as it came time to head back to recover, I noticed that I did not have quite as much fuel as I would have liked.

Then, like the hand of God, an S-3 *Viking* checks in on his way back from a flight to Okinawa. Upon check-in, he said, "I have about 3,000 pounds of extra fuel to give if anybody needs some." I immediately responded, "This is *Vigilante* Two, I would be happy to take a drink." We spotted the tanker overhead *Connie* and initiated a rendezvous. Trots did not need any gas, so he just stayed on the port wing of the tanker, while I dropped back and plugged in. After completing the refueling, I moved up on the tanker's starboard wing and signaled him to reel in his refueling hose and basket. Once it was safely reeled in, I gave him a thumbs up. Now with plenty of fuel, I lit my afterburners, pulled the nose straight up and commenced a couple of vertical rolls.

Meanwhile, in the right seat of the S-3 cockpit, RADM Bordy watched as I lit the afterburners and climbed away. He said to the pilot, whom I believe was the VS-38 XO, CDR Wilburn, "Do you know who is flying that *Hornet?*" He replied, "Yes sir, that is the skipper." "Well, you tell CDR Trotter to stop wasting my gas showing off in full afterburner." "Ah, sir, it is not CDR Trotter. It is Captain Rud." The admiral than looked down at *Connie* steaming with the numerous ships in the exercise, turned to the CAG, who was in the jump seat of the same S-3, and said, "Admiral, CAG and the captain of the ship all airborne; who in the hell is in charge down there?" The quick-thinking S-3 pilot responded, "Well admiral, I would guess it is the navigator, at least on the bridge."

Just prior to starting the next launch, the Air Boss gave us clearance for a bow to stern flyby. True to his word, Trots got us to 400 knots and began a constant, positive-G, tight-arcing turn to basically wrap us around the ship in what was probably about a 70-degree angle of bank. I moved in close enough to where many observers thought that it was a single aircraft flyby. What I had not mentioned to Trots was anything about what altitude the flyby would be flown. Well, although I saw nothing but his aircraft, we were—according to those observing—at flight deck level! Thanks, Tom Trotter, for making my last flight in the Navy so memorable. Especially for the 50-year-old geezer that I was!

My last carrier-arrested landing in an F/A-18C Hornet *aboard* Constellation.

Following completion of my last trap, the yellow shirt directed me to the farthest forward parking position on the bow of the ship. He then passed me off to this rookie trainee, who directed me to the dreaded, nose-over-the-water, reverse-slash, parking spot. Once I was out of the cockpit, the flight deck crew immediately surrounded me, and, as is the custom for a final flight, I was summarily hosed down with what I recall was some really cold water.

The next morning, prior to the change-of-command ceremony, the flight deck crew strapped my flight boots to the catapult, along with those of Joe Rixey, who was also leaving the ship. They then fired both sets of boots far out into the Pacific Ocean.

It was over. I would never fly again as a naval aviator. I finished with 5,600 hours of flight time and 786 arrested landings. These landings included perfect ones, good ones, average ones, below average ones and some awful ones. Most of the awful ones were in the early years, and I learned from those so that most of my later career landings were good ones. The day traps were all a blast, and the night ones never got easy—or fun for that matter. The same goes for the catapult shots: day fun, night not.

As I boarded the C-2 COD to leave the ship and my beloved Navy, I thought to myself, "I really have been the luckiest man in the world!"

Joe Rixey and me getting ready to catapult our flight boots into the Pacific Ocean.

Epilogue

My Luck Continued, as I managed to land a marketing position with the McDonnell Douglas Company and went to work for them while I was still on terminal leave in the Navy. Instead of going into detail about the eventual 14-year career that followed with McDonnell Douglas and the Boeing Company (this just might deserve another memoir), let me just say that it was almost as much fun as my 28 years in the Navy.

Personally, I lost my wife, Barbara Carroll Rud, to breast cancer in 2005. Although I never expected to fall in love again, through the help of mutual friends, I met, fell in love with and married Carol MacDonald Simmons in 2009. I knew Carol's late husband, Carl, who was an A-7E and F/A-18 pilot, but did not know Carol. Carol knew my late wife from school related activity in Lemoore, but did not know me. Since, between us, we now had five children, we decided that the best course of action to get their approval for our marriage was to invite them to an all-inclusive-destination wedding at a beautiful resort in Cancun/Playa del Carmen, Mexico. Once they strapped on those purple wristbands, the party began, and the Rud-Simmons adventure was underway.

What an amazing adventure it has been. Carol is my best friend and confidant. We are blessed and living life to its fullest. I married into a large, loving and welcoming family. Carol has two amazing daughters, Erin and Stacey, both with great families of their own. She comes from a wonderful family of seven children, the MacDonald Clan. She describes herself as the perfectly well-adjusted, middle-child. Since I only had the one sister, who was four years older than me, I particularly enjoy participating in the MacDonald family gatherings, which are welcoming and loads of fun.

The Rud–Simmons combined family. From left to right: Stacey, Erin, Carol, Gil, Valerie, Sally, and Ryan.

I did not realize that there could be such a thing as a perfect match of a man and a woman. Carol and I together make that perfect match. Once again, I am the luckiest man in the world.

Although I did not fly much immediately following my retirement from the Navy, a fellow Boeing employee, Jim McCarthy, who owned an A-36 Bonanza and was also a Certified Flight Instructor, got me back into flying. I eventually purchased a Cirrus SR-20 and then an SR-22, which I flew extensively (1,200 hours), while carrying out my duties as the head of Boeing's Navy and Marine Corps Marketing Field Offices.

I retired from Boeing in 2009. Carol and I settled in the San Diego area. We both stay very busy volunteering (I am a docent on the USS *Midway* Museum), traveling, golfing, chasing the now nine grandkids, and staying involved in the *Blue Angel* Foundation Wounded Veterans Program led by Mike "Manny" Campbell, who was #8 in 1987–88. Bernie Willett, an honorary *Blue Angel*, and former fund-raising executive for American Airlines ran a Celebrity Golf event for the Susan Komen Foundation for many years. Several of our 1986–88 *Blue Angel* team members, including Mike and I, served as staff for this event. Following his retirement from American, Bernie now runs a similar tournament for the *Blue Angel* Foundation. The proceeds of this tournament and several other fund-raising events, all go toward a treatment protocol for veterans with Traumatic Brain Injury (TBI) and/or Post Traumatic Stress Disorder (PTSD). We are losing these combat heroes to suicide at the rate of twenty each day. So far, this protocol has been very successful in deterring this unacceptable loss of Americas finest. *Please consider supporting the* Blue Angel *Foundation effort.*

In 2013 I decided that I needed a new aviation challenge. Having regularly embarrassed myself attempting to land Fugie's Aeronca Champ, I decided to sell the Cirrus and purchase a backcountry capable tail dragger. I found a used 2008 Aviat Husky A1-C that just so happened to have the perfect paint job. With the 31-inch bush wheels, I can land just about anywhere, and I have done just that. I also use the Husky to fly folks who donate to the *Blue Angel* Foundation, as well as Young Eagles for the local chapter of the Experimental Aircraft Association (EAA). When I am in North Dakota with my Husky, I hang out with the "Hatton Flying Circus," which includes my good friends, Eddie and Ethan Grindeland, Keith "VOR" Thorsgaard, Steve Hilstad and other backcountry/tail dragger pilots.

With my beloved Husky.

With Carol in front of the last F/A-18 Hornet flown by the Blues *prior to transitioning to the* Super Hornet. *The aircraft is in the Smithsonian — Udvar Hazy Museum of flight and has my name on one side of the cockpit as the first* Hornet *Flight Leader, and Brian Kesselring's name on the other as the last flight leader of the 34-year* Hornet *era. Ironically, both of us are North Dakota natives.*

* * *

FAMILY SERVICE: I am very proud of our families' participation in the military. My cousin Don Leland was a Marine A-6 *Intruder* and OV-10 *Bronco* pilot in Vietnam, and my cousin Jim Rud was an officer in the Army Corps of Engineers during the same time frame. In addition to my daughter, Valerie's, career as a naval aviator, her husband, Travis Overstreet, also served 22 years as an E-2 *Hawkeye* pilot and now is a United Airlines 737 Captain. My son Ryan served as a Cavalry Recon Scout in the California Army National Guard. My nephew, Jim Kringlie, did two tours in Iraq, one of which was outside the wire as a

Cavalry Recon Scout Sniper. Jim's wife Sheila is a nurse in the North Dakota National Guard. Jim's oldest son, Michael, is on active duty as an Army Ranger, and Jim's daughter, Lindsey, is a member of the National Guard. His youngest son, Jack, just enlisted in the Army National Guard and will go on active duty upon his graduation from high school. My nephew Tristan, Sarah's son, deployed to Afghanistan as an active duty Army Medic and is now deployed once again as a member of the National Guard. Sarah's husband Greg also completed a combat tour in Iraq. When it comes to serving our country, we have certainly done our part.

Acknowledgments

I Take this Opportunity to thank my friend Dan Beintema for his immense patience in the early editing of my mediocre work. I can tell you that the initial set of pages that he sent back to me looked like my first graded college-freshman English paper. They were covered in red-ink corrections, minus only the C– grade I remember getting on that first paper. By the way, I was thrilled to get a C– as most of my classmates got Fs.

I would also like to thank my long-time friend and fellow Naval Aviator, Denny Irelan, for his mentoring and professional editing contribution.

Thanks to Ryan Nothcraft and Nick Veronico for Beta reading the manuscript and professionally scanning the photos, many of which were in marginal condition.

To the publishers, Mick and Diane Prodger, thanks for your patience, guidance and superb professionalism. If someone reading this book is looking for a publisher, I strongly recommend Elm Grove Publishing.

Finally, a special thank you to my wife, Carol. Babe, without your support and encouragement, I would never have even attempted to write *From the Prairie to the Pacific*. I love you so much.

Index

Printed in the USA
CPSIA information can be obtained
at www.ICGtesting.com
CBHW060844040224
3974CB00002BA/3